Structural Integrity
Monitoring

Structural Integrity Monitoring

R. A. Collacott
PhD, BSc(Eng), C.Eng, F.I.Diag.E, F.I.Mech.E,
F.I.Mar.E.

London New York
Chapman and Hall

First published in 1985 by
Chapman and Hall Ltd
11 New Fetter Lane, London EC4P 4EE
Published in the USA by
Chapman and Hall
29 West 35th Street, New York NY 10001

Printed in Great Britain at the
University Press, Cambridge

ISBN 0 412 21920 4

British Library Cataloguing in Publication Data

Collacott, R. A.
  Structural integrity monitoring.
  1. Structural stability   2. Reliability
  (Engineering)
  I. Title
  624.1'71   TA656
  ISBN 0-412-21920-4

Library of Congress Cataloging in Publication Data
  Collacott, Ralph A. (Ralph Albert), 1918–
  Structural integrity monitoring

  Bibliography: p.
  Includes index.
  1. Structural stability.   2. Structural failures.
I. Title.
TA656.C65   1985   624.1'76'0287   85-7778
ISBN 0-412-21920-4

Contents

Preface

Today's astronauts who jet at fantastic speeds enjoy a remarkable confidence in the structural integrity of the spacecraft in which their mission is performed; internal systems may malfunction but, in general, the structure remains secure. It has not always been so. Not many years ago early shuttle flights were imperilled by the possible 'rip-off' of insulation tiles – special heat shields to protect the spacecraft in the friction heating zone during re-entry. Ground-based monitors through spacecraft CCTV inspection were able to appraise the condition of the exterior and advise the astronauts accordingly. Engineering technology has gone a long way since the early industrial disasters – and yet disasters still occur.

This book aims to bring the management consequences of disasters and the wide range of existing expertise to the attention of all concerned with the integrity of structures. Through the use of high technology it has been possible to evolve a very wide range of monitoring equipment which can be applied to solve nearly every conceivable problem ... and even to conceive the probable problems themselves. With the high technology typical of the nuclear power, chemical and aviation industries it has not been sufficient simply to wait to 'learn' from horrendous disasters; techniques have had to be evolved for appraising the criticality of the schemes themselves and means introduced for monitoring such conditions.

The ability to monitor deteriorating conditions in mechanical plant has been well established with mechanisms; it can also be used more extensively for structures.

The author became involved in diagnostic matters in the early 1970s and formed the UK Mechanical Health Monitoring Group and Condition Monitoring Association as a means of helping organizations to protect their assets by condition monitoring (sometimes called 'mechanical health monitoring'). It was found that, all over the world, skilled engineers had faced and overcome problems of monitoring and had solved a local problem; yet there the matter had lain. The author's organization sought out these experts and invited them, through a series of seminars and publications, to bring their expertise to the attention of others who might be struggling with the same difficulties. From there, a request arose for a professional organization which would recognize the special skills needed for the diagnosis of deterioration and the identification of faults; this led the author to found the Institution of Diagnostic Engineers, which in a very short time has established a considerable international membership. Through research, the preparation of a regular series of newsletters and a bi-monthly journal '*Diagnostic Engineering*', the author has acquired a vast amount of information which has been distilled and presented in this book. The scheme has been to present the subject through four themes:

Theme 1
(Chapter 1) Why worry about structural integrity?
A chapter dealing with the fact that disasters do occur and that it is no use thinking they will 'happen to the other man' or that your 'insurance deals with it all'. If you are responsible for structures you have worries – Union Carbide, Bhopal, India 1984.

Theme 2
(Chapter 2) How to start structural integrity monitoring
This chapter introduces the managerial factors: determining the structure or parts of a structure most at risk; appraising the manner/mode by which failure may occur

Theme 3
(Chapters 3–9) An in-depth study of major methods

Theme 4
(Chapters 10–12) Decision-making: understanding the nature of failures and preparing decision-making information. In other words these chapters aim to help the reader through an understanding of stresses and strains of cracks and crack growth to acquire the knowledge needed to make shut-down decisions optimally – neither too early or (worse still) too late.

It has been held by the author for some time that the activities of a maintenance organization could benefit by being built up in two sections:

1. A diagnostic centre.
2. A task force.

This book is aimed at those who would work in the diagnostic centre. The task force, with its appeal to craft skills and methods organization, is already well provided with technical literature. The diagnostic centre would need to interpret information derived from its monitors and obtained either from its own staff resources or through persons with a regular daily association with the maintained structure.

Members of a diagnostic centre require a deep insight into the functions of each structure throughout its life – much the same as medical physicians minister to the human frame. The task force members require a good understanding of the daily corrective needs – much the same as nurses provide for the sick and injured.

To make this book effective as a manual, references have been included to sources of supply and to firms who manufacture or supply equipment. The inclusion of such names does not imply any endorsement or recommendation for such equipment, instruments or devices; nor have I or my Institution any known commitment towards such organizations. We do have members among such firms and support the contention that firms concerned with the supply of diagnostic equipment should employ staff who are members of the Institution to so imply a continuing education and training of such persons in the many different aspects of diagnostic engineering and integrity assurance. We advise buyers to enquire whether the staff of their suppliers are such members.

Leicester, England   R. A. Collacott
1985

1
Hazards and failures

1.1 Introduction

Hazards exist at all times. Failures result from many factors but are an inevitable termination of functional life. Catastrophies do occur.

Corporate management is often unaware of catastrophies averted by the skill of their plant engineers. It should always be remembered that catastrophies can and do occur, and that when they take place they are followed by financial, political, social and legal consequences which no amount of insurance cover can (or should) protect.

A review of some notable faults, failures and catastrophies is beneficial for the purposes of:

(1) Reminding doubting executives of their responsibilities.
(2) Advising administrative engineers of the need to document and monitor fault occurrences.
(3) Showing that experts exist whose studies enable them to make a substantial contribution to the assurance of structural integrity.

1.2 Failure causes

All failures exhibit three different types of weakness:

(1) The technical fault through which a part fails because of:

    (i) overstressing because the load was greater than expected (wrong specification); or
    (ii) overstressing because the part was not strong enough for the particular load (wrong design); or
    (iii) the part was not manufactured or fitted according to the design (wrong construction).

(2) The organizational fault – because the project coordinator had not established means of avoiding 1(i)–(iii).

1

(3) An educational aspect – the responsible person had not been trained to avoid the mistake. The engineering designer needs to anticipate potential risk from the operation of his plant, system or structure under the conditions of 'reasonable use'.

1.3 Reasonable use

It is possible that designers should extend the range of man–machine interaction (cybernetics) to consider the effects of unfavourable maintenance, deliberate tampering, vandalism and possible sabotage. The knowledge and experience of operators should be exploited. Thus the senior officials of the National Coal Board did not envisage the possibility of waste tips constituting a major hazard, and those with an awareness of the dangers had difficulty in having their case heard. Staff at a Steel works knew that water in contact with molten steel in a confined space had catastrophic consequences, yet this was allowed to occur.

1.4 Operators

The abilities of persons employed to operate on or with structures are as important as the technical aspects. Mental ability, dexterity and physiological characteristics should be considered.

Thus colour recognition may be defective. Colour identification is vitally important with electrical wiring, yet it is estimated that in the UK more than two million men and 150 000 women have colour vision deficiencies, principally red and green. This defect arises from faulty alignment of the eye's optical system. A green-defective person would confuse the steam and water code for pipelines as set out in BS1710 (green = water; silver-grey = steam). It is reported that when the atomic chemist John Dalton presented a paper (in 1794) on colour blindness he made the mistake of wearing a scarlet cloak (forbidden by Quakers) in the belief that it was a sombre black; he was colour blind and to him blocd was a 'bottle-green colour'.

1.5 Domino effect

1.5.1 Mississauga train crash

In part because of the enormous growth of handling operations involving flammable liquids, the potential risk to the public of catastrophic explosions involving such liquids is increasing. Thus in Mississauga, Ontario, Canada, the derailment of a chemical train in the middle of the town caused a major incident. The accident illustrates the domino effect, where a small initial problem grows into a major disaster. Of the 106 rail cars making up the train, about a quarter were derailed, among them eleven propane, three toluene and three styrene tankers, and one containing liquid chlorine.

The initial result of the crash was a 'bleve' from one of the propane tankers followed by prolonged and intense burning of the propane and other failed tankers. A second explosion soon after, which was seen 50 miles away across Lake Ontario, badly damaged other tankers, including the 90 tonne capacity chlorine tanker.

The subsequent release of chlorine gas and the problem of the 20 tonnes of liquid chlorine remaining in the damaged tanker prompted wholesale evacuation of the heavily built-up area. Some 7000 people were moved immediately after the accident in the early hours of a Sunday morning, and 24 hours later a total of 240 000 had been evacuated. By this time the propane was burning in a controlled way but the chlorine tanker was leaking through a large hole. This hole was eventually sealed, but not before 150 000 people, who had returned to their homes, had to be re-evacuated when some of the released chlorine reacted with water and sent clouds of hydrochloric acid into the atmosphere. Although major casualties were avoided, the scale of the evacuation operations can only have heightened that particular population's perception of risk.

1.5.2 Cascade failure of electricity pylons

Part of a paper entitled 'Static and dynamic behaviour of mechanical components associated with electrical transmission lines' describes a further example of the domino effect [1].

With electricity pylons a progressive failure propagates from one span to the next. Several very long cascades have occurred, most notably in Denmark (50 km of a 150 kV line) and in Wisconsin (125 km of a 345 kV line). Towers have usually failed as a result of inadequate longitudinal strength, although vertical and transverse cascades have also occurred. In general, failure of any along-line component leads to a release of the gravitational and strain energy of the conductor; the result is an impact load at the attachment points of the conductor, followed by a residual static load if the tower survives.

An overhead transmission line is an interactive system consisting of cables, insulator strings, towers and foundations. The forces acting on each component depend upon climatic loads and on the configuration and mechanical properties of the system as a whole. Transmission towers must be designed for static and dynamic conductor loads that can act in different combinations of the vertical, transverse and longitudinal (i.e. along-line) directions.

Recommendations for representative load combinations are given in codes and tower design guides. Such recommendations are based mainly on the performance of conventional lines that use self-supporting towers. Line loads are increasing in magnitude owing to the use of larger conductors and conductor bundles in high-density power corridors. Consequently, heavier-duty supporting structures are needed that are both economical and aesthetically acceptable.

The most severe impact is caused by a broken wire or phase. Although this impact can initiate a failure of several adjacent towers, three factors can alleviate impact loads as the failure propagates. They are:

(1) Energy dissipation at a previous tower.
(2) A graduated release of energy in the conductor span. The structural deflection and insulator swing at the previous tower will transfer some of the energy of the conductor along the line before the collapse of the previous tower releases the remaining energy.
(3) Frictional drag of conductors along the ground.

1.6 Structural weakness: Severn Bridge

An editorial in the journal *Corrosion Prevention & Control* [2] refers to a number of reports which have been published concerning the deterioration of the Severn Bridge, which carries the M4 motorway between England and Wales. Corrosion and fatigue have been two of the main causes of suspension cable damage and cracks to welds in the steel deck structure. The latest interim report by consultants Flint & Neil Partnership (F & N) expresses serious doubts regarding the ability of the structure to meet safety criteria under the most severe appraisal loadings.

(1) The steel towers are inadequte for supporting the main suspension cables.
(2) The innovatory inclined hangers may have to be replaced by traditional vertical hangers owing to fatigue and corrosion effects.
(3) Strengthening of the deck itself is required.
(4) One of the main suspension cables is showing signs of 'distress'.

Corrosion and fatigue of hangers and support cables have plagued the bridge for several years, and more recently cracking of steel deck box welds has occurred. The blame has largely been laid on increased traffic flow and the incidence of heavy vehicles.

1.7 Tay Bridge disaster

On 28 December,1879 a train set out to cross the great new Tay Bridge. The swaying tail-lamps passed into the distance then vanished in a spray of sparks as the two-mile sweep of latticed ironwork collapsed into the water 60 feet below. It took with it a railway engine, six coaches and 75 men, women and children. All of them died in the black abyss.

1.8 Los Alfraques camping site disaster

Propylene was being carried in a road tanker trailer near a camping site at 2.30 in the afternoon of Tuesday 11 July 1978. Unknown to the driver, his trailer leaked. Eyewitnesses reported seeing a huge white cloud of gas sweeping over the camp. The cloud caught fire and exploded, resulting in a tragedy of 101 instant deaths and a further 43 dying later from burns. A three-inch crack was suspected as the cause of the disaster.

1.9 Chlorine freight train disaster

Chlorine gas killed at least six people and sent 34 others to hospital after a freight train was derailed near Youngstown, Florida. Several cars of the train left the rails and plunged into a swamp. At least two tanks containing the poisonous gas commenced to leak [3].

Some of the victims were overcome by fumes while driving through the area. Several were found in or beside their cars on a nearby highway.

Some cars apparently stalled and ran into ditches when the chlorine gas cut off the oxygen needed for engine combustion. As the motorists panicked and fled into the swamp, many were quickly overcome by the deadly yellow-green cloud.

It was the second accident involving a derailed freight train in America's south in the two days of 25 and 26 February 1978.

1.10 Typhoon disaster

Sometime during the week commencing 8 September 1980 it would seem that the 91 000-tonne bulk carrier *Derbyshire* was lost in typhoon 'Orchid'. The freight carrier was carrying a cargo of iron ore from the USA to Japan when struck. Samples from oil slicks found in the vicinity of its last known position were analysed to provide some confirmation of the sinking.

1.11 Typhoon protection

At Kowloon Yard, Hunghom, of the Hong Kong United Dockyards, the maritime manager is responsible for safety when a typhoon strikes. Very considerable damage was done in the harbour when typhoon 'Rose' struck Hong Kong in 1971.

Whenever bad weather is expected in Hong Kong, the Royal Observatory and other sources supply information, carefully monitoring the storm's progress and direction. Production managers are kept informed and all essential communications equipment is checked.

When it is known that the island is in the direct path of a typhoon, all ships moored at the dockyard under repair are made ready for the onslaught:

(1) All vessels not in the safety of dry dock are towed out to typhoon buoys in the harbour. This cuts down the risk of both the vessel and the dockside wall being damaged more than is necessary. All cables, pipes, hoses and electrical wires are removed from the ships as well as staging, gangways and fenders. Berthing gangs are on standby to let go.

(2) Ships in dry dock when the typhoon strikes are protected by a wreckage chain which is stretched across the dockyard entrance, and all vulnerable openings in the vessel are protected. A pump man will stand by to deal with high tides or heavy rains which might flood the dock.

(3) If a major conversion is being carried out at the time, the vessel will be docked for the duration of the storm (if possible), or it will be taken to a typhoon buoy in the harbour. Failing that it will be anchored at Kellet Bank. It is safer to take a ship into the open sea in foul weather, eliminating the risk of collision if the ship breaks its moorings.

(4) Giant crane barges proceed to typhoon shelters. Other cranes from 20 to 100 tonnes in size are anchored and their jibs secured.

1.12 Bridge failures

Most bridges will survive several hundred years, but incidents have occurred in which failure has resulted within months of their opening [4]. Some of these incidents, in alphabetical order, are as follows:

Ashtabula Bridge, Ohio – crossed a steep gorge near the shores of Lake Erie. In December 1877 the bridge collapsed as the New York train was crossing; 90 people were killed. The accident was due to the failure of a cast iron Howe truss.

Chain Pier, Brighton, England – destroyed by aerodynamic forced vibration on 30 November 1836 after two previous similar failures during storms (described by John Scott Russell, *Transactions of the Royal Scottish Society of Arts,* vol. I, 1841 'On the vibration of suspension bridges and other structures; and the means of preventing injury from the cause').

Chester Bridge, Illinois – a continuous bridge over the Mississippi, built in 1942, fell into the river in 1944 after a violent windstorm. Failed due to aerodynamic instability.

Covington Bridge near Newport, USA – a small bridge spanning the Licking River completed in 1853 but collapsed two weeks after it was opened while a drove of cattle was crossing.

Frankenthal Bridge, Germany – spanning the Rhine. Originally started in 1939 with work interrupted by World War II. The second span was wrecked in a windstorm, and 36 people lost their lives.

Liard River Bridge, Alaskan Highway – wire breaking caused by vibrations.

Peace River Bridge, Alaska, USA – completed in 8 months with towers built through ice several feet thick in temperature of $-40\,°F$. In 1947 engineers worked under 7 ft thick ice to build steel cofferdam and save the bridge, where floodwaters had undermined a 250 ft tall tower which leaned 14 inches out of line. Aerodynamic vibration caused strands of cable wire to break under fatigue.

Rialto Bridge, Venice – in 1450 a crowd of Venetians pressed its way on to the wooden bridge in order to watch the entry of Emperor Frederick III along the Grand Canal. The parapet timbers gave way and it collapsed.

Sando Bridge, Angerman River, Sweden – begun in 1938, it collapsed in 1939 with the loss of 18 lives when false work failed during pouring of concrete.

Saulte Ste Marie Bridge – a Canadian Pacific railroad bridge built in 1941 over the ship canal. Of double-leaf bascules design, it collapsed on 7 October 1941 when a freight train was on it, and 2 men were killed. Apparently the two leaves failed to latch together properly. Repaired, it is still functioning.

Tacoma Narrows Bridge, Puget Sound, USA – a suspension bridge with a span of 2800 ft (900 m). On 7 November 1940 (4 months 6 days after the official opening) the oscillations of the bridge during a gale increased to destructive amplitude until the main span broke up, ripping loose from the cables and crashing into the waters of the Puget Sound. The characteristic twisting/writhing action (which was fortuitously recorded on cine film) is characteristic of aerodynamic instability.

Wheeling Suspension Bridge, spanning the Ohio River – completed in 1849 but destroyed by the wind during a storm on 17 May 1854. Described as twisting and turning as it vibrated.

Williamsburge Bridge, 10 November 1903 – fire started in small workshed near cable saddles on lip of Manhattan Tower, 333 ft in air. Temporary footbridge and supporting cable destroyed.

1.13 Earthquakes

Shaking danger to buildings may be ascribed broadly to the exaggerated structural strain which arises when the natural period of a structure coincides or resonates with the period of shock wave.

Many factors determine the area affected by an earthquake – magnitude, geological conditions, depth of focus, etc. – but in general short-period waves are absorbed very quickly, so that at long distances from the epicentre of an event only the longer-period waves remain. These resonate with the taller, longer-period buildings[5].

1.13.1 Earthquakes in Britain

Shocks occurred in Britain during early 1984 but on 26 December 1979 the strongest earthquake in Britain for nearly 50 years occurred over an area from Carlisle in the north-west of England to Aberdeen in the north-east of Scotland. It registered five on the Richter scale – the same as that recorded in June 1931 on the Dogger Bank in the North Sea.

Mr Graham Neilson, a seismologist at the Institute of Geological Sciences, Edinburgh, said that though the earthquake was small on the world scale, it was the worst recorded in the Carlisle area for 200 years.

The earthquake lasted about 15 seconds, and was followed by two smaller tremors and shock waves. The epicentre appeared to be at Longtown, 10 miles north of Carlisle, at the head of the Solway Firth. People reported a sound like a plane crashing; others thought that the nuclear power station at Chapel Cross near Annan on the Scottish border had blown up.

Much attention in recent years has been focused on the Stoke-on-Trent, Staffs, area which has had hundreds of tremors, most of them insignificant.

1.14 Tunnels

Fortunately, few serious problems arise with the many tunnels which exist. Tunnels have been dug for railways (including underground systems), for sewers, hydro-electric power, mining and roadways. Military offensive and defensive tunnels are plentiful. Some difficulties reported by Pequignot [8] in the use of tunnels are as follows:

Bolsover (railway) Tunnel, near Chesterfield – suffered continuous deterioration due to coal workings; a decision to mine a new seam below the tunnel led to its closure in 1951.

Bo-peep Tunnel, near Hastings [9] – unsuspected geological fault found after construction.

Clifton Hall (railway) Tunnel – 28 April 1953, severe collapse in a filled-in construction shaft. Constructed in 1849 through ground which later suffered coal-mining subsidence such that houses collapsed and several deaths occurred.

Dudley Tunnel – linking the Dudley Canal south of Birmingham with the

Birmingham Canal. Nine men were killed and 18 seriously injured during construction. Mining subsidence made substantial thickening necessary.

Holland Tunnel, New York, USA – a lorry carrying 12 tonnes of highly inflammable carbon disulphide was involved in an accident on 13 May 1949 while entering the tunnel in contravention of rules. No lives were lost in the ensuing explosion and fire but damage amounting to $1 000 000 was incurred. About 500 ft (170 m) of the reinforced concrete ceiling collapsed.

Hoover Dam – one of the spillway tunnels was badly damaged by cavitation caused by a small hump in the lining.

Polhill (railway) Tunnel, near Sevenoaks – water seeping through natural chalk side walls carried soot and slurry into the ballast.

San Joaquin Valley, California, USA – 21 July 1952, severely damaged by earthquake disrupting tunnels on the Southern Pacific Railroad.

Severn Tunnel [10] – considerably flooded in 1879 by the inpouring of water. To shut off a door in a headwall a diver walked underwater a distance of 1000 ft (310 m) from the bottom of the shaft using a then-new self-contained type of diving-dress.

Thames Tunnel – original shaft by Vazey and Trevithick in 1807 was broken-into by the Thames. Attempt by Brunel using a shield also failed by water break-in. On resumption of construction, workmen fell senseless due to presence of sulphuretted hydrogen gas.

Woodhead New Tunnel [11] – completed on 16 May 1951 after 2 years construction but 3 weeks later (8 June) a section collapsed which took 6 months to clear.

1.15 Failure of dams [12]

In a study *History of Dams* by N. Smith [12] events at various years were shown:

Year (AD)	Event
942 1241	Nahr Isa, Baghdad. Failed twice. Causes unknown
1191	Grenoble dam, River Drac, France. Earth embankment collapsed
1226–42	Sassanian Beldai dam, Baghdad. Neglected by the authorities responsible for the Euphrates–Tigris–Nahrwan irrigation system and allowed to silt up
1284	Saveh, SW Tehran. Dam never filled. Built on sand/gravel river alluvium. Inflowing water found own way out

1305	Anio Novus, Rome. Stones were removed by two monks in an effort to lower the level of the lake
1626	San Idefonse dam, Potosi, La Paz. Built to power grinding wheels for the silver mines. Failed on 15 March, possibly due to undermining of foundations. Within 2 hours 4000 people killed and many grinding mills destroyed. This loss led to the collapse of silver mining in the locality
1709	Bazacle dam, River Garonne, France. Made of earth, wood and stones yet lasted 700 years until completely destroyed by an exceptional flood
1799	Leicester Navigation dam, built to take water from Blackbrook reservoir to the canal at Loughborough. In February, heavy rain and melting snow overloaded the spillway and burst a hole in the dam
1799	Guadarrama, River Tagus, Spain. When construction had reached height of 57 metres there was a heavy fall of rain which soaked the clay which swelled and overturned the front wall
1802	Peuntes dam, Segura, Spain. Pressure of 154 ft depth of water build-up blew a hole underneath dam. Whole reservoir of 2000 million gallons discharged. 608 people drowned (including director of works) and 809 houses destroyed
1877	South Fork dam, Johnstown, Penn, USA. Built 1839 as a canal supply but later abandoned and taken over by a fishing club who raised the level to 75 ft. Unprecedented rainfall overfilled the dam, spillway was inadequate and whole structure collapsed
1881	Habra dam, Algeria. Suspiciously pervious since it was built, the rise in water level was thought to produce an uplift on the water face. This encouraged tension cracks to form which reduced the effective thickness and created high compressive stresses at the air face as a result of which the dam collapsed
1882	Ponte Alto dam, River Fersina, Italy hit by a flood of water when another dam further up the river failed and released 45 million cubic feet of water. The dam survived but a thin arch dam was built the following year to relieve pressure on this dam
1885	Gran Cheurfas, Spain. Partial collapse because although designed as a profile of equal resistance it was subjected to tension on its water face
1895	Bouzey dam, near Epinal, France. Built 1700 ft long and 49 ft high it suffered continual severe leakage, cracks developed and a 600 ft length of wall collapsed
1923	Gleno dam, Bergamese Alps, Italy collapsed on 1 December due to high shear stresses and bad workmanship
1924	Cowlyd dam, Wales. On New Year's Eve the dam overflowed and

a large section of the air face was washed away. The concrete core-wall remained intact and managed to support the rest of the dam

1925 Eigian dam, Wales. Built on the slopes of Carnedd Llewelyn in 1908 it impounded 160 million cubic feet of water to drive Pelton wheels for the Dolgarrog power station and works of the British Aluminium Company. On the evening of 2 November the dam failed apparently as the result of 'piping'. Subsequent investigations showed that the concrete was of poor quality and the foundations inferior. Where failure started the concrete base was only 2 ft below the top of the clay. Leaking water had formed 'pipes'. A previously dry summer had dried these out and the structure weakened. Flood water swept into another dam, the Coedty dam and swept this away as well. Sixteen people died in Dolgarrog – a total which might have been higher except for the fact that most of the village were away attending a film show at the time

1925 Skelmorlie reservoir, Scotland. A small dam failed and discharged 800 000 cubic feet of water. This disaster had a curious origin: the drains of a nearby quarry which emptied into the reservoir became blocked. When heavy rain filled the quarry with water the increase in pressure suddenly cleared the choked drains which flooded the small reservoir below. The small overflow was not able to cope, resulting in the dam being toppled over and washed away

1928 St Francis dam, Los Angeles, USA completely wiped out when foundations washed away

1957 Loughborough, Leicestershire. Concrete dam built in 1906 cracked and damaged by an earthquake

---

1.16 Structural and foundation failures

A whole range of troubles can arise which in turn cause financial loss and litigation. Such problems have been described in depth both from the technical and legal aspects in 34 case studies by Le Patner and Johnson [13] – a happy combination of practising US attorney and a consulting engineer both specializing in structural problems. A few of the cases are summarized in Table 1.1.

1.17 Accident data

Records are available from various sources dealing with accidents in mines, aircraft crashes, fires, explosions, railway accidents and lists of other situations in which structural failures have occurred.

Table 1.1

Facility	Problem	Comments
Bridge	Cracking and distortion of welded girders	Liquid damages of $500 000 assessed by owner against contractor after litigation based on; (a) arbitrary and capricious rejection of steel girders (b) improper rejection of contractor's request to repair bridges by rewelding (c) failure of prime and subcontract engineers to pass on to the owner the contractor's request for a time extension in order to complete the project (d) faulty design of the proposed bridges which caused unexpected and extensive costs in connection with pouring the concrete deck The contractor recovered a substantial sum from the owner, the fabricator and the consultants who inspected the fabrication
Bridge	(a) Structural failure of bearing seats (b) Excessive deflection	The designer failed to take into consideration how the structure would react to temperature changes: ignorance of local conditions is not a defence against a claim of professional malfeasance It was mentioned that in order to fund remedial action it was necessary to impose a local tax on motorists. When coupled with the inconvenience of the repair work this did not help in selecting an unbiased jury
Septic system	Failure to function	The subsoil did not allow soak-away. Lack of adequate information provided by the soils engineer was behind the problem
Hospital	Excessive settlement of foundations	A deficient soils report had been relied upon by the structural engineer and contractor. Costs of $400 000 involved
Sludge digester tanks/sewage treatment plant	Cracking and spalling of corbel supports for floating roofs of tanks	Repair costs of $129 000. Some dispute as to the information offered to the designer

Table 1.1 *continued*

Facility	Problem	Comments
Industrial building	Collapse of roof during heavy rainstorm	Losses of $3 million incurred following a storm of such severity that according to the local weather history such an occurrence would be likely once every 50 to 300 years. Insurer sued for repair costs (owner did not join in law suit) against architect and contractor who pleaded 'Act of God' Settlement of $500 000 reached with the architect contributing $350 000 and the contractor $150 000
Grandstand	Collapse of roof	Structural engineer failed to evaluate stresses that would be imposed. Proven damages for repair exceeded $350 000
Warehouse	Excessive flexibility and cracking of framed first floor	Designer had considered only the static loads and omitted the concentrated (movable) load of 5000 lb weight from fork lift trucks. Demand for $94 000 damages settled by arbitration at $83 000

Boiler explosions dominated disasters in the early days of the Industrial Revolution. Between 1816 and 1848 over 233 steam boats used in America suffered boiler explosions resulting in 2560 deaths [6, 7]. These early failures prompted the enactment of pressure vessel codes such as ASME III and BS 5500.

Dusts have been responsible for many explosions. Structural integrity itself cannot be monitored to anticipate such situations but the actual dust concentration can be monitored. A grain elevator explosion at Galveston Wharves, USA in 1977, devastated a wide area and led to the introduction of stringent scales by the US Occupational Safety and Health Administration (OSHA).

In the UK the Flixborough disaster of 1974 arose from the use of a temporary by-pass pipe of inadequate design; this led directly to a massive explosion and the deaths of 28 employees.

Perusal of the reports of Courts of Inquiry, insurance records, Government records world-wide, reveals that structural accidents continue to occur – and while they decrease in frequency, those failures which do occur, increase in severity.

As a consequence of this increasing severity, the question arises 'What is a satisfactory accident rate?'. In an ideal world the answer to such a question must obviously be 'none', but experience and such quantities as actuarial figures advise that while accidents do occur, there are rates of occurrence which can be considered as 'acceptable'.

1.18 Accident rate philosophy

To apply the techniques described in this book it must be recognized that accident prevention/structural integrity monitoring is an essential function of modern design and management.

The standards and concepts for accident rate philosophies have been established by such high-risk industries as the nuclear, chemical, power and aviation industries.

In the aviation industry accidents due to the failure of the aircraft structure have seldom exceeded 1% or 2% of the total fatal accidents in terms of aircraft flying hours. Before World War II the structural accident rate of acceptable military aircraft was in the region of 1 per 10^7 flying hours. Under war conditions for some aircraft this rose to about 5 in 10^7 hours, a rate which pilots and crews began to regard as dangerous. Changes to aircraft structure were demanded by aircrew even though the risk of death due to other engineering causes or to their own errors was much greater and risk of death due to military action even greater still.

Civil airline experience, mainly post-war, has confirmed this tendency for personnel to react against structural failure very strongly as compared with other causes of death. In airline operation it is believed that the structural accident rate should not exceed 1 in 10^7 hours. Thus for every 100 aircraft built, assuming a life for each of 20 000 flying hours, if even one of these aircraft fails due to some structural collapse, the structural accident rate will be 5 in 10^7 hours and so likely to be regarded as excessive. During World War II the number of aircraft constructed to one design could well have been 1000 and the life of each (due mainly to military action) only about 1000 hours. If only one of these 1000 aircraft failed structurally, the structural accident rate could have appeared as 10 in 10^7 hours – quite unacceptable for military pilots. Thus for aeroplanes, structural safety must be expressed in terms of an accident rate for the type in flying hours.

1.19 Risk analysis: repeated loads

Structural failure as a consequence of load repetitions has long been a recognized hazard. In mechanical engineering, from the days of Wohler onwards, materials (mostly steels) have been selected which have definite endurance limits. Such endurance limits are expressed as ranges of stress below which no fatigue failures can occur. Until about 1935 metals used in aircraft had fatigue endurance limits (or their equivalent at, say, 10^6 cycles) that were a substantial fraction of their ultimate strength; yet the flying lives of those aeroplanes were usually less than 1000 hours, very seldom more than 5000 hours. Towards the end of World War II some bombers achieved over 1500 flying hours; some of these had wing spans and joints of newer high strength aluminium alloys with 'endurance limits' at the level of earlier alloys and so corresponding to unusually low fractions of their ultimate strengths. A number of fatigue cracks and some complete collapses resulted.

References

1 Trainor, P. G. S., Popplewell, N., Shah, A. H. and Pinkney, R. B. (1983), 'State and dynamic behaviour of mechanical components associated with electrical transmission lines', *The Shock & Vibration Dig.* **15**, No. 6, April, 29: Shock & Vibration Information Centre, Code 5804, Naval Research Laboratory, Washington DC 20375, USA.
2 Halliday, R. (1983), *Corrosion Prevention & Control.* **30**, No. 1, February, 1.
3 Miller, H. (1978), 'Six die as poison gas cloud spills over US Highway', *Daily Telegraph*, 27 February.
4 Steinman, D. B. and Watson, S. R. (1957), *Bridges and their Builders*, Dover, New York.
5 Bannister, J. E. (1976), 'Earthquakes', *Br. Engng Bull.*, **10**, No.1, 10–13.
6 Collier, J. G. (1983), 'Reliability problem of leak transfer equipment' *Atom*, 322 August, 172–8.
7 Institution of Chemical Engineers *The Assessment of Major Hazards*, EFCE Event No. 272, Symposium Series No. 71, Pergamon Press, Oxford.
8 Pequignot, C. A. (1963), *Tunnels and Tunnelling*, Hutchinson, London.
9 Campion, F. E. (1951), 'Part reconstruction of Bo-Peep Tunnel at St Leonards-on-Sea', *J. Inst. Civil. Eng.*, **36**, 52.
10 Walker, T. A. (1891), *The Severn Tunnel: Its Construction and Difficulties*, Bentley, London.
11 Scott, P. A. and Campbell, J. I. (1954), 'Woodhead New Tunnel: construction of three-mile main double-line railway tunnel', *Proc. Inst. Civil Eng.*, **3**, (Part 1), 506.
12 Smith, N. (1982), *A History of Dams*, Peter Davies, London.
13 Le Patner, B. B. and Johnson, S. M. (1982), *Structural and Foundation Failures*, McGraw-Hill, New York.

Further reading

Acker, H. G. (1953), 'Review of welded ship failures', Ship structure Committee Report, Serial No. SSC-63, National Academy of Sciences, National Research Council, Washington DC, 15 Dec.
——(1967), 'Collapse of US 35 Highway Bridge, Point Pleasant, West Virginia', NTSB Report No. NTSB-HAR-71-1, 15 Dec.
Blake, M. G. (1959), *Roman Construction in Italy*, Carnegie Institution Publication No. 616, Washington.
Diesendorf, W. (1962), 'The Snowy Mountains Scheme', Horowitz Publications, Sydney, Australia.
Guthrie Brown (1964), 'Discussion on dam disasters', *Proc. Inst. Civil Eng.*, **27**.
Hartley, H. A. (1956), *Famous Bridges and Tunnels of the World*, Muller, London.

Jaeger, C. (1963), 'The Malpasset Report', *Water Power*, **15.**

Lane, E. W. (1935), 'Security from under-seepage masonry dams on earth foundations', *Trans. Am. Soc. Civil Engrs*, **100.**

Madison, R. B. and Irwin, G. R. (1971), 'Fracture analysis of King's Bridge, Melbourne', *J. Struct. Div. ASCE* 97, No. ST9, Sept.

——(1958), *The Work in Compressed Air Special Regulations*, HMSO, London.

Rolt, L. T. C. (1955), *Red For Danger: a History of Railway Accidents and Railway Safety Precautions*, Bodley Head, London.

Rothe, J. P. (1968), 'Fill a dam, start an earthquake', *New Scientist*, **39**, No. 605, 11 July, 75–8.

Sloan, C. L. (1957), 'Cofferdam problems plague Harvey Tunnel builders', *Civil Engng, NY*, **29,** 870.

—— (1954), 'Ministry of Transport Accident Report: Watford, February 3, 1954, British Railways/London Midland Region', *Railway Gazette*, **101,** July–Dec., 246.

Wade, L. A. B. (1908–9), 'Concrete and masonry dam construction in New South Wales', *Min. Proc. ICE*, **CLXXVIII.**

—— (1964), *World Register of Large Dams*, ICOLD (International Commission on Large Dams) 4 Vols, Paris.

2
Integrity assurance techniques

2.1 Introduction

Structures are more likely to collapse as a consequence of gross failure or through fatigue, than as a result of wear. For this reason, structural integrity monitoring is more likely to use physical effects than the chemical phenomena associated with wear monitoring. It is just possible that wear-out failures may dominate some structures and for this reason (and in anticipation of other, possibly less usual, failure modes) this chapter seeks to outline the whole range of monitoring techniques.

2.2 Monitoring

As a principle this involves: (a) regular and consistent measurement of a parameter; (b) recording and comparing changes against an as-new datum reference.

Methods of measurement can be:

(1) Semi-skilled – recording of a temperature, pressure or overall vibration;
(2) Skilled – frequency analysis of vibrations, logging of performance data, visual inspection, strain gauge measurement;
(3) Highly skilled – pattern recognition, ultrasonics, ferrography, holography, acoustic emission.

The general effect is that the simpler (and less skilled) the technique, the cruder must the data be and accordingly the less sensitive to change: the less skilled the techniques, the less will be the 'lead-time' warning of developing disaster.

2.3 Primary and secondary effects

It is possible to assess the various techniques into either:

(1) Monitoring of the primary purpose – thus a pipeline is required to convey a fluid without leaks and without loss of pressure; hence for this, leak monitoring and differential pressure would be primary monitoring techniques.
(2) Secondary monitoring – whereby consequences from the use of the structure or machine set up phenomena which are secondary to its direct purpose. For example, stresses set up in a pipeline are secondary (a consequence of pressure), and thus stress monitoring by strain gauges or acoustic emission might be regarded as secondary techniques.
(3) Tertiary monitoring – as for example, inspection for cracks or wastage.

With machinery, performance monitoring involves the measurement of many parameters, evaluated by computer and (on the basis of theoretical assessments) made comparable with various defect states. A typical installation on the Alaskan Oil Pipeline cost over $\frac{1}{2}$m for the monitoring of turbines and pumps. Measurement of flow parameters, Reynolds number etc., could be incorporated easily to monitor actual pipe integrity.

2.4 Bridges: temperature and movement instrumentation

Thermal expansion is significant with bridges and causes movement which (if restricted) can cause damage. Seven bridges were instrumented by the UK Transport and Road Research Laboratory [1] with a view to determining their responses to the thermal environment, specifically to measure:

(1) The distribution of temperature throughout the cross-section of the deck;
(2) The temperature of the surrounding air;
(3) The movement of the deck;
(4) The flow of heat through the deck;
(5) The amount of solar radiation incident on the deck.

The instrumentation used on each of the bridges was as follows:

2.4.1 The Medway Bridge [2]

Fig. 2.1 shows the layout of instruments used on this bridge. Transducers were fixed across the expansion gaps of the bridge using either mountings on wooden blocks or brass tubes (Fig. 2.2). The wooden blocks were made from laminated wood and were waterproofed to eliminate spurious movement due to the warping of the wood. On all sites, the transducers were located at the position of the

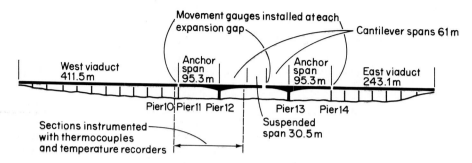

Figure 2.1 *Medway Bridge: side elevation showing position of instrumentation.*

neutral axis of the bridge wherever possible. Where this was not possible a correction was made to the measurements recorded to compensate for this.

Thermocouples were distributed sparsely because the details of the instrumentation had not been completed before the construction of the bridge was started. This also meant that most of the thermocouples in the viaduct section had to be placed by drilling and cementing. In addition, two thermocouples were suspended beneath the structure to measure the shade temperature. As well as these thermocouples suspended beneath the bridge, the shade temperature was also recorded on a thermohygrograph inside a Stevenson screen, but these records were unsatisfactory.

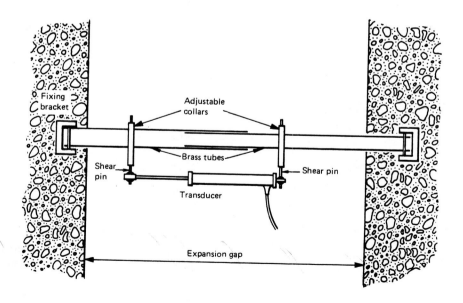

Figure 2.2 *Medway Bridge: movement transducer.*

2.4.2 Hammersmith Flyover [3]

Twenty-eight thermocouples were installed in the cross-section in the positions shown in Fig. 2.3 and one further thermocouple was suspended beneath the structure to measure shade temperature. As at the Medway Bridge, shade temperature was also recorded by a thermohygrograph in a Stevenson screen, but this was eventually discarded. All the thermocouple leads were brought to the inside of the middle box section where they were connected to a potentiometric recorder.

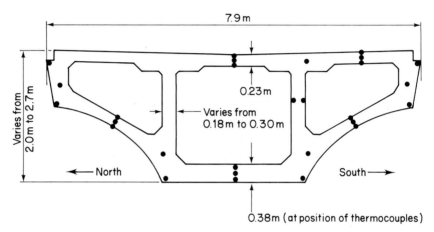

Figure 2.3 *Hammersmith Flyover: distribution of thermocouples in a beam segment.*

2.4.3 The Beachley Viaduct/Wye Bridge

Thirteen thermocouples were attached to the steel throughout the cross-section of the bridge near the expansion joint at the positions shown in Fig. 2.4. One thermocouple was suspended beneath the bridge to measure shade temperature. Shade temperature was also measured using a thermohygrograph inside a Stevenson screen, but, as at previous sites, the records from this were discarded.

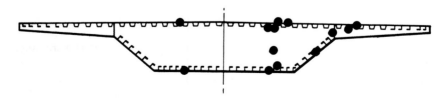

Figure 2.4 *Beachley Viaduct/Wye Bridge: thermocouple positions.*

2.4.4 The Coldra Viaduct [4]

Eight thermocouples were cast into one of the beams and three into the road slab in the positions shown in Fig. 2.5. The movement of the structure was measured by means of two 100 mm transducers mounted on wooden blocks connected across the expansion gap on the north and south sides of the east

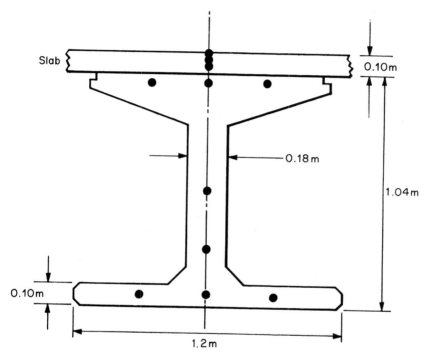

Figure 2.5 *Coldra Viaduct: thermocouple positions in beam and stab.*

abutment. The thermocouple and transducer cables were run along inside the cellular structure to one of the columns where they were led through the ducts down to a site hut and connected to a potentiometric recorder.

2.4.5 The Mancunian Way [5]

The positions of the thermocouples are shown in Fig. 2.6. The leads were taken through a duct and run inside the hollow box section of the spine beam to the west abutment. The movement of the western part of the structure was measured using two 203 mm transducers situated inside the west abutment. One was mounted between the top of one of the southern columns and the bottom of the spine; the other was mounted between the north wall of the abutment and the inner face of the anchorage block at the position of the neutral axis. Four heat flow meters were installed in the structure.

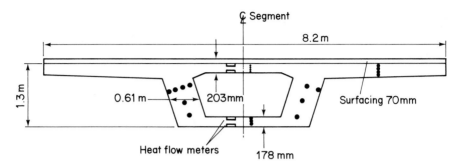

Figure 2.6 *Mancunian Way: thermocouple and heat flow meter positions in a typical two-lane box section.*

2.4.6 Adur Bridge – main viaduct

Twenty thermocouples and two heat flow meters were installed in the eastern part of the main viaduct at the positions shown in Fig. 2.7. Two more thermocouples were suspended beneath the bridge to measure the shade temperature.

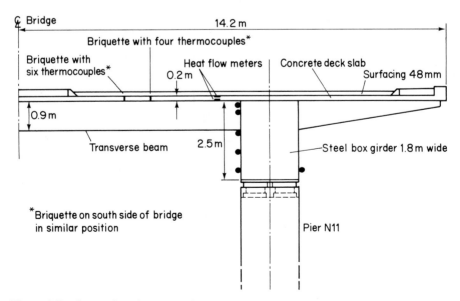

Figure 2.7 *Adur Bridge: thermocouple and heat flow meter positions (cross-section of main viaduct).*

Two 203 mm transducers were fitted, using wooden block type mountings, between the base of each of the steel box girders and the east abutment, and parallel to a straight line drawn between the fixed point and the expansion joint. They are not at the position of the neutral axis. The extension leads from the transducers were cast into the deck slab from the expansion joint to the top of pier N11. The leads from the thermocouples and heat flow meters were also taken to the top of this pier. From here all the leads are ducted down the pier to a site hut.

2.4.7 Adur Bridge – slip road

There are 36 thermocouples cast into the slip road and one placed at the top of column NE13 to measure shade temperatures. The positions of the thermo-couples in the cross-section are shown in Fig. 2.8. The cables were taken through a duct down column NE13 to a site hut. Two heat flow meters were cast into the slip road.

Two transducers were installed between the slip road and the column beneath the main viaduct end expansion gap. One was located at the bottom of the spine beam and the other was fixed to the underside of the north cantilever as near as possible to the position of the neutral axis. These, like the transducers on the main viaducts, were mounted on wooden blocks. Both transducers were fixed in line with the fixed point and the expansion joint. The curvature of the slip road is such that the transducers were offset about 45° from the direction of the slip road at the expansion joint.

Experience for more than 9 years shows that copper–Constantan thermo-couples and linear resistance potentiometers are effective and reliable; they provide valuable results [6, 7, 8].

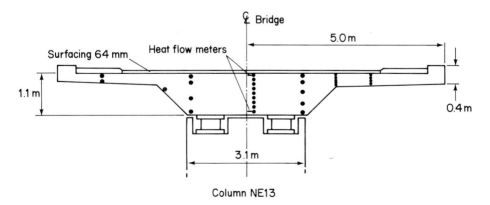

Figure 2.8 *Adur Bridge (slip road): thermocouple and heat flow meter positions.*

2.5 Primary, secondary and tertiary techniques

The measuring abilities of monitoring methods are related to the phenomena which they are sensing. They can be considered in terms of:

Primary techniques – which sense parameters related to the primary purpose for which the monitor is needed. Thus if a pipeline is involved, the primary purpose would be: (a) to prevent leaks; (b) to encourage pumping by reducing blockage or fouling (Fig. 2. 9).

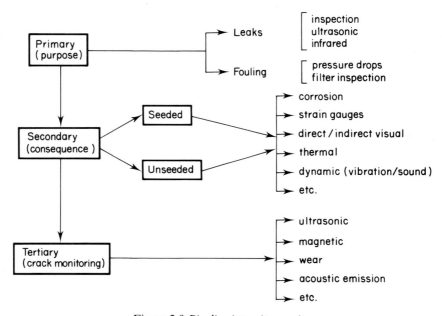

Figure 2.9 *Pipeline integrity monitors.*

Secondary techniques – use signals which are generated as a consequence of a primary fault and are related to the failure mode.

Tertiary techniques – directly measure one significant failure parameter related to the secondary failure. Thus for leakage from a pipeline the size of the crack may be the most relevant tertiary technique.

2.6 Discriminatory sensitivity

The complex factors leading to the loss of strength in a building structure are indicated in Fig. 2.10. Even the factors relating to wear and its physical effects are complicated, as shown in Fig. 2.11.

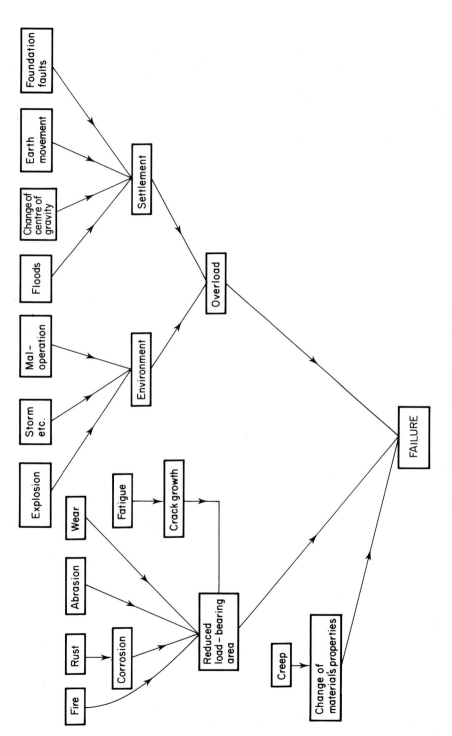

Figure 2.10 *Sources of failure in a building structure.*

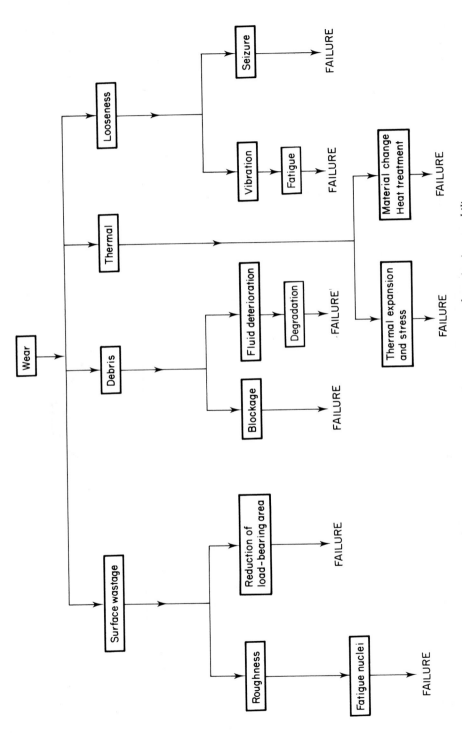

Figure 2.11 *Wear: its physical consequences and monitoring capabilities.*

In selecting the most appropriate form of integrity monitoring it would be appropriate initially to select one technique for advanced, early warning and revert to another technique when deterioration has advanced. Some idea of the discriminatory sensitivity of various methods throughout the life of an offshore rig are shown in Fig. 2.12 and the sensitivity of some structural techniques in Fig 2.13.

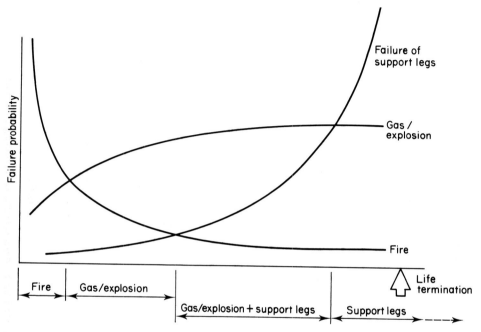

Figure 2.12 *Significance of different failure modes and monitoring methods throughout whole life: offshore production platform.*

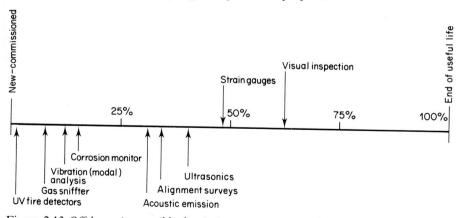

Figure 2.13 *Offshore rig: possible discriminatory sensitivity of various sensors. Note that the mode under surveillance may change in significance.*

Thus the executive concerned with the whole-life integrity assurance of such a structure should consider the needs for providing for the installation of all these diagnostic facilities when the rig is first built. Some of these facilities will have their beneficial sensitivities only during the critical period towards the end of the rig's working life when hazards will be at their greatest.

2.7 Monitoring methods: some techniques

A brief summary of other techniques, without comment or appraisal, is as follows (alphabetical order):

Accelerometers – piezoelectric devices used to measure vibration, recording instantaneous vibrational acceleration.

Airborne pollution monitoring – infrared cameras mounted for sideways photography based on a system developed by Svenska Rymbaoleget, Stockholm.

Anisotropic etched silicon crystals, with integral transducer circuitry – used for highly sensitive pressure sensors [9,10].

Borescope – a rigid viewing system of lenses (originally used for the internal inspection of the bores of guns).

Bronchofibrescope – a fibrescope, originally designed for medical operations, which includes manipulators which can use small tools, scrapers, pliers. It has four ways of angulation and some upward/downward, left/right orientation.

Catalytic combustion analyser – incorporates Pellement detectors and a balanced four-element bridge to measure air pollutants.

Charge amplifiers – used in conjunction with sensors based on the accumulation of electric charge.

Corrosion monitoring – see detailed description.

Cyclonic debris monitor – precipitates debris particles by cyclonic action with a means of measuring quantity and size [11].

Data couplers – such as the Acurex system [12], to communicate via radio transmission from strain gauges on a rotating shaft.

Depletion rates – coolant inhibitors. Regular testing to BS 3150 to monitor the amount of Na MBT (sodium mercaptobenzothiazole) present to check the effectiveness of antifreeze in engine coolant systems.

DEPOTMAIDS – a technique used by the US Army, Latterkenny Depot, Chambersing, Pa, in conjunction with a dynamonitor test cell to evaluate defects in tank engines.

Dust analysis: infrared spectroscopy – which uses the refraction of infrared light to identify and quantify a sample of dust [13].

Dust monitor: quartz crystal microbalance – captures particles and assesses concentrations and accumulations by sensing the impact on a retentive target [14].

Electron microscopes – from specially formed replicas of materials it is possible to examine surface structures by means of the high magnifications (up to $2000 \times$) achievable by this type of microscope [15].

Explosives detector – functions by detecting the presence of vapours given off when explosive materials are sublimating.

Fatigue load meter – an instrument which records the number of stress or load excursions beyond the normal working limit and cumulatively states the amount of 'safe life' remaining in the structure.

Fibrescopes – an optical system of inspection whereby light is reflected internally along the inside of a coated glass fibre. Can see round bends.

Flammable/toxic gas detector – uses a silicon semiconductor gas sensor to detect and measure hydrogen sulphide, chlorinated organics, silanes, silicone vapours, halogenated compounds, leaded gasoline vapours [16].

Flare gas monitor – automatic and continuous monitoring of O_2 levels in flare stacks to prevent the dangerous conditons which can be set up if air enters at the base of a stack or through any vessel connected to a flare stack system [17].

Flash-point – In the UK the Pensky–Martens apparatus is used with oils to evaluate fuel contamination as a monitor of combustion.

Frequency-modulated telemetry – a means of providing contactless communication between sensors on a rotating shaft and the surrounding signal processor [18].

H_2S gas monitoring – a diffusion adsorption sensor in which conductivity change is proportional to the hydrogen sulphide concentration. H_2S is a colourless gas, smelling of rotten eggs, which can anaesthetize a person's smell and is lethal at quite low concentrations.

High-speed motion photography – ultra-high speed photographs when projected at slow speed reveal faults in mechanisms.

Image intensifiers – an electro-optical intensification of the contrasts between the greys of viewed surfaces. This facilitates the identification of cracks.

Infrared optic thermal monitors – whereby a photon detector can be buried in the structure and thermal effects transmitted to the electronic console through optical fibres [19].

Infrared pollution monitor – uses two similar Nichrome filaments as the source of infrared radiation; one beam passes through a reference cell, the other through polluted air. Produces a signal measuring the concentration of selected pollutants such as CO, CO_2, NO_x, SO_2.

Kurtosis meter – a method of evaluating bearing damage by statistical appraisal of the roughness of the action of the bearing.

Load cells – whereby compressive effects are measured by means of a strain gauge incorporated in the cell and integrated to provide a direct force measurement.

Microprocessor-based coordinate inspection – a survey theodolite associated with a microprocessor which evaluates coordinates from multiple locations in terms of a reference plane. Used underground and above ground.

Particle sizing – HIAC. Uses the principle of light obscuration to measure the concentration of particles in a fluid (air, liquid, gas).

Patch tests – a simple technique for the visual evaluation of contamination levels in fluid systems (including bacteriological contamination) and can be applied to ASTM Method D2276, MIL-H-5606, MIL-H-6083, [20].

Peak strain indicator – as developed by the British Hovercraft Corporation, a strain transducer is incorporated in their TM 14A peak strain indicator which displays a steady maximum reading until reset.

pH measurement – electrodes, such as the Beckmann range, to measure the acidity/alkalinity of wetted systems in relation to the possible corrosive activity.

Piston-ring micro-seizure monitor – a surface sensor developed by Burmeister and Wain to measure microscopic surface seizure between piston rings and cylinder walls.

Pressure-decay leak detector – monitors the rate of change of pressure with time $(\mathrm{d}p/\mathrm{d}t)$ when a vessel is subjected to pressure [21].

Radio-link telemeters – miniature radio transmitters coupled to strain gauges, thermocouples and vibration transducers to transmit signals from rotating shafts to external recorders.

Romeo 'Pix' smoke detection – based on the principle of light obscuration with calibrations in terms of Ringelmann numbers [22].

Routine lubricating oil analysis – a 'used oil' analysis programme which is able to:
(a) detect abnormal wear;
(b) ascertain the causes of such abnormal wear;
(c) determine lubricant serviceability and oil change periods.

Satellite telemetry – signal communication through the use of industrial satellites [23].

Seismic pick-up – a vibration transducer which produces a signal proportional to instantaneous vibratory velocity.

Shock pulse monitoring – a technique of special value to the monitoring of rolling element bearings by recording the shock pulses set up by defective bearings.

Signal amplifier – a device incorporated in electronic circuits (especially piezoelectric vibration transducers) to amplify the measured signal.

Slenderscope – a very slender borescope which can even be used to inspect the internal profile of welds on small bore hydraulic tubing.

SPATE and TSA – *stress pattern analysis by thermal emission*, and *thermoelastic stress analysis* are non-contacting methods of surface stress determination developed for the British Admiralty Surface Weapons Establishment, Portsmouth by the Scientific Instruments Research Association (SIRA), Chislehurst, Kent, BR7 5EH, England.

SPECTRASPAN – a spectrometer incorporating a microprocessor for the rapid analysis and evaluation of wear debris.

Standard time division multiplexing – a system using CMOS analogue switches and LSI logic to provide a multi-channel shaft telemetry system [24].

Stereoscopic photography – a method of photography which when used with a viewer makes three-dimensional effects for better appraisal [25].

Strain indicators – a wide range of specially developed measuring devices [26] such as the Whittlemore strain gauge, the Berry gauge, the Olsen gauge, the Porter–Lipp strain gauge, the Huggenberger strain gauge, the Marten-mirror extensometer, the Tuckerman optical strain gauge, the Vose interferometer strain gauge, and the pneumatic extensometer.

Strippable magnetic film – a self-curing rubber with a dispersion of iron oxide particles which forms a replica of the original surface; used for crack monitoring [27].

Telephone telemetry – the use of the public service as by the 'Teltel' system [28] to transmit signals automatically from remote sensors to a central control position.

Temperature indicators – phosphor compounds available as crayons, paints or labels [29] which when applied to a surface change colour if the temperature exceeds that for which the indicator was designed. Originally a Rolls-Royce patent, No. 1 029 605.

TEMPLUG temperature sensors – a range of special alloys whose hardnesses alter at defined temperatures [30].

Thermographic scanning – heat pictures made of surfaces. Used to indicate thermal barriers and applied to the detection of faulty insulation, defects in reinforced concrete, failure of switchgear, leakage from joints, etc.

Thin-layer activation – developed by the UK Atomic Energy Authority, Harwell to measure changes in surface layers due to erosion or corrosion.

Torque transducer – based on strain gauge networks to evaluate surface shear stress and thus deduce the applied torque.

Total acid number (TAN) – measures the total acids present in an oil (preferably less than 2·5 mg KOH/g) and indicates proneness to sludge and lacquer formation.

Total base number (TBN) – indicates an oil's ability to neutralize acids caused by a high sulphur content in diesel fuels.

Tourmaline underwater pressure transducers – used to sense pressure variation under waves. Developed by the AWRE Foulness, Essex using sandwiches of piezoelectric tourmaline crystals [31].

Ultrasonic camera – the Polaroid sonic focusing system [32] incorporates both a transmitter and a receiver.

Ultrasonic diffraction – a technique developed by the UK Atomic Energy Authority, Harwell to detect subsurface flaws in structures even with the surface painted.

Ultrasonic goniometry – measures surface stresses on the basis of changing ultrasonic velocities at a critical angle [33].

Underwater CCTV – with appropriate lighting, reveals gross faults to depths of possibly 120 m (400 ft) [34].

Vibrating wire (acoustic) strain gauges – originated in 1958 and applied extensively from 1968, these are buried in concrete structures and function on the basis of the changing frequency of the wire when stretched due to movement of the structure [35].

Vibration frequency analysers – resolve the 'spectra' of a vibration into its intensities at different frequencies.

Vibration meter – an accelerometer or a seismic transducer integral with a meter which then measures the 'average' vibration.

Viscosity measurement – with engines, lubricating oil viscosity may increase due to oxidation and/or carbon from piston-ring blow-by, or it may decrease due to fuel dilution owing to combustion defects.

Water leak detectors – uses microphones and headphones to listen to the sounds of leaking water [36].

Waveform analyser – records the quasi-sinusoidal form of a vibration, usually by means of an oscilloscope or recorder, or by means of a mechanical pen.

Xeroradiography – in which selenium-coated aluminium plates provide 'dry' photographs sensitized by an electrostatic charge.

X-ray fluorescence wear monitor – uses the energy of radiation from X-ray irradiated liquid to measure ultra-low concentrations of impurity.

X-ray radiography – based on the penetration of X-rays through optically opaque materials. Penetration influenced by internal cavities and faults.

2.8 Wear monitoring

Techniques have been developed to monitor the rate of wear in lubricated systems. The wearing components can be identified through the composition of the wear particles and the wear rate can be measured by quantifying the amounts of such chemicals present. Such monitoring techniques include the following (all in active use, particularly where machinery is involved).

2.8.1 Magnetic plug inspection

A magnetic chip is inserted in a plug which is screwed into the pipe system such that the magnetic chip attracts ferrous and other debris. The shape of the debris is examined under a microscope or magnifying glass and its quantity can be determined by the use of a debris tester such as the type DTI.

2.8.2 Spectrometric analysis

The chemical constituents of a contaminated fluid can be identified from their luminious spectrum. A sample is burnt in either a gas flame (atomic emission) or in a high voltage electric arc (electronic emission) and the 'rainbow' produced by light passing through a prism identifies the constituents present – these may be ions from wear products, silicon from the ingress of dust, sodium from salt-water ingress, etc.

2.8.3 Chromatography

This is a physical technique whereby contaminations are separated as a consequence of their size in relation to the interstices of the substance in which they are placed; typically a 'blotter' test uses the spread of sample from the blot on the surface of an absorbent paper such that the contaminants separate out in a series of rings.

2.8.4 Ferrography

In this technique a sample of fluid is pumped to flow down a glass slide and comes under the influence of a progressively increasing magnetic field. This separates particles according to their size; they can then be assessed in terms of their quantity. The source of these particles is determined from their coloration when viewed under a bichromatic microscope.

2.8.5 X-ray fluorescence analysis

A technique of very high detection sensitivity used to identify very small traces of elements. A specimen is exposed to a source of radiation such that contaminants in the oil will emit X-rays. The rate of X-ray emission and its spectrum identify the type and quantity of pollutant present.

2.8.6 Photoluminescence

A method used to identify structural changes in a fluid such as those due to electrolysis from corrosive activity. The principle is based on ultraviolet absorptiometry.

2.8.7 Ultraviolet absorptiometry

This is used to determine changes in basic fluids such as straight and additive lubricating oils as a result of oxidation; the principle involves irradiation of a sample with UV light and the measurement of the radiations given off by the sample.

2.8.8 Scanning electron microscopy

Frequently used in the examination of contamination products, this technique reveals the shape of significant constituents from which the deterioration can be assessed.

2.8.9 Capacitance oil debris monitor

In this method the quantity of debris which has accumulated and the rate at which it is being produced are measured by collecting the particles in the annular space between the plates of a cylindrical capacitor: total values and changes of capacitance provide analogues for quantity and rate of debris.

2.8.10 Other techniques

For special purposes, methods which may be used (and are mostly refinements of the foregoing) include: mass spectrometry; induced radioactive seeding; electron probe microanalysis; electron spin resonance spectroscopy.

2.9 Temperature monitoring: pyrometry, infrared thermography

Such a very wide range of applications exists for modern sensitive thermographic techniques that it is possible to single out only one or two special applications. Infrared methods can be used to deal with ultra-low temperature effects, e.g. leakage from a cold store; or they can be applied with equal effectiveness in furnaces.

Thermographic surveys to detect lining deterioration in torpedo ladles in steel mills enabled information to be obtained which extended their use. One particular ladle's life was extended from 72 000 to 143 000 tonnes with an estimated saving in bricking costs alone of the order of £2000 per annum for every 1% increase in the average lining life [37].

By examining black and white thermograms it is possible to identify dirt-precipitated and congested pipes. A built-up composite grey thermogram of a gas main to a blast furnace showed cold dark areas in the heat picture which represented a build-up of dust. This blocking-up of the gas mains severely jeopardized production; there was also the added hazard that the gas main might have collapsed under the weight of dust with obvious disastrous consequences.

2.10 Temperature measurement: infrared

A hand-held infrared thermometer by the name of 'Heat Spy' [38] reacts to thermal radiation to provide a digital read-out. Operation consists of simply pointing the thermometer at a distant object and pulling the trigger.

Operation of this thermometer is based on the fact that infrared energy is emitted by all solid objects above absolute zero. The amount of energy emitted is proportional to the body or target temperature. By means of fixed focus optics Heat Spy collects this energy onto a sensitive detector and reads out directly in degrees Celsius. Operation is fast because IR energy travels at the speed of light, and the detector has a very low mass. The time constant is 0·1 second, about ten times faster than conventional contact methods. Measurements are displayed in less than 1 second. Heat Spy is the fastest thermometer available. The latest technology is utilized: a vacuum-deposited ultra-stable sensor with advanced CMOS electronics, precise digital readouts and linearized output, allowing use with direct reading percentage recorders.

Purposes for which this thermometer has been used include a wide range of predictive maintenance measurements and such specific objectives as:

- Measuring printing web/drier temperature during set-up to adjust quickly for paper weight, coverage and moisture.
- Instantly checking die face temperature without stopping production.
- Measuring the temperature of moving hot rubber sheet being rolled and cooled.
- Monitoring the temperature of compressor valve head in machinery preventive maintenance programme.
- Temperature measuring of most surfaces up to 600 °C, including glass and infrared heated surfaces.
- Measuring continuously moving materials such as plastic film, extrusions, pulp and paper, textiles, rubber, coatings and painting.
- Checking rotating machinery, switchgear transformers for overheating.
- Preventive maintenance of steam traps and lines, insulation and boilers for energy conservation.
- Measuring temperatures up to 1700 °C with a spot size of 76 mm focused at 6 metres.
- Surveillance of distant surfaces in foundries, steel mills, furnace tubes, forging, welding and kilns.
- Measuring metal and ceramic parts through glass ports.
- Measuring objects as small as 2.5 mm diameter with target defined by light beam.
- Detecting hot spots of electronic components and switchgear.
- Measuring temperatures of small extrusions, moulds, laminations, seats, brazed joint tubing and V-belts.
- Identifying faulty or obsolete insulation – both thermal and electrical.

2.11 Temperature monitoring: other applications

2.11.1 Temperature/life integration

In an effort to monitor cyclic fatigue, particularly in relation to gas turbine monitoring, a value for the expired life of an engine can be obtained by accumulating data on the time spent at each elevated operating temperature, the integral being weighted to take account of the proportionally greater ageing effect of higher temperatures. Attempts are being made at several centres to measure blade temperature directly by optical pyrometry. Two-colour pyrometry may be more successful than brightness pyrometry which can be affected by attenuation caused by particulate matter.

2.11.2 Stockpile/tip temperature sensing probe

Undetected, uncontrolled combustion in coal stockpiles and colliery spoil heaps can cause dangerous and costly fires. Scientists at the National Coal Board's Laboratories [39] have designed a probe which enables the internal temperature of stockpiles and tips to be monitored. This consists of a robust, bullet-shaped heat-conducting shell containing a silicon diode or thermistor.

In use, the probe is inserted into the stockpile or tip on the end of a hollow pipe. Electrical leads attached to the probe pass through the pipe to a meter calibrated to read directly the temperature of the probe, and compensated for variations in ambient temperature. The probe may either be withdrawn, using the pipe, after a measurement has been taken or left in position – the pipe being removed – so that the temperature can be checked regularly.

The probe has been used extensively in the NCB's North Nottinghamshire Area, both in stockpiles and in colliery spoil heaps, where it has proved a valuable aid to the control of combustion.

It could also be used in other fields – for example the storage of grain or chemicals.

2.11.3 Steam-trap failure

A steam trap which blows steam is not doing its proper job of clearing condensate without losing steam, and it can waste prodigious amounts of money and energy. (One continuously leaking half-inch trap can mean the loss of the order of £1000 a year in wasted energy.) Often trap failures are not detected. There are several methods of monitoring steam traps.

Stethoscopes may be used on steam systems but, as with listening to the human body, there may be a lot of extraneous noise, and an expert is required to interpret the sounds.

Ultrasonic leak detectors are better at pin-pointing noises. But again, expert interpretation of the evidence is required. There is only a slight difference between a trap functioning properly and one leaking slightly. A trap passing condensate at one pressure can produce the same 'read-out' as a leaking one at a different pressure.

A third method embraces surface pyrometers, temperature-sensitive crayons and the like. Measuring the inlet temperature of a mechanical trap blowing steam is unlikely to show any variation from a trap in good order. With a really accurate measuring device, and exact knowledge of the pressure immediately before the trap, it may be possible to detect leaking thermostatic traps.

Temperature sensing on the discharge side is no more useful. In most cases temperature is dictated by the pressure in the condensate line, and in a mixture of good and bad traps discharging into the same system all will indicate the same temperature. In effect, temperature sensing is practical only to identify serious water-logging – which veteran engineers can spot anyway by putting a cautious hand on the trap!

SPIRA-TEC uses vortex forces from induced spiral flow to enable people with no steam skills to check a trap positively without expensive or bulky equipment, delay, trouble or doubt [40]. A steady green light appears when the trap is functioning correctly. A continuous red one reveals some fault. This device can easily detect leaks as small as that produced by a bit of grit 12 thousandths of an inch (300 μm) thick in a valve orifice.

2.12 Ullage monitoring

Changes in the volumetric capacity of tanks holding volatile liquids may incur losses which can only be settled by ullage monitoring to assess the volumetric change (shrinkage). This is especially critical in connection with the metering of liquid fuels.

A portable shrinkage meter [41] is approximately 24 inches (610 mm) high and weighs around 50 lb (25 kg); it is simple to use and can give results within two hours of the sample being taken.

2.13 Hull stress monitor

A device has been developed by Mitsui Shipbuilding & Engineering Co. Ltd to monitor the stress created by waves on a ship's hull. It is of particular significance in preventing possible dangers on very large cargo-carrying vessels. The first unit has been installed on the 270 000 dwt tanker *Navigation Monitor* owned by Exxon, USA.

Hazards which a ship may encounter are monitored by a continuous watch on the hull strength under various weather and sea conditions. This is done by a sensor which is affixed to a specific position in the midship section of the hull, where the hull strength as well as degree of strain exerted on the hull are most critical and apparent [42]. Strain detected is converted into a strain value to represent the condition of the hull and is shown on a display unit in the wheelhouse. If any danger is indicated an alarm warns the crew in charge. The device also makes forecasts on possible dangers to the ship if she still continues to sail under the condition perceived by the sensor. The sensor incorporates an explosion-proof stress gauge and the display unit is equipped with a signal change unit, root-mean-square processor, meter, indicator and printer.

2.14 Pipeline pigs: maintenance and inspection vehicles

Crack monitors and detectors can be passed through pipelines using 'pigs' and 'spheres' of a type developed from those originated for pipeline descaling and cleaning [43]. Such pigs may be self-powered and position-controlled or they may be propelled by the pipeline fluid itself. Tracking of GD pigs is possible by means of signallers (Fig. 2.14) in which the pig or sphere actuates a flexible trigger which extends into the pipeline.

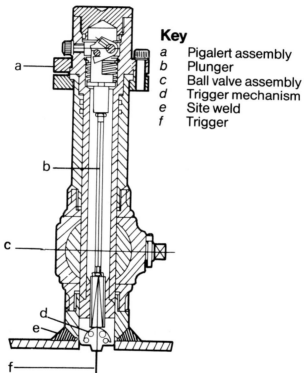

Key

a	Pigalert assembly
b	Plunger
c	Ball valve assembly
d	Trigger mechanism
e	Site weld
f	Trigger

Figure 2.14 *Type 'TB' Pigalert®. (Reproduced by permission of GD Engineering Ltd).*

2.15 Remote maintenance vehicle (RMV)

The RMV robot is designed to be used by Shell when producing oil in waters too deep for conventional production systems. The RMV robot will do away with the need for divers to maintain the production site.

When required, the RMV will be carried out into the North Sea by a semi-submersible vessel. A buoy mounted on the un-manned craft (UMC) is then released by remote control. This buoy floats to the surface trailing a line, which the RMV uses to winch itself down to the UMC. When it reaches the bottom the robot latches itself to a track that runs through the UMC. Using its 45 kW electric motor the 236 ft (72 m) long, 226 ft (69 m) high robot then propels itself along the track until it is opposite the valve that needs replacing. All the RMV's operations are controlled by an electrical umbilical cord running from the mother ship.

Shell's robot is equipped with two tools for maintenance tasks on the UMC. One is a large socket wrench used for removing and installing valves, the other a forklift with dual socket wrenches to replace control modules. Both tools have the ability to lift an object 6·84 ft (3 m) and have a 3·42 ft (1·5 m) reach. All the RMV's operations are monitored by television cameras, and spares are carried on board the robot in special racks. Once it has completed its task, the RMV releases itself from the track and floats back to the surface.

2.16 Condition monitoring in electrical power generating stations

The prevention of unscheduled down-time arising from failures in generating plant is being carried out by various organizations world-wide by the use of condition monitoring techniques. Plant monitoring may be either: (a) static non-destructive testing carried out when the plant is shut down; or (b) those carried out while the plant is running. A typical schedule of techniques is listed in Table 2.1.

2.17 Continuous multiparameter recorders

With the probable failure modes established there appears to be no technical reason why any structure should not be continuously monitored to assure its continuing integrity; with in-depth interpretation of the recent consumer and safety legislation there appears to be no reason why a public authority should not be called upon to prove the continuing integrity of its structures. A precedent exists with aircraft flight recorders; there are proposals to introduce ship's condition recorders.

From the early days of aviation when manuscript notes were made by the pilot or navigator, in-flight recording has developed to its present stage involving highly complex instrumentation systems, recording in some cases millions of discrete measurements throughout the period of a single flight.

Early records were used for the purposes of flight test development and more recently for airworthiness data which subsequently expanded into their use for performance monitoring and accident investigation. Legislation requiring the carrying of reliable crash-resistant data recorders on large civil aircraft came

Table 2.1 Some condition monitoring techniques used in electrical power stations

Plant type	Condition monitoring techniques
Rotating machinery in general	Vibration monitoring Performance analysis
Electrical rotating and stationary plant	Insulation condition
Steam turbines	Vibration monitoring Performance analysis
Gas turbines	Comprehensive performance trend analysis Spectrographic analysis of lubricating oils Acoustic and vibration monitoring
Alternators	Insulation conditions Leak monitoring from coolant system Vibration monitoring Performance analysis
Steam condensers	Leak monitoring
Boilers and boiler plant	Flame monitoring Mill fire detection Leak monitoring Smoke and flue gas composition monitoring
Reactors	Burst fuel element detection Graphite monitoring Boiler tube leak detection Corrosion monitoring
Transformers	Buchholz gas analysis Transformer oil monitoring Insulation monitoring
Transmission system	Detection and recording of fast transients Infrared detection of faulty joints

into force in the USA in August 1958, closely followed by similar requirements in France and Australia, and in 1965 in the UK [44].

Experience shows that flight recorders are valuable in the analysis of two very general accident categories:

(1) Where some form of aircraft or system malfunction has been the primary accident cause. This category may extend from, for example, a simple electrical or instrument system failure to the in-flight disintegration of the whole aircraft structure. In such cases wreckage analysis will probably play the principle role in attempting to establish the primary failure. The data recorder serves a very useful purpose in this type of accident, however, by supplying supporting data such as time, height, velocity and acceleration factors.

(2) Where the accident has an operational cause, that is, it is one in which no

engineering defect or deterioration in aircraft performance has occurred. Examples of this are certain handling accidents resulting in loss of control and those due to weather or navigational factors. In such an accident wreckage analysis may provide data concerning final configuration and broad details of impact attitude and speed. Beyond this the information derived from the wreckage will probably be largely negative. In this type of accident an accident data recorder may well play the predominant role by providing a time history of the principal manoeuvres throughout the flight.

The common parameters recorded are: pressure altitude, indicated airspeed, aircraft heading, normal acceleration. Other information which is mandatory in some countries is: pitch angle (UK), radio beacon interception data (France).

Experience which has been accumulated in the extraction and evaluation of accident data records suggests that more performance monitoring data as a mandatory requirement would assist prevention and improve flight safety. Protected 'voice recorders' have been used in Australia and the USA for a number of years, they have facilities for bulk erasure following an uneventful flight.

Three basic methods of recording are used in the present accident data recording systems. They are:

(1) Oscillographic engraving on metal foil. In this type of recorder either aluminium or high nickel content steel foil is used as the record medium on which styli engrave, in analogue form, either continuous or sampled traces representing each parameter.
(2) Photographic recording. With this method the displacement of light beams, via galvanometer-controlled mirrors, produces traces (also in analogue form) on photosensitized paper.
(3) Electromagnetic recording. In these systems both analogue and digital recording techniques are used. Of the former, frequency modulation methods are normally employed using plastic-based magnetic tape as the recording medium. The digital type of recording systems currently use pulse code modulation techniques to produce digital data on stainless steel wire. Developments in electromagnetic recording media include the use of stainless steel tape, primarily because of its good survival and data packing properties.

Most of the present types of recorder are of the fixed duration type involving the removal of the record and where required, erasure of the data at specified intervals; a few are of the continuous loop type where a constant amount of data is stored and continuous erasure of the 'first recorded' data takes place on the aircraft. The continuous loop types of recording systems have a facility for parallel recording of the data on a second non-protected recorder from which the recording medium may be readily removed for replay and storage. Means are provided on all types to ensure compliance with statutory data storage periods.

Electromagnetic recording systems generally include such subsystems as: sensors, transducers, signal conditioners, multiplexers, modulators, analogue-to-digital converters, stabilized power supply, recording stages.

With digital recorders the binary system of digital coding is used although recording techniques vary and differing formats are employed for word sequences and frame synchronization.

2.18 Integrity monitors: their own reliability – integrity testing

Apart from the need to ensure that only integrity monitors of the highest quality are used, it is invaluable to have a system of self-integrity testing incorporated in the monitor itself. Electronic circuits and systems used in computers can (and do) include means for self-integrity testing. Two different systems are available: built-in test equipment (BITE) and self-test.

BITE implies a periodic automatically-initiated test sequence to check to a high degree of confidence the equipment's operational capacity without hindering its normal function. BITE must generally be incorporated in the integral system design, possibly using microprocessors; it is rarely added-on.

Self-test is initiated and monitored by the user. It is an abnormal test configuration added to the system. In an analysis of computer-based ship automation systems the computing requirements relevant to the fault condition monitoring applications of a system are given in Table 2.2 [45].

Table 2.2

Application	Program size (16 bit words)	Program loading (add equivalents/s)
Machinery surveillance (diesel)	8850	17500
Performance and condition monitoring (diesel)	8050	7200
Ship motion stress analysis	1500	500
System self-test	600	500–1000

2.19 Electronic systems test: built-in test equipment

The correct functioning of marine electronic control equipment can be checked by means of BITE which identifies defective printed circuit modules so that they can be replaced by new modules. A typical BITE is produced by Hawker Siddeley Dynamics Engineering Ltd [46].

The system to be tested is connected to the BITE through an interface unit. This is essentially a routeing system such that the stimulus and measurement devices can be connected to the system under test. The operator controls and monitors the test sequence via the control and display panel. Control signals are fed to the central test sequence and signals are fed back to the display panel. The operator is informed by the display of the result of the test and, if required, the most probable corrective action.

The BITE system self-test facility may be exercised periodically to ensure that it is functional prior to the testing of a suspect system. The system to be tested is connected to the test system and identified by coding of the connectors. Three lines coded in binary are used for identification. The insertion of the connector mutes the system, taking it off-line.

2.20 Intrinsic safety

Equipment which is likely to be used in hazardous areas should incorporate protective devices. A hazardous area is defined as either:

(1) an area in which a gas/air mixture may be present; or
(2) an area in which a gas/air mixture has a degree of probability of being present such that an introduced electronic device may be a source of ignition.

Ignitable fibres and dusts may be included in such a definition.

Locations in which such hazards may exist (set out in detail by Code NFPA-70-1981 by the National Fire Protection Association, Quincy, MA 02269, USA) are:

(500–4) Class I   flammable gases or vapours in sufficient quantities (with air) to produce explosive or ignitable mixtures
(500–5) Class II   combustible dust may be present
(500–6) Class III ignitable fibres or flyings present but not suspended in air

In the UK three zones are defined by BS 5345 as follows:

Zone 0 explosive gas/air mixture is continuously present or present for long periods
Zone 1 explosive gas/air mixture likely to occur in normal operation
Zone 2 explosive gas/air mixture, if it occurs, will exist only for a short time

Apparatus to be used in such hazard zones must be proved incapable of generating such energy as to ignite the combustibles. Two categories of intrinsic safety recognized in Europe (BS 5501), are briefly:

Ex ia electrical apparatus incapable of causing ignition in normal operation with a single fault and with any combination of two faults
Ex ib electrical apparatus incapable of causing ignition with a single fault

To achieve such objectives safety circuits need to be incorporated into the electronics of all equipment. A Zener safety barrier is the most common device for this pupose. Limitation of both current and voltage are the objectives of this barrier, shown in Fig. 2.15, whereby the diodes act in a regulatory manner (BS 9000).

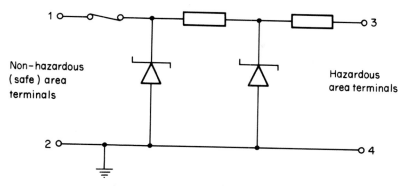

Figure 2.15 *The Zener safety barrier.*

2.21 Residual stresses

Stresses which remain in components following production processes may be so high that the total of operating and residual stress exceeds the material's strength. Account must be taken of residual stresses when examining failed components as a basis for redesign and as a criterion for the introduction of condition monitoring techniques. Knott expresses this point succinctly [47]: 'residual stresses remaining in metal plates or sections after forming operations give rise to one major uncertainty.'

2.21.1 Classification of residual stresses

A residual stress may be defined as 'a stress system within a material which is not subjected to load or temperature gradients yet remains in internal equilibrium'. Such stresses may be macroscopic or microscopic.

Macroscopic residual stresses are typical of production influences. Typically, they arise from:

● forced alignment of parts during assembly
● loads causing non-uniform plastic flow or creep
● metal forming
● non-uniform temperature changes large enough to cause plastic deformation or creep
● volume changes due to metallurgical or chemical processes.

Such effects most frequently arise as a consequence of defective heat treatment, welding, cold forming, casting and their control is an example of condition monitoring applied to production quality control.

Microscopic residual stresses are of limited location, as for example parts of a material where a homogeneous strain continuum is not possible. On the smallest scale, microresidual stress may arise from misfitted atoms and individual dislocations or local inhomogeneities. On a larger scale, residual stresses arise from dislocation build-ups, kink boundaries, deformation-induced tilt boundaries, deformation twins, etc. Such configurations are important in plastic deformation, crack initiation in brittle fracture, fatigue crack growth, and in the Bauschinger effect (an anisotropy of work hardening) [48,49]. Such materials experience a drag effect on the inhomogeneity during deformation; when this ceases, residual stresses are set up.

2.21.2 Residual stress measurement

Two methods are prominent in the techniques for measuring residual stress:

(1) Mechanical methods based on the measurement of strains observed during the removal of material.
(2) Observations of elastic strain in the crystal lattice determined by X-ray diffraction methods.

Mechanical methods consist of measuring changes in shape due to the removal of metal such that from a knowledge of the mechanics of deformation it is possible to calculate what the stresses must have been in the metal that has been removed. When a series of layers is removed, the removal of one layer redistributes the stress in the remainder of the part, and this must be taken into account in calculating the original stress.

Uniaxial stresses, for example in a rail or pipe, can be adequately assessed by measuring changes in length. Biaxial stresses, for example in a sheet, require strain measurements made from a coupon cut from the sheet.

One of the most widely used techniques involves the measurement of strain around a centre hole, either by placing individual strain gauges around the hole or by adjacent positioning. Values for the residual stress have been established by Osgood [50] and Heindlhofer [51], and for other techniques have been summarized by the American Society of Metals [52].

Because they are non-destructive and can be applied to structures in use, X-ray diffraction methods have credibility as an operating control method for appraising the condition of structures [53]. By taking diffraction patterns of two sets of planes, one parallel to the surface and one making a steep angle with the surface, the stress in the x direction can be determined. The interplanar spacings are determined from the Bragg equation and differences between these spacings and the stress-free value give the strains in the other planes. From the transformation rule for components of strain, a relation can be obtained between the two measured strains and the component of normal strain lying in the plane. Combination with the elastic stress–strain relation, assuming the material to be isotropic, then yields a value for the residual stress. After repeating the process twice more to obtain three normal components lying in the plane of the plate, the entire state of stress can be determined.

(a) Centre-hole drilling technique

This technique, developed by Beaney and Procter [54], involves cutting a hole 1·5 mm diameter and 1·5 mm deep through the centre of a three-element strain gauge rosette attached to the surface of the material. As a hole is cut stresses in the material relax and changes in strain are recorded, from which the principal stresses may be calculated. The method is considered accurate to $\pm 20 \, \text{MN/m}^2$.

(b) Sentinel holes

Local changes in geometry can also be used to redistribute stresses and hence by measuring deformation to obtain the residual stress. Two methods use this principle: (a) the trepanning method; and (b) the hole-drilling method.

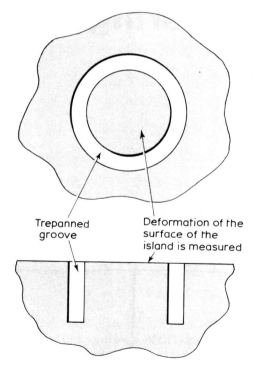

Trepanned groove

Deformation of the surface of the island is measured

Figure 2.16 *Small island trepanned from a surface with a resultant deformation from residual stresses.*

In the first of these a small island is formed, and the strain at the centre of the island measured (Fig. 2.16).This method leaves a hole about 10 mm diameter and 10 mm deep which is too damaging on operational plant. The second method uses a central hole which is only 2 mm diameter, 1·8 mm deep: small enough to be ignored on heavy plant (if necessary, the damaged area can be dressed smooth by grinding in order to remove the hole).

A hole drilled in a uniformly stressed structure reduces the radial stress at the edge of the hole to zero; stresses will be redistributed around the hole. Radial stress in the direction of the applied load will be of the form shown in Fig. 2.17.

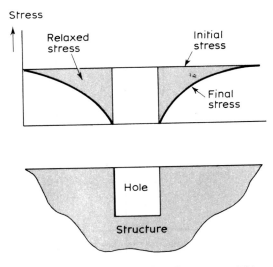

Figure 2.17 *Radial stresses at the edge of a hole reduce to zero. This causes redistribution of stresses and surface deformation near the hole.*

Deformations associated with this reduction in stress are clearly related to the relaxation in stress at the edge of the hole, that is, the applied stress.

A purpose-made strain-gauge rosette (Fig. 2.18) measures these deformations. It is constructed from a very thin metal etched into a wire grid with a resistance of 120 ohms and mounted on a thin backing material.

Residual stresses can be calculated from these measured strains using empirical formulae in which the relaxed strains are proportional to a calibration constant. This constant depends upon the geometry of the hole and the position of the strain gauge with respect to the hole.

Applications research [55] indicates that in using this technique hole diameter and location relative to strain gauge are very important. Although the obvious way to make the holes is to use drills or milling cutters these produce some strain due to machining stresses. The nearer the hole is to the gauge, the greater this machining strain becomes; thus a compromise has to be made between using a large hole for greater sensitivity and accuracy, and a small hole to cut down the machining-induced strains.

Consideration of other methods such as spark erosion, lasers, electrochemical machining [56–62] led to an evaluation of an air-abrasive system [63] in which a stream of air containing fine abrasive particles is directed against the workpiece. The abrasive particles, normally 50 μm particles of aluminium oxide, chip away microscopic particles of the workpiece material.

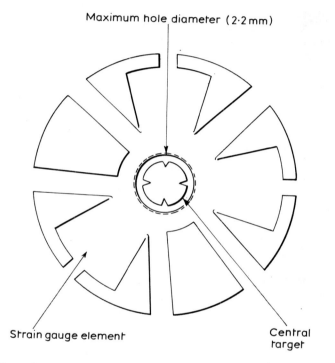

Figure 2.18 *Strain-gauge rosette bonded to the surface to be trepanned.*

This equipment in its finally developed form has been used to make measurements on glass, ceramics and plastics. The equipment is manufactured by a CEGB licensee as a single package containing an air-abrasive machine, the drilling and optical devices. Although the air-abrasive machine is basically a standard item, it has been modified to include power supplies for the drilling device and illumination within the optics, and to semi-automate the drilling operation.

The equipment is now used throughout the CEGB to further research in the subject of residual stresses and to assist with solving operational plant problems.

References and list of companies

1 Mortlock, J. D. (1974), 'The instrumentation of bridges for the measurement of temperature and movement', TRRL Report 641, Transport and Road Research Laboratory, Crowthorne Berks, UK.

2 Kerensky, O. A. and Little, G. (1964), 'Medway Bridge: design', *Proc. Instn Civil Engrs*, **29**, September, 19–52.

3 Rawlingson, Sir J. and Stott, P. F. (1962), 'The Hammersmith Flyover', *Proc. Instn Civil Engrs*, **23**, December, 565–600.

4 (1966), 'Erection of the Wye Viaduct', *The Engineer*, **221**, 10 June, No. 5759, 888–91.

5 Bingham, T. G. and Lee, D. J. (1969), 'The Mancunian Way elevated road structure', *Proc. Instn Civil Engrs*, **42**, April, 459–92.

6 Emerson, M. (1969), 'Bridge temperatures and movements in the British Isles', TRRL Report LR 228, Transport and Road Research Laboratory, Crowthorne, Berks, UK.

7 Capps, M. W. R. (1965), 'Temperature movements in the Medway Bridge – interim report' Road Research Laboratory Note No. LN/914/MWRC, Harmondsworth, September.

8 Capps, M. W. R. (1968), 'The thermal behaviour of the Beachley Viaduct/Wye Bridge', TRRL Report LR 234, Transport and Road Research Laboratory, Crowthorne, Berks, UK.

9 Wittier, R. M., Endevco Corporation, Rancho Viejo Road, San Juan Capistrano, California 92675, USA (Tel. (714) 493-8181).

10 Wilner, L. B. (1977), 'A diffused silicon pressure transducer with stress concentrated at transverse gages', *Proc. 23rd ISA Symp.*, May.

11 Tauba, T. (President), Technical Development Company, 24 East Glenolden Avenue, Glenolden, Pa 19036, USA (Tel. (215) 583-9400).

12 'Acurex Autodata', Acurex Corporation, Autodata Division, 485 Clyde Avenue, Mount View, California 94042, USA (Tel. (415) 964-3200, Telex 34-6491).

13 Harris, G. W. and Revell, G. S. (1981), 'Environmental monitoring 15', *Health & Safety at Work*, January, 52–6.

14 'Real-time particulate measurement', Celesco Industries Inc., Costa Mesa, California 92626, USA

15 Sira Institute Ltd, South Hill, Chislehurst, Kent, BR7 5EH, UK.

16 'SEMA Gas Detection Systems', Sema Electronics Ltd, Unit 32, Dundonald Camp, Irvine, Ayrshire, KA11 5BJ, UK (Tel. 0294-311252).

17 National Air Oil Burner Company, Inc., 1284 East Sedgley Avenue, Philadelphia, Pa 19134, USA.

18 Experimental and Electronic Laboratories, British Hovercraft Corporation, East Cowes, Isle of Wight, UK.

19 Vanzetti Infrared and Computer Systems Inc., 607 Neponset Street, Canton, Massachusetts 0201, USA.

20 Millipore Filter Color Standards Publications Department, Power Generation Division, Babcock and Wilcox, Barberton, Ohio, USA.

21 AI Industrial, London Road, Pampisford, Cambridge, CB2 4EF, UK (Tel. 0223-834420).

22 Henderson, R. M. P., Rojon Technical Services, 1 Croxley Close, St Pauls Cray, Orpington, Kent, BR5 3DD, UK (Tel. 01-302 1898). Also, Reliance Instrument Corporation, 164 Garibaldi Avenue, Lodi, NJ 07644, USA (Tel. (201) 778-2293).

23 (1980), 'Vessel monitoring by satellite–telex link' Ref. ASK 7681-007 *Ship & Boat International*, July/August, p. 23 (Pax Marine Press AB S-150 30 Mariefred, Sweden (Tel. (0159) 106 15).

24 Astech Electronics Ltd, 73 Castle Street, Farnham, Surrey, UK (Tel. 02572-725585).
25 Vanzetti Infrared and Computer Systems Inc., 607 Neponset Street, Canton, Massachusetts 0201, USA.
26 Prewitt, R. H. (1979), 'Mechanical strain indicators and recorders', *Transducer Technology*, November, 12–15. Also, Prewitt Associates Inc., Lexington, Kentucky, USA.
27 'Eddyprobe Mk II', Inspection Instruments (NDT) Ltd, 32 Duncan Terrace, London N1 8BR, UK.
28 Dynamic Logic Ltd, Doncastle House, Doncastle Road, Bracknell, Berks, UK (Tel. Bracknell 51915).
29 'Thermindox', Synthetic & Industrial Finishes Ltd, Imperial Works, Imperial Way, Watford, Herts, UK (Tel. Watford 28363).
30 Belcher, P. R. and Wilson, R. W. (1966), 'Templugs', *The Engineer* **221**, 305.
31 The Meclec Company, No. 7 Unit, Star Lane, Great Wakering, Essex, SS3 0PJ, UK.
32 Polaroid (UK) Inc., St Albans, Herts, UK.
33 Andrews, K. W. and Keightley, R. L. (1978), 'An ultrasonic goniometer for surface stress measurement', *Ultrasonics*, Sept., 205–9.
34 Kinergetics Inc., 6029 Reseda Blvd, Tarzana, CA 91356, USA. Also, 161 Clifton Road, Aberdeen, UK (Tel. 0224-876513).
35 Tyler, R. G. (1974), 'Measuring bending moments', *Tunnels & Tunnelling*, September.
— Tyler, R. G. (1968), 'Development in the measurement of strain and stress in concrete bridge structures', TRRL Report No. 189, Transport and Road Research Laboratory, Crowthorne, Berks, UK.
36 'Water leak detectors', Fuji Sangyo Co. Ltd, 1-11 Izumi-cho, Kanda, Chiyoda-ku, Tokyo (Tel. (03) 862-3196). Also Subtronic Ltd, 1 Lucas House, Craven Road, Rugby, UK (Tel. 0788-70241).
37 Green, L. (1977), 'The role of thermography', *National Conference on Condition Monitoring, NTC*, 15/16 February, 22–34.
38 Wahl, W., 'Heat Spy', Wahl International Ltd, 5750 Hannum Avenue, Calver City, California 90230, USA.
39 NCB East Midlands Laboratory, Mansfield, Notts., UK (ref: Mr T. Whitelam/Mr D. Holdam).
40 Spirax Sarco Ltd, Charlton House, Cheltenham, GL53 8ER, UK (Tel. 0242-21361).
41 Caleb Brett, marine bulk oil surveyors, cargo superintendents, analytical and testing laboratories, tank calibrators, consultant petroleum chemists, Wellheads Crescent, Dyce Industrial Park, Aberdeen, AB2 0GA, UK (Tel. 0224-723242).
42 (1975), 'Hull stress monitor for VLCCs', *Marine Engineering Review*, June, 66.
43 GD Engineering Ltd, Worksop, Notts, S80 2PY, UK.
44 Feltham, R. G. (1970), 'The role of flight recording in aircraft accident investigation and accident prevention', *J. R. Aero. Soc.*, **74**, No. 7, 573–6.
45 House, J. H., Harding, E. J. and Lucke, P. R. (1975), 'Technical aspects of computer-based ship automation systems', Royal Institution of Naval Architects, 11 March, Paper 4.
46 (1974), 'Trouble-shooting marine control systems', *Marine Engineering Review*, November.
47 Knott, J. F. (1973), *Fundamentals of Fracture Mechanics*, Butterworth, Guildford.
48 Bauschinger, J. (1886), 'Changes of elastic limit and hardness of iron and steel through extension and compression during heating and cooling and cycling', *Mitteilung aus dem Mechanisch*, Technischen Laboratorium, Technische Hochschule, Munchen 13, Part 5, 31.
49 McClintock, F. A. and Argon, A. S. (1966), *Mechanical Behaviour of Materials* Addison-Wesley.

50 Osgood, W. R. (Ed.) (1954), *Residual Stresses in Metals and Metal Construction,* prepared for Ship Structure Committee of National Academy of Sciences, National Research Council, Reinhold Publishing Corp., New York.
51 Heindlhofer, K. (1948), *Evaluation of Residual Stress,* McGraw-Hill, New York.
52 (1948), *Metals Handbook Supplement* (1955), American Society of Metals.
53 Doig, P. (1978), 'Residual stress determination by X-ray goniometry', UK Mechanical Health Monitoring Group, Leicester, September.
54 Beaney, E. M. and Procter, E. (1974), 'A critical evaluation of the centre hole drilling technique for the measurement of residual stress', *Strain,* **10,** 7.
55 Procter, E. and Beaney, E. M. (1981), 'A technique for measuring residual stresses', CEGB Research Report No. 12, July, 3–10.
56 Savin, G. N. (1961), *Stress Concentration Around Holes,* Pergamon, Oxford.
57 Kelsey, R. A. (1956), 'Measuring non-uniform residual stresses by the hole-drilling method', *Proc. Soc. Experimental Stress Analysis,* **14,** 181.
58 Rendler, N. J. and Vigness, I. (1966), 'Hole-drilling strain-gauge method of measuring residual stresses', *Exptl Mechanics,* **6,** 577.
59 Soete, W. and Vancombrugge, R. (1950), 'An industrial method for the determination of residual stresses', *Proc. Soc. Exptl Stress Analysis,* **8,** 17.
60 Bush, A. J. and Kromer, F. J. (1973), 'Simplification of the hole-drilling method of residual-stress measurement', ISA Trans., **12,** 249.
61 Boiten, R. G. and Ten Cate, W. (1952), 'A routine method for the measurement of residual stresses in plates', *Appl. Sci. Res.* A, **3,** 317.
62 Beaney, E. M. and Procter, E. (1972), 'A critical evaluation of the centre-hole technique for the measurement of residual stresses' (CEGB Report RD/B/N2492.
63 Beaney, E. M. (1980), 'The air-abrasive centre-hole technique for the measurement of residual stresses', CEGB Report RD/B/N4806.

3
Crack monitoring: electronic methods

3.1 Introduction

Cracks alone do not necessarily present a hazard or imply a gross loss of structural integrity, it is the loss of load-bearing capability which needs to be appraised. This is a combination of the stress concentration within the defective region (the load/shape effect) and the reduction of load-carrying section (area and shape) resulting from the crack.

In practice, a defect may comprise several small cracks or a multitude of cavities. In a region under intense stress even small cracks are significant—they can quickly 'grow' larger and initiate a failure. In a region of low stress intensity even quite large cracks can be tolerated; they either grow slowly or may even be dormant.

Crack monitoring therefore involves both the task of identifying the location of defects and the measurement of their size.

Cracks may appear on the surface or may be hidden within the material itself. While most of the techniques originating from non-destructive testing (NDT) methods can identify cracks, it is their regular measurement and growth-rate assessment which distinguishes quality-assurance NDT from integrity-appraising condition monitoring, i.e. measuring not only the apparent surface dimensions but also the more important dimensions of intrasurface penetration.

In general, crack acceptability can be determined from such size–time curves under critical operating conditions more detailed understanding of cracks, crack growth rates and regional stresses is necessary.

3.2 Crack monitoring techniques

Two distinct stages apply to crack growth monitoring: (1) location/identification of the crack or defects; and (2) measurement and trend recording.

Crack or defect recording may be either: (a) direct visual, with or without aids (seeding) such as dyes; or (b) by secondary effects, such as magnetic effects or acoustic phenomena (wave propagation within the material).

3.3 Sonic 'ring' tests

The ring of a metal when struck is an old technique for detecting gross flaws. The note emitted from a steel specimen containing a flaw sounds dull and harsh compared with that emitted from a flawless material. A well-known example of the application of this technique is the wheel-tapping test for cracks in railway rolling stock wheels. Devices such as stethoscopes, microphones and electronic amplifiers have been used to improve the listening sensitivity of such tests. While these tests provide reasonable results with welds and similar joins they are of limited value with more complicated faults.

3.4 Ultrasonics

Acoustic excitations at frequencies of 18 kHz and above which are beyond the hearing range of the normal human ear are called ultrasonic waves (see Section 3.4.2). By generating and detecting ultrasonic waves small material defects can be found since the wavelength of an ultrasonic wave is of the same order of magnitude as the size of the defect. Because of their good elastic properties most metals readily transmit ultrasonic vibrations which are scattered or reflected by flaws due to acoustic mismatch (see Section 3.4.9).

Table 3.1

Material	Wave velocity (mm/s $\times$ 10^5)		
	Longitudinal	Shear	Rayleigh
Aluminium	62.7	31.0	28.0
Brass	47.0	21.4	19.3
Copper	46.3	21.3	19.1
Lead	19.5	6.4	5.8
Steel	57.5	30.9	27.9

Waves which propagate over the surface of a solid whose thickness perpendicular to the surface is large compared with the wavelength of the waves are known as Rayleigh waves. They have a velocity less than that of waves which propagate through the material (see Section 3.4.5). For metals this velocity is approximately 90% of the shear wave velocity. Typical values for ultrasonic velocities are given in Table 3.1.

3.4.1 The nature of ultrasound

Orderly oscillations generated by a crystal in contact with the outside surface of a medium produce an interaction with the particles of the medium and the transmission of vibrations. Continuous oscillatory generation causes continuous wave propagation. Intermittent (discontinuous) oscillatory generation produces pulsed wave propagation. Oscillations may have amplitudes of less than 10^{-8} mm.

3.4.2 Ultrasonic frequency

The human ear may detect sounds in the frequency range 20 Hz to 18 000 Hz (18 kHz). Frequencies above 18 kHz are regarded as being 'ultrasonic'. Diagnostic ultrasound is normally generated at frequencies of 1 to 15 MHz where

$$1 \text{ MHz} = 1\,000\,000 \text{ Hz} = 60\,000\,000 \text{ cycles/minute}$$

Ultrasound frequency is measured by converting the mechanical oscillations into electrical voltage and metered electronically.

3.4.3 Crystal ultrasound generators

Typical crystal ultrasound generators are based on the piezoelectric effect where an applied charge alters the size of the crystal. Thus ultrasonic transducers take the form of small cylindrical tubes about 15 mm diameter with a disc-shaped crystal slice at one end (Fig. 3.1). By applying a fluctuating electrical voltage to the crystal slice a pulsed vibration may be generated.

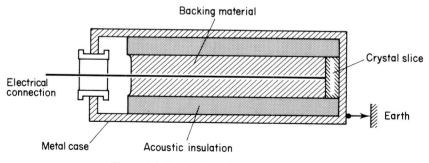

Figure 3.1 *Basis of an ultrasound generator.*

Ultrasound beams have the greatest frequency at their axis of propagation and this reduces at the edge. Such beams may be focused by using concave crystals or by using the acoustic equivalent of an optical lens in front of a flat crystal. A focal length of 5 mm produces a powerful effect, average focal lengths are 30 mm.

3.4.4 Interface interrogation

Interface interrogation involves three stages:

(1) Detection – when an interface reflects part of an incident beam.
(2) Location – by noting the direction of the beam and timing the interval for an echo to return.
(3) Identification – by the height and shape of the echo.

Ultrasound is reflected (echoed) in the same way as audible sound, the amount of vibrational energy reflected depending on the difference in acoustic impedance between the transmitting media and the interface boundary (see Section 3.4.9). Echoes from flat surfaces such as A and B in Fig. 3.2 will be strong while those from C and D will be spread out and weaker.

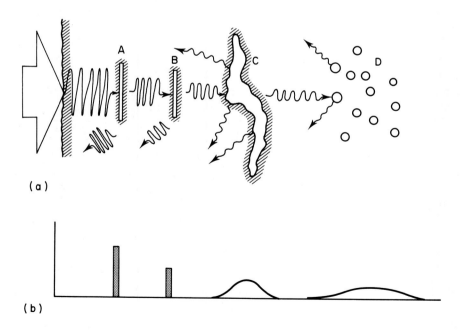

(a)

(b)

Figure 3.2 *Interrogation of flaws: (a) effect of defect; (b) echo display.*

Localization of a defect is undertaken by the simple expedient of relating velocity and time, thus

$$\text{depth to defect} = \text{ultrasound velocity} \times \text{transmission time}$$

Ultrasound velocity varies according to the medium as in Table 3.2. Transmission time is one-half of the overall generation–echo–reflection time.

Table 3.2

Material	Acoustic impedance (g/cm^2 s)	Velocity (m/s)
Air (NTP)	0.0004×10^5	330
Water	1.48×10^5	1480

General interface identification may be established from the height, length and shape of the echo. Such reflections are displayed by cathode ray tube scanning with the echo fixed as necessary by a storage screen.

3.4.5 Wave phenomena

Ultrasonic waves can combine in similar manners as for other wave phenomena.

Interference between waves at similar frequencies may produce alterations to waveform according to their phase, based on the principle of superimposition. For waves at different frequencies the resulting waveform is not purely sinusoidal although when the frequencies are only slightly different the combination produces a beat frequency, involved in the Doppler technique.

Standing waves may be created by propagating identical waves from opposite directions. At the point where those waves meet the pressure vibrations are enhanced, they become a 'standing' wave for which the intensity of ultrasound is increased (Fig. 3.3). Although standing waves are less important when pulse techniques are used, they may be generated accidently by reflections and thereby produce spurious signals.

Resonance may arise when surfaces are separated by a distance equal to a complete number of half-wavelengths. Waves transmitted to and reflected from such surfaces produce a train of enhanced standing waves.

High-frequency mechanical waves may have such physical characteristics as:

● shear waves – in which vibration occurs at right angles to the direction of wave travel, similar to those on the surface of the water.
● Rayleigh waves – transverse waves, similar to shear waves, used to detect surface flaws.
● Lamb waves – as with flexural vibrations of sheets of material.

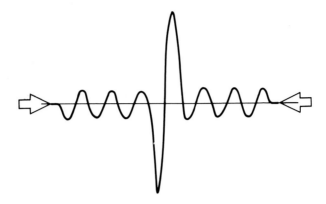

Figure 3.3 *Standing wave produced by ultrasound.*

Waves may change from one mode to another and thereby produce spurious results. For example, when a longitudinal wave strikes a solid surface at an angle both longitudinal and transverse waves will be set up.

3.4.6 Gain

Echoes produce small voltages which must be amplified for adequate display. Gain control governs the amount of amplification. An amplifier with a linear response increases the size of both large and small signals. The result is that large echoes become excessively amplified, for which reason logarithmic amplifiers are used as they overload large echo signals to a smaller extent. A decibel notation is a convenient label for the gain control, an input signal V_0 amplified to V_1 with a gain ratio of $\dfrac{V_1}{V_0}$ being expressed as

$$\text{Gain (dB)} = 20 \log_{10} \frac{V_1}{V_0}$$

On this basis, for a gain of 40 dB, the amplification is $10^2 = 100$, while for a gain of 60 dB the amplification is $10^3 = 1000$.

Electronic attenuators may be used to amplify echoes, they adjust the voltage level used to excite the transducer.

Deep echoes need more amplification than echoes from near-surface interfaces. This can be achieved by swept gain or depth-varied gain, carried out by changing the amplifier gain at an appropriate rate after each instant of transmissions.

3.4.7 Pulse repetition frequency (PRF)

This is the rate at which an ultrasonic pulse is transmitted. A machine may have a fixed PRF or it may be varied, a range from 500 to 3000 pulses being normal. The lower value is fixed by the persistence of vision, the higher value by the die-out time of a pulse; at 200 pulses/s persistence of vision would be 500 μs.

3.4.8 Echo gating

Selected echoes at a particular depth may be studied in more detail by gating them electronically. After the whole echo pattern has been examined the layer of interest is marked off by electronic markers (Fig. 3.4 (a)). A gating switch then expands the marked-out display (Fig. 3.4. (b)) so that the details of this layer can be studied.

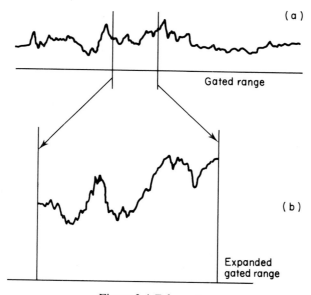

Figure 3.4 *Echo gating.*

3.4.9 Acoustic impedance

Acoustic impedance is the product of the density of the transmitting medium and the velocity of the ultrasonic wave. Thus an ultrasonic pulse encountering a flaw experiences a substantial change of impedance arising from the change in density at the interface. The strength of an echo can be shown to be represented by the amplitude of the pressure wave reflected, given by

$$\text{Pressure amplitude ratio} = \frac{Z_1 - Z_2}{Z_1 + Z_2}$$

where

Z_1 = acoustic impedance of incident wave
    = $\rho_1 V_1$
Z_2 = acoustic impedance of echo
    = $\rho_2 V_2$
ρ_1 = density of incident wave
ρ_2 = density of echo
V_1 = velocity of incident wave
V_2 = velocity of echo

3.4.10 Attenuation

Transmitted ultrasound and reflected echoes are reduced in intensity (attenuated) by such processes as reflection, scattering, refraction, absorption and wavefront divergence. Typically, increased attenuation with increased frequency implies a high absorption rate; a very divergent wavefront results from a small defect. Attenuation may be expressed in terms of decibels per millimetre, being the logarithmic pressure change per unit length. Such a characteristic provides a further method for the measurement of thickness.

3.4.11 Doppler ultrasonic diagnosis

Reflected continuous ultrasonic waves to which are added or subtracted the echoes from a moving source form the basis of this technique. With a source and detector placed close together, the source emitting at frequency f_0 and the discontinuity moving at velocity u, there will be a shift in frequency which will be registered by the detector. Such a 'Doppler shift' may be used to measure motions within a moving body, for example changes in a crack in a machine component.

3.5 Ultrasonic testing techniques

Techniques available for electronic testing include pulse echo, transmission, resonance, frequency modulation and acoustic imaging.

3.5.1 Pulse echo technique

A piezoelectric crystal coupled to a surface by means of a liquid medium is commonly used to generate ultrasonic waves. Three main types of ultrasonic wave may be generated: (a) longitudinal; (b) shear (or transverse); and (c) surface. Longitudinal waves arise when the crystal is applied parallel to the surface, if the angle of incidence of the crystal is 30° about 98% of the energy becomes shear.

The basic testing technique is that of pulse echo. The transmitter emits an ultrasonic pulse into the material under test and the echo is picked up and displayed on a cathode ray tube. The echo is a combination of return pulse from the opposite side of the specimen and from an intervening defect. The travel time of the reflected wave, which appears as a pulse to the side of the original pulse, is determined and its position determines the location of a defect. An approximate idea of the size, shape and orientation of the defect can be obtained by checking the specimen from another location or surface.

3.5.2 Transmission technique

This technique uses continuous waves from one transducer which pass right through the test piece. Flaws tend to reduce the amount of energy reaching the receiver and by this means their presence can be detected. This was the original ultrasonic testing technique but has been replaced owing to problems of modulation associated with standing waves which can be created in the test piece, causing false readings to be obtained.

3.5.3 Resonance technique

This consists of moving the transmitter over the surface and observing the transmitted signal. Resonance in the absence of flaws will keep the transmitted signal strong. Flaws will cause the transmitted signal to weaken or disappear. Such a method is highly suitable for the testing of thin specimens as a development of the transmission technique; it is widely used as a means of testing the bond between thin surfaces.

3.5.4 Frequency modulation technique

Only one transducer is used to send ultrasonic waves continuously at changing radio frequencies. An echo returns at the frequency of the early original transmission and thus interrupts the new changed frequency. By measuring the phase between frequencies the location of the defect can be determined.

3.6 Defect sizing: time delay diffracted ultrasound

Crack depth gauging is a major requirement. Hitherto the stumbling block has been the interpretation of discovered defects in terms of size and orientation. Significant errors and misleading results often arise because the amplitude of ultrasound echoes is affected by defect roughness, transparency or orientation as well as material attenuation and coupling efficiency.

Ultrasonic bulk wave diffraction is claimed to be able to size a subsurface flaw with an accuracy greater than $\pm$ 0.5 mm, typical accuracy being within 0.3 mm.

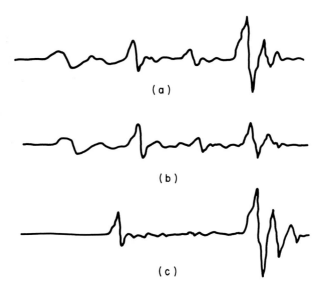

Figure 3.5 *Typical ultrasonic responses to cracks: (a) internal crack; (b) crack open to surface opposite probes; (c) crack open to same surface as probes.*

The system uses two probes set astride a detected flaw or crack. The probes receive echoes from the backwall of the specimen and from the tips of the defect plus a signal arising from a lateral wave passing directly between the two probes. The lateral and backwall echo waves help to define a time field corresponding to plate thickness. If a subsurface crack is present, two diffracted echoes will be observed in the time field (Fig. 3.5). If the crack breaks surface, one diffraction echo will be observed (Fig. 3.5 and 3.6). The user will detect which surface is broken since either the lateral wave or backwall echo signal will be lost. Usually probes are separated by a distance of four times the plate thickness so that measurements are not disturbed by lateral position of the crack tip or by probe misplacement.

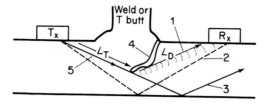

Figure 3.6 *Diffracted ultrasound used in crack depth measurement: (1) path of diffracted ultrasound; (2) path of weak component of reflected ultrasound; (3) path of reflected ultrasound; (4) crack; (5) centre-line of beam.*

Time domain analysis relies on the diffraction of ultrasound from the tip of the crack (Fig. 3.7). The first signal to reach the receiving probe arises from the diffracted component; the transit time of this component depends on the depth of the crack. The technique has the following advantages:

(1) The ultrasonic wave passes through the body of the material and would not be affected by rough weld surfaces or the attachment of other members.
(2) Longitudinal waves may be used which ensure a rapid transit time and thus little interference from other ultrasonic modes or reflectors.
(3) The parameter measured is the *depth of the crack tip* below the surface on which the probes rest.
(4) The estimate of crack depth is little affected by small lateral movements of the probe system or unknown changes in crack direction.

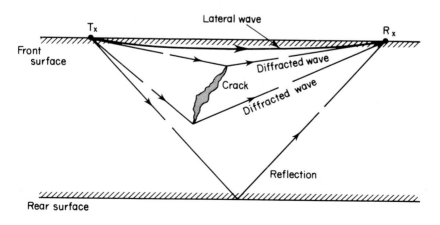

Figure 3.7 *Ray paths of diffracted ultrasound.*

During the evaluation of this technique [14] a standard fatigue specimen 19 mm thick mild steel plate approximately 305 × 102 mm had a stress raiser, a slit 6 mm deep machined at the centre of the long axis of the plate. This slit could either be left in as 'part of the crack' or the slit could be removed by reducing the thickness of the plate, leaving a tight natural crack. Fig. 3.8 is interesting in showing two estimates of crack depth before and after this machining process. Apart from the consequent reduction in depth it is clear that the two estimates, necessarily obtained in differing testing geometries, agree to a high degree of accuracy.

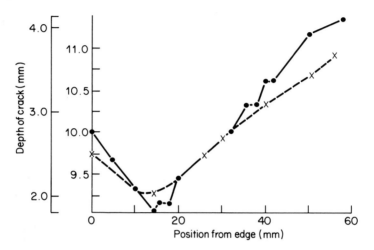

Figure 3.8 *Crack profile determination.*

3.7 Ultrasonic ranging: sonar focusing

Precise distance readings while moving are made possible by the Polaroid Sonar Focusing System [1] (Fig. 3.9).

For SX-70 autofocus cameras, distance is measured by the direct use of ultrasonic waves. There are no moving parts or optical elements in the range-finding system which comprises an electrostatic transducer – to transmit and receive sound signals – a crystal oscillator clock, a return signal detector, an accumulator and focus motor.

As millisecond-long 'blips' are transmitted, the oscillator clock starts to count by emitting regularly-spaced timing signals to the accumulator until the echo is picked up by the transducer. As Polaroid has developed the device, the focusing range has been divided into 128 depth-of-field zones which are matched with the same number of sites in the accumulator.

Each of these is sequentially activated by the timing signals. On receipt of the first echo, the detector signals the counter to stop, determines the travel time of

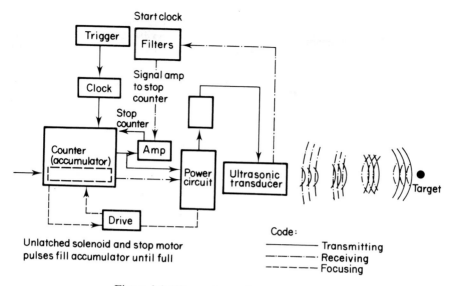

Figure 3.9 *Ultrasonic ranging: sonar focusing.*

the signal, activates the lens rotation motor by data from the accumulator and produces a precise focus. As the sonar system saturates the 'target' area with ultrasonic energy it is extremely accurate and effective. Independent assessment claims that no alternative system available today can match the zone focus capability of this range-finding technique.

Linked to a small speech synthesizer instead of a lens rotation motor, this principle can be used to 'announce' the precise proximity of an approached object.

3.8 Ultrasonic triagulation fault location

Sound waves can be used in much the same way that electromagnetic waves are used to detect and pinpoint flaws in materials.

In contrast to existing ultrasonic techniques which require the transducer to be located fairly near the suspected crack, triangulation can detect a flaw as long as it is within the area being scanned by the sound signals. A system devised by the George C. Marshall Center, Huntsville, Alabama, USA was used to detect flaws on the Saturn rocket case.

Four barium titanate crystals, one of which was used as a transmitter, were arranged in a rectangular pattern near the weld area being examined. Each crystal was able to move round a quadrant, which coupled a waveguide to the plate being tested. This gave a null and well-defined ultrasonic beam which radiated through the metal. The beam was resonant in the thickness of the plate, and thus was capable of travelling through it, as if the plate were a waveguide. Reflected energy from any flaws was picked up by one or more waveguide detectors. The return signal was then processed by a computer that directed the sensor beams so they intersected at the flaw. This point was either displayed on a cathode-ray tube or on a digital readout. The amplitude of the return signal determined whether the flaw should be passed or failed; this was done by automatic signalling equipment.

3.9 Scanning systems

Much of the work involved in structural monitoring has a counterpart in medical and surgical diagnosis, where inspection, measurement and monitoring involve similar (sometimes identical) equipment. Scanning systems are now evident in both medical and engineering applications.

The simplest one-dimensional system, known as A-scanning, uses a transmitter which doubles as a receiver and is pointed in a determined direction. Surfaces intersected by the beam set up echoes for which the length of their vertical trace represents the interface size. Horizontal distance from the origin measures time and is therefore proportional to depth.

A pictorial representation of an interface is obtained by B-scanning. Reflections produce only bright spots on the time axis so that by orbiting the transmitter an orbit of reflection vectors traces out the profile of interfaces.

3.9.1 A-scan instrumentation

The components of a system which will scan a complete surface in the form of an A-scan are shown in Fig. 3.10 and operate as follows:

(1) Start pulse generator – Triggers off the system by producing an electronic pulse which is fed simultaneously to components (2), (3), (4).
(2) High-voltage pulse generator – Excites the transducer crystal with a pulse of possibly 3 kV amplitude and $0.5\,\mu s$ duration and produces a short burst of mechanical vibration in the crystal, thereby generating an ultrasonic pulse to the test body.
(3) Swept gain amplifier – Adjusted to obtain the maximum response from the applied pulse.
(4) Radio frequency amplifier – Performs the first stage of amplification for frequencies in the MHz range. The output from (3) produces an increase in the gain at suitable depths.
(5) Demodulator – Smooths the signals into voltage pulses of a non-oscillatory (rectified) shape.
(6) Video amplifier – Large-voltage amplification of the signals so that they show on the Y-plate display of the cathode-ray tube.

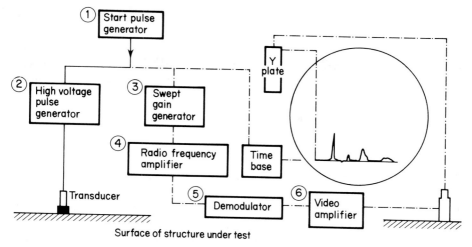

Figure 3.10 *A-scan instrumentation.*

3.9.2 B-scan instrumentation

Echo signals to produce B-scan images are presented as bright spots at appropriate points on a line. The scanning can be carried out by effectively six scanning actions as shown in Fig. 3.11

(a) simple linear scanning.
(b) simple sector scanning.
(c) simple radial scanning.
(d) simple contact scanning.
(e) compound linear scanning.
(f) compound contact scanning.

Developments from simple linear scanning arise because many interface boundaries do not present themselves at suitable angles to the beam. Consequently the beam must be submitted at larger angles by rocking the transducer, this may be performed manually or by using a mechanical linkage.

The component parts of a B-scan machine are similar to those for A-scan with the addition of a coordinate measuring system and a generator for registering these X, Y coordinates.

3.10 Pipes: ultrasonic testing

Longitudinal and circumferential defects in pipes may be examined simultaneously and locations marked automatically on the pipe surface by paint sprays [2] using the following ultrasonic testing technique.

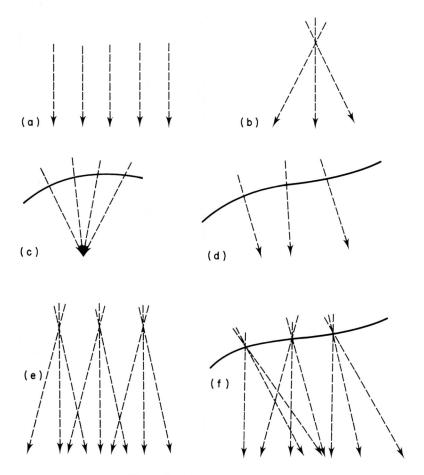

Figure 3.11 *Six scanning actions.*

The pipe is rotated by drive rolls chain-driven by a 7 kW motor with a rotational speed continuously variable to give a range of pipe surface speeds from 0 to 0.5 m/s. Adjustable end location rollers prevent 'creep' of the pipe during rotation. A tray between the rails collects the surplus coupling medium from the ultrasonic probes which is filtered and recirculated. The ultrasonic probe complex is mounted on a positioner with the probes supported by a multiple pivot nylon wheel roller system which 'floats' on the pipe surface and automatically maintains the probe in correct alignment with the surface even if the pipe is bent or oval.

Irrigated, 15 mm diameter, 45°, $2\frac{1}{2}$ MHz shear-wave probes detect longitudinal and circumferential defects. They are positioned 0.75 mm from the pipe with couplant fed just in advance of each probe. A compression wave

5 MHz jet probe measures thickness. Monitors respond to defects in excess of a pre-set level, trigger paint spray location markers and set off a klaxon horn.

3.11 Bolt strain measurement

Ultrasonic crack detection incorporating precision timing measurements has been modified for the measurement of bolt strain (elongation) [3].

If a sound pulse is sent into the head of the bolt at an angle, the individual threads of the bolt will all reflect part of the incident pulse as though they were small cracks. Accurate timing of the intervals between successive echoes may then be related directly to thread pitch. Currently, timing may be achieved to an accuracy of $\pm$ 4 ns, representing an extension of 12 μm in thread pitch or about 0.5% strain in a 18 mm bolt (see also Section 12.5.4).

3.12 Boiler tube thickness probe

Instead of evaluating metal wastage in marine boiler tubes by removing selected tubes and assessing condition from the frequency and size of the pits, an ultrasonic probe technique has been developed by Peters, O'Malie and Greenwood of the Royal Canadian Navy Defence Research Establishment [4]. This method was developed specially to deal with the superheater tubes of the Babcock and Wilcox X100 boiler which are difficult to reach from the combustion area of the boiler.

Two ultrasonic units were built, one utilizing a contact probe and one utilizing an immersion probe. The ultrasonic contact probe used a 5 MHz transducer (Automation Industries SFZ 57A2216 5.0/.375) incorporated into the probe; this transducer was chosen to suit a Sperry UCD battery operated reflectoscope. A lucite or nylon shoe which tends to wear into the profile of the tube, and can easily be replaced when it wears out, was fitted onto the transducer.

Alcohol coupling was maintained through a small white plastic tube and the alcohol inserted by means of a large hypodermic syringe. The alcohol supply could be attached to either of two receptacles on the probe so that the couplant runs down onto the probe–tube interface when used in either the port or starboard boilers.

The special feature of the probe is that it can measure tube wall thicknesses at the bends and around the bends, as well as along the straight lengths. A flexible steel spring, to which the probe is attached, keeps the transducer at right angles to the tube wall at all locations. The probe is attached to a long flexible steel wand which enables the operator to manipulate the probe from the end; that is, the probe can be rotated, pushed or pulled while in the tube.

3.13 Ultrasonic inspection case studies

3.13.1 International inspection of boiler tubes

This system was originally developed for Burmah-Castrol at Ellesmere Port, England, who were having trouble in inspecting their boilers for tube thinning. External thickness calibration was impossible because of the boiler multi-row tube arrangement. A system was developed whereby a number of ultrasonic probes are mounted in a dolly carriage drawn through the tube bore by a wire. Tube thickness can be measured on a 'go–no go' basis or as a direct readout.

3.13.2 Underwater structural nodes

In-service inspection for fatigue cracks on complex weld formations such as those on oil rig nodes is necessary. A system has been designed such that this requires minimum diver participation. In fact all that a diver has to do is clean the joint to be examined and place an inspection unit in position. Any subsequent skilled interpretation is carried out by non-diving NDT personnel, either in the relative comfort of a submersible or above the water line.

This system features custom-built probes arranged along either side of the weld. The individual probe angles are chosen and the stand-back distance calculated to allow the beam to pick up defects on any of the fission faces or vertical cracks. The probes are arranged on a flexible belt and are switched sequentially. The signals are analysed in two separate gates so that it is possible to clear long lengths of weld quickly and concentrate on doubtful areas for more detailed analysis. The resultant data can be processed in a microcomputer and presented either as a printout or in A- or B-scan form.

3.14 Eddy current testing

3.14.1 Basic principles

Eddy current testing is based on the principle of electromagnetic induction and is concerned with the magnetic field generated by a coil carrying an alternating current. The magnetic field interacts with a test object brought near to the coil, and the effect on the field depends, *inter alia*, on the electrical and magnetic properties of the test object.

The effect is shown in the vector diagrams of Figs 3.12 and 3.13. In Fig. 3.12 the empty coil is supplied with an alternating current which creates a flux φ_0. The voltage V_0 lags this flux by 90°.

When a test object is introduced into the coil the flux induces eddy currents to flow in the test object which themselves create a flux φ_1 in vector opposition to

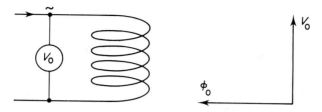

Figure 3.12 *Flux in empty coil.*

φ_0, as in Fig. 3.13. This combines with the original flux to produce a resultant flux vector φ_R as shown. The resultant voltage across the coil now lags this vector by 90° and is shown as V_R. The vector voltage V_1 due to the eddy currents in the test object can be found by vector addition of V_0 and V_R. The simple diagrams assume zero coil resistance. The dotted line is the locus of the resultant voltage V_R across the test coil caused by variation in the characteristics of the test object within the coil.

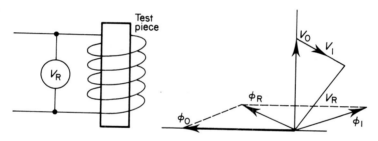

Figure 3.13 *Flux due to presence of test piece.*

3.14.2 Simple instrument

The resultant voltage V_R is used to measure the eddy current effect in simple instruments. This can be manipulated to produce a meter reading indicative of the desired property or feature under investigation. In such instruments the phase and amplitude variations exist in combination, and variation of one may mask the variation of the other.

In the Phasec instrument [5] the complex response of the coil is analysed by a technique known as impedance analysis. In this the vector voltage V_R is analysed into two components in quadrature, one representing the resistive (real) component R, and the other the reactive (imaginary) component X of the impedance. These components are converted to horizontal and vertical deflection voltages, and used to draw a vector diagram on the storage cathode-ray tube display, or for chart or magnetic recording puposes.

The display can thus give a true record of the variations of V_R both in phase and amplitude (also time in changing situations). It permits one to differentiate between effects which on simpler types of instrument may appear identical.

Phase rotation circuitry makes it possible to change the orientation of the display. Thus when using a probe coil with the Phasec, the lift-off signal, created by moving the probe away from the metal surface, may be aligned according to convention in the negative horizontal direction.

The precise effect of a particular test object on the Phasec system is strongly dependent on the construction of the coil, its relationship to the test object, and particularly, the frequency of the current in the test coil. In the Phasec system maximum freedom is given to the user to determine the precise frequency of operation from a continuously variable range between 0.5 kHz and 2.5 MHz, and to use any design of probe or coil, absolute or differential, permitted by the design of the a.c. bridge.

3.14.3 Wire rope testing

Mines rely heavily on the wire ropes that are used to hoist cages; in a multistorey building the lift depends upon the integrity of the lifting cable; many ships and offshore platforms rely upon the moorings using wire ropes.

Failure of such cables arises from breaks of strands within the rope. Such breaks and local changes in cross-sectional area may be detected by measurement of the magnetic anomaly fields produced by these flaws as the rope passes through the instrument.

Detection is usually achieved using one or more induction coils, mounted on a collar around the rope in the centre of the instrument, and arranged to detect the rate of change of the radial or longitudinal magnetic field on the rope surface. Thus a rope flaw passes through the instrument, the induction coils detect the rate of change of the radial or longitudinal component of the magnetic field anomaly (which is seen on the rope surface) due to the flaw. Only localized flaws in the rope are detected by this system and the amplitude of the signal obtained is proportional to the speed of the rope.

A development in this technique reported by Marchent [6] is featured in patent application no. 41530/77, October 1977 to provide a direct indication of the steel cross-sectional area of the rope, even for a rope that is stationary. Also, the use of 'Hall effect' magnetic sensors in the central collar for broken wire detection, produces a signal output which is not dependent on rope speed. Thus the instrument can be moved slowly along the rope while still enabling broken wires and other rope flaws to be detected. In this way splices, wear, corrosion, broken wires, missing wires and rope distortion can be detected. Also any regular occurrence of rope flaws along the rope length can be seen and may indicate if a part of the rope system (for example a particular pulley wheel) is causing excessive rope deterioration.

The rope within the instrument is magnetized along its length and 'Hall effect' magnetic sensors under the pole pieces enable an indication of steel cross-sectional area of the rope to be obtained. Also magnetic sensors in a central collar enable broken wires and localized flaws in the rope to be detected.

An electromagnet system (Fig. 3.14) or a permanent magnet system (Fig. 3.15) could be used to magnetize the rope. In the case of an electromagnet the

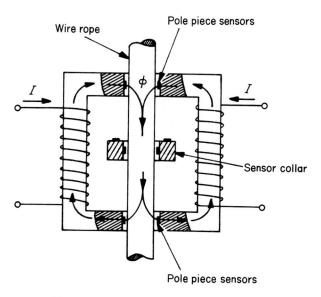

Figure 3.14 *Electromagnetic cable detector.*

instrument could consist of a number of yoke bars (usually two) with a coil, carrying a d.c. current, wound on each bar. The magnetic flux from these bars is coupled into the rope via the pole pieces so that a magnetic flux passes along the rope between the pole pieces of the magnet.

Broken wires are detected by 'Hall effect' magnetic sensors mounted in a collar around the rope in the centre of the instrument (see Fig. 3.16). If a broken wire is present then a magnetic fringing field will appear on the surface of the rope over this break. The form of the fringing field can be obtained from consideration of the magnetic flux along a single broken wire with a gap as shown in Figs 3.17 and 3.18.

The magnetic flux in the wire will have to pass over the gap. This will produce a magnetic fringing field, the form of which will be similar to that obtained from a magnetic dipole. The form of the fringing field detected by both radial and longitudinal magnetic sensors in the collar can be derived easily as shown in Figs 3.17 and 3.18 for both short and long break lengths. If the broken wire is hidden within the rope some attenuation of the fringing field levels is obtained,

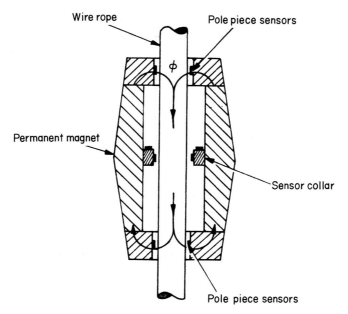

Figure 3.15 *Permanent magnet cable detector.*

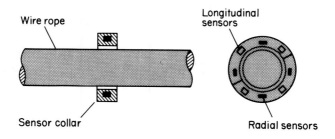

Figure 3.16 *Sensor collar for broken wire detector.*

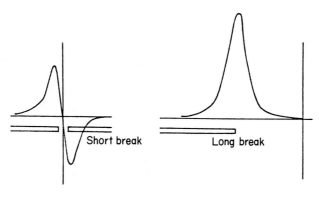

Figure 3.17 *Transient signal – broken wire; detected by radial magnetic sensors on collar.*

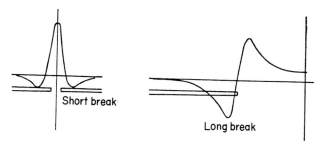

Figure 3.18 *Transient signal – broken wire; detected by longitudinal magnetic sensor on collar.*

but if the rope magnetization is approaching magnetic saturation this attenuation is not excessive.

In practice, the 'Hall effect' radial magnetic sensors are usually found to be sufficient for broken wire detection. They have the advantage that the signal output is not affected by the speed of the rope. The outputs from a number of sensors equally spaced around the rope circumference, are summed and a.c. coupled with a long time-constant, say five seconds, to eliminate sensor drift. This output may also be passed through a low-pass filter to reduce magnetic noise detected from the rope. The output is thus an analogue signal with transient peaks (of the form shown in Fig 3.17) to indicate the presence of broken or missing wires. This output may be applied to a threshold detection circuit to provide an alarm.

3.14.4 Eddy current tester (computerized): CANSCAN specification

Following an awareness of the dominant criticality of steam generator tubes arising from their extensiveness, frequency and cost [7], Ontario Hydro in 1974 adopted an AECL-RC target specification [8] for the remotely controlled eddy current inspection of the Pickering nuclear generating station in Canada.

Each of the eight Pickering reactors has twelve steam generators located in two opposing rows of six inside each reactor containment vessel. Tube details are as shown in Table 3.2.

The limiting tube bend radius was determined by the mechanical dimensions of the eddy current probe, the cable connector and the drive cable. A prototype semi-automatic inspection system was built to assess the problems likely to be encountered for an eddy current inspection involving hundreds of tubes in an active environment. This system used a remotely operated probe drive with the controls and instrumentation outside the reactor containment vessel. The maximum inspection rate achieved was 4 tubes per hour. A test crew dose penalty of 4.7 man-rem was incurred in this inspection of 51 tubes.

Table 3.2 Pickering Nuclear Power Station: steam generator
details

Number of reactors	4
Steam generators per reactor	12
Tube material	Monel 400
Tubes/steam generator	2600
Average tube length	18.5 m
Tube inner diameter	10.2 mm
Tube outer diameter	12.7 mm
Limiting tube bend radius	97.6 mm ($3\frac{27}{32}$ in)
Maximum possible number of full length tube scans	2237 (86% total tubes)
Partial scans (straight portions only)	229
Probable no. untested tubes	134 (5% total tubes)

CANSCAN specification

Duty cycle

(a)	simultaneous tube scans	2
(b)	probe inspection speed	0.5 m/s (30 m/min)
(c)	scan time (2 tubes)	80 s
(d)	tubesheet walker transfer time	20 s
(e)	target inspection time (time to scan steam generator with 2600 tubes; 0.75 load factor)	48 h

Defect detection

The minimum detectable defects would be:

(a)	tube through-wall hole	0.3 mm diameter
(b)	eccentric external wall fretting wear	0.1 mm

The system electrical signal-to-noise ratio for a 1.6 mm (1/16 in) diameter through hole would be greater than 75 : 1.

3.14.5 Tube thickness monitor

A technique which is particularly applicable to the monitoring of heat exchanger wall thickness uses an eddy-current technique with a data storage method using a computer to establish changes in thickness. The equipment, as described by Boogard [9] was developed from other established principles [10–12].

Thickness measurement is carried out with an internal bobbin type eddy current probe of absolute coil configuration. With an applied frequency of 1 kHz the instrument has about the same sensitivity for internal and external wall thickness variations, which means that a real wall thickness measurement is obtained.

The probe is pulled at a speed of 32 m/min and the signal recorded on an oscillographic recorder. Fig. 3.19 shows the record of one such measurement. Tubesheets (1), baffles (2) and the reference tube (3) show up clearly. A distinct wall thinning occurs at the centre, a length of about 1 m. Measurements are accurate and reproducible to within 0.05 mm.

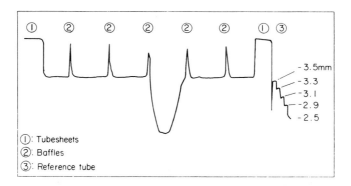

Figure 3.19 *Heat exchanger tube: wall thickness record by eddy current probe. (1) tubesheets; (2) baffles; (3) reference tube.*

Initial records for a heat exchanger tube in 'new' condition, reported by Boogard, showed a substantial thinning over a length of 1 m in a 6 m tube. The wall had a thickness of 2.1 mm instead of the required 3.5 mm. Although new and unused the tube had to be plugged.

3.14.6 Computerized data processor

Very long plots would be needed with direct recording; to select the relevant data quickly (at least within the duration of a plant stop) and arrange it in a simplified form for further interpretation a computer is used.

Based specifically on the detection of corrosive thinning, use is made of the observation that corrosion always proceeds in the area where it originally started, so that a place of minimum wall thickness in a tube will never shift. Moreover, in view of strength aspects, the value of the minimum wall thickness is the criterion for rejection.

A punched card recording the minimum wall thickness, tube identification number and test data is read into the computer and stored in the computer memory, together with the data obtained from previous measurements. This procedure is shown in Fig. 3.20. A computer program is applied to check the data for validity before storage. Thereafter it is possible to compile and print out several surveys:

● a list of tube identification numbers, arranged according to minimum thickness in 0.05 mm increments;
● a survey of wall thickness decrease for each tube over the period since the previous test, arranged in 0.05 mm increments; the average decrease over this period is also calculated;
● a survey of wall thickness and wall thickness decrease values found in all measurements, arranged according to tube identification number; this survey contains the complete (reduced) information on the heat exchanger tubes.

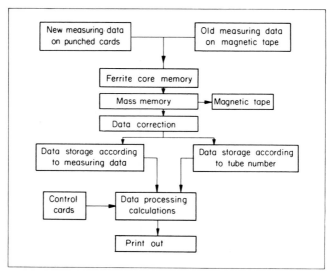

Figure 3.20 *Data processing: heat exchanger tube thickness measurement.*

3.14.7 Eddy current tester (computerized) case study: corrosion distribution

Corrosion in a heat exchanger depends on:

(a) Liquid temperature.
(b) The liquid distribution such as the amount of liquid passing through each tube in unit time. Fouling causes the liquid distributing system to become irregular, which is reflected in variations in the degree of corrosive attack.

Two different corrosive distributions are shown in Figs 3.21 and 3.22.

Central corrosion is shown in Fig. 3.21. When the liquid distribution system was examined more closely it appeared that fouling was particularly strong near the circumference. This means that the innermost tubes had been subjected to a relatively high liquid load, so that the corrosion rate was highest here. The liquid distribution system had to be cleaned.

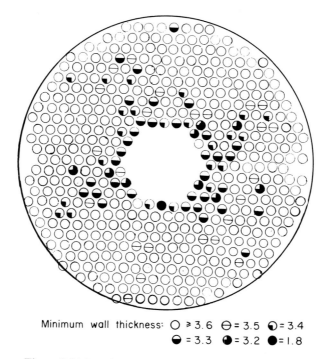

Minimum wall thickness: ○ ≥ 3.6 ⊖ = 3.5 ◐ = 3.4
◑ = 3.3 ◕ = 3.2 ● = 1.8

Figure 3.21 *Local corrosion due to irregular liquid supply.*

Offset corrosion in two groups is shown in Fig. 3.22. In this tube plan the open circles represent the tubes in which corrosion was least severe. They are found to be just opposite the steam inlet, where the temperature is lowest. The full circles represent tubes having the severest attack. They are grouped together in an oblique band. There was no question of fouling. The cause was a slightly less corrosive-resistant charge of tubes used in this configuration when the heat exchanger was constructed.

3.15 Crack growth monitoring: pressure vessels

Models which have been suggested as the basis for life assessment relate crack length, growth rate and life to such parameters as:

- net section stress [13].
- equivalent stress [14].
- crack opening displacement (COD) [15].
- reference stress [16].
- limit load [17].

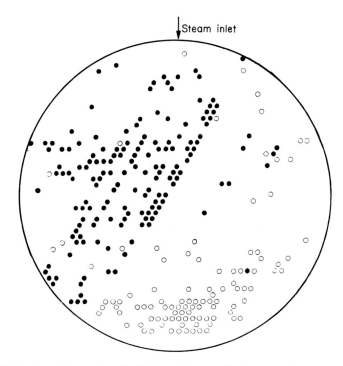

Figure 3.22 *Regions of exceptionally high and abnormally low corrosion rate as determined by eddy current monitoring:* ●, *high corrosion rate;* ○, *low corrosion rate.*

To relate plant and laboratory conditions cracks were monitored [18] at a pressure vessel testing facility of the Central Electricity Generating Board's Marchwood Engineering Laboratories (MEL) in which full scale tests up to 800 °C and 1700 bar (170 MPa) can be performed. Crack depth measurements were made continuously using: (a) electropotential drop methods; and (b) capacitance strain gauge monitoring. Both methods were regarded as being reliable and giving more than 1500 hours of operation at 565 °C.

In practice, changes in the potential across the defect are used with either experimental or theoretical calibration curves to give the crack length. Theoretical calibration curves for uniform current fields and fields produced by application of the current at a point are shown in Fig. 3.23.

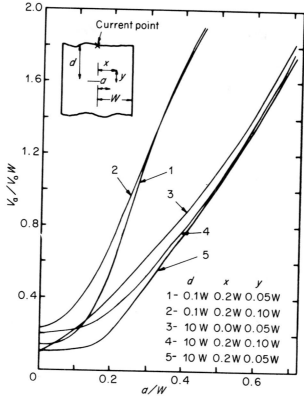

Figure 3.23 *Potential drop calibration curve:* $V_0 = $ *potential gradient for uncracked specimen;* $V_a = $ *measured potential at points x, y; a = crack length.*

3.16 Crack depth measurement

3.16.1 Capacitance strain gauges

Capacitance-type strain gauges developed at CERL [19] are suitable for bridging defects and monitoring COD. Such gauges have a low drift because of their low compliance and have a temperature maximum well above 600 °C, the limit of the BLH gauges.

The CERL/Planer gauge shown in Fig. 3.24 consists of two compliant arches 20 mm long, spot welded together and to the surface to be gauged. The high compliance of the arches results in the two small mounting spot welds being lightly stressed and the two capacitor plates separate when the structure is strained.

The resulting capacitance change varies nonlinearly with strain, over a strain range of 10 000 μm/m. It can readily be measured by connecting each plate by a

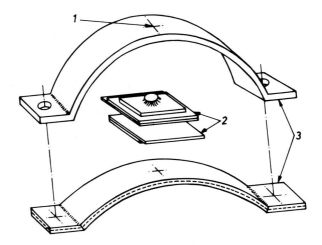

Figure 3.24 *Exploded view of capacitance strain gauge: 1, crown of arch; 2, capacitor plates; 3, foot of arch.*

screened lead to a Blumlein transformer bridge (Fig. 3.25). This bridge has the ability to measure accurately a small capacitance in the presence of very much larger cable capacitances to earth. A range of such bridges and their associated multichannel switching units is made for site use within this gauge. They are capable of operating with up to 100 m of polythene-insulated coaxial cables and 10 m of high temperature 3 mm diameter mineral-insulated stainless steel sheathed cable. The latter should be fitted with special seals at the hot ends.

Since the capacitance strain gauge, unlike a resistance strain gauge, is attached only at two points 20 mm apart it is not destroyed by the presence of a crack underneath it. Examination of micrographs of such cracks shows widths at the surface (not all such cracks are at the surface) of about 15 μm. As a displacement this would produce an apparent strain of the order of 750 μm/m from a capacitance strain gauge, so that a change of up to this figure could be expected from a gauge straddling a crack.

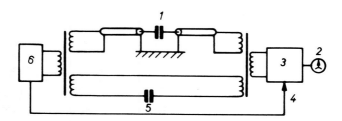

Figure 3.25 *Capacitance bridge circuit: 1, unknown capacitance; 2, output; 3, detector; 4, phase reference; 5, standard capacitance; 6, oscillator.*

The capacitance strain gauge first applied in this manner was on the Cr–Mo–V main steam pipe of a ten-year-old 120 MW steam turbine in the CEGB south-east region. Circumferential cracks were discovered in the HAZ adjacent to a weld between the 216 mm diameter main steam pipe and a strainer during routine nondestructive testing. They are shown diagrammatically in Fig. 3.26. The diametrically opposite positions of the two cracks and the layout of the pipework suggested that the cracking was caused by bending moments in the pipe.

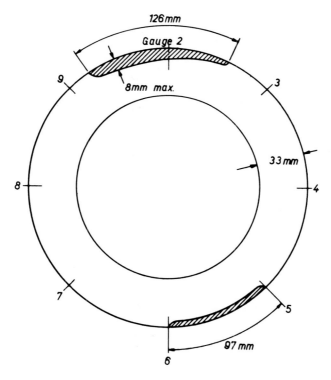

Figure 3.26 *Circumferential cracks of main steam pipe: determined by capacitance strain gauge.*

3.16.2 Electropotential drop method

An electric current flowing through a material produces a potential field and, if the current application points are sufficiently spaced, the potential gradient will be uniform over most of the material.

The presence of a crack, or any discontinuity in this field, will disrupt the potential in a manner which is a function of the crack dimensions. This factor forms the basis of the electropotential drop technique. Experimental measurement of the potential drop across a crack is compared with that across an uncracked area, and changes in this relationship can be related to crack growth [18].

In applying this technique to metallic materials the problem of measuring very low potential drop values, resulting from low electrical resistance, is encountered. However, appropriate solutions are available depending on whether an alternating or direct current system is used.

(a) Alternating current system

A constant a.c. supply with a frequency in the range 5 Hz to 20 kHz is commonly used. To allow measurement of very low potential levels, the potential is increased either by using high currents or by amplification of the output potential. The existence of very stable a.c. amplifiers has resulted in amplification being the preferred system with a.c. supplied as, in principle, quite sophisticated phase sensitive or phase lock loop detectors can be used to extract the signal from background noise. However, problems can arise experimentally as a result of cable interaction effects and the complexity of the circuits used. In this context, the mutual interference between current and potential leads and the very intense fields generated by furnace currents are particularly troublesome.

(b) Direct current system

In this system the experimental arrangement is identical to that of the a.c. system. However, high stability, fast response, d.c. amplifiers, although available, are relatively expensive. Consequently, alternative methods are employed to increase the potential to directly measurable values by increasing the current density. This can be achieved by using higher currents or by moving the current application points closer to the crack or, alternatively, through a combination of these actions.

Long-term operational experience gained during uniaxial and pressure vessel testing has indicated that the d.c. technique is easy to handle and reliable. For these reasons, and because of its relatively low cost, the d.c. technique was selected for this work.

References and list of companies

1 Polaroid (UK) Inc., St Albans, Herts, UK.
2 'Ultrasonoscope pipe tester type TG-150', Ultrasonoscope Company (London) Ltd, Sudbourne Road, London SW2, UK (Tel. 01-274 4041).
3 Tombs, F. (1979), '25 years of nuclear power' *Atom*, No. 272, June, 146–9.
4 Peters, B. F., O'Malia, J. A. and Greenwood, B. W. (1969), 'Ultrasonic evaluation of tube wastage in marine boilers', *Trans. IERE London, Joint Conference on Industrial Ultrasonics*, pp 177–84.

5 Hocking Electronics Ltd, 40 London Road, St Albans, Herts AL1 1NG, UK.

6 Marchent, G. R. (1978), 'An instrument for the non-destructive testing of wire ropes', *Systems Technology,* August, No. 29, 26–32.

7 Stevens-Guille, P. D. (1973), 'Steam generator tube failures: a world survey of water-cooled nuclear reactors to the end of 1971', AECL Report No. AECL-4449.

8 Wells, N. S. (1981), 'An automated eddy current in-service inspection system for nuclear steam generator tubing', Report AECL-7406.

9 Boogard, N. J. (1978), 'Measurement of heat exchanger tube thickness' *NDT International,* January, 123–5.

10 Cecco, V. S. and Bax, C. R. (1975), 'Eddy current in-situ inspection of ferromagnetic monel tubes' *Mater. Evaluation* January, 1–4.

11 Dau, G. J. and Libby, H. L. (1974), 'Steam generator and condenser tubing non-destructive inspection by the eddy current methods', *Inst. Mech. Engrs Conf. Publ.* 8.

12 Stumm, W. (1976), 'The adaptation of eddy-current test methods to the multiple problems in a reactor' *Materialprufung,* **18,** No. 4, 109–14.

13 Harrison, C. B. and Sandor, G. N. (1971), *Eng. Fracture Mech.* **3** (3), 403.

14 Haigh, J. R. (1974), PhD Thesis, (CNAA) London.

15 Formby, C. L. (1972), CEGB Report RD/B/R2067.

16 Williams, J. A. and Price, A. T. (1974), CEGB Report R/M/R205.

17 Goodall, L. W. and Townley, C. H. A. (1974), CEGB Report.

18 Coleman, M. C., Fidler, R. and Williams, J. A. (1976), 'Crack growth monitoring in pressure vessels under creep conditions', *The Detection & Measurement of Cracks,* The Welding Institute, Cambridge.

19 Noltingk, B. E., Owen, C. K. V. and O'Neill, P. C. (1972), CEGB Report RD/L/R1748.

4

Acoustic emission (stress wave) monitoring

4.1 Introduction

A technique of considerable potential has become available through research into fibre dislocations and the signals arising from the internal re-distributions in an over-strained material. Several applications for this technique have already been successfully established. Some case examples are given in this chapter. Other applications relating to leak monitoring are given in Chapter 6.

The subject of stress wave emission analysis as known today really began with the PhD thesis of Kaiser [1]. This dealt with the use of electronic amplification to listen to materials under varying loads. He studied a wide range of materials, and found (among other things) that if the load were removed and re-applied, very little noise was emitted until the original load was exceeded. This became known as the 'Kaiser effect'. The phenomenon of 'tin cry' is well known, it is the audible manifestation of a form of acoustic emission during crystelline changes in stressed tin. Dunegan and others in the USA showed that high frequencies (up to 3–5 MHz) could be used, and this permitted the technique to be used in noisy environments, since ambient and machinery noise levels are rarely troublesome at such frequencies.

In the application of this phenomenon the terms acoustic emission (AE) and stress wave emission (SWE) are interchangeable although strictly AE refers to the response at the detector while SWE is the originating source.

4.2 Acoustic emission – how it works

When a crack or other defect grows, i.e. gets longer or larger, it releases bursts of energy in the form of stress waves called 'acoustic emission'. Even if the defect is a microscopic crack, pressure or strain on the structure will make it

86

generate stress waves. The stress waves can be detected by piezoelectric sensors arranged in an array around the structure.

Acoustic emission (AE) is different from other methods in that the energy comes from the material itself. As a result, AE is more sensitive to growing defects and less reliant on operator interpretation than methods that require external stimulation, such as ultrasonics and radiography.

A simple AE system consists of sensors, preamplifiers, signal conditioners and counters. The piezoelectric sensors – so sensitive they can detect material movement on a molecular scale – are placed at strategic locations on the structure to be tested. Sensor signals are picked up by the preamplifiers, converted to digital pulses, and counted. An experienced AE test operator can use counts from several sensors to detect the presence, location and severity of a flaw.

An advanced AE system provides the same data as a simple AE system, but uses a minicomputer to calculate flaw location. It locates the source by triangulation, using data from three or more sensors. The minicomputer also analyses the stress waves and tells the operator a great deal about the type and severity of a flaw.

4.2.1 Stress wave emission sources

Arrington [2] indicated that the stress waves generated by material deformation, degradation and flaw growth arose from a driving force due to the strain energy stored in the material. These emissions are generally broadband transients at their source but the waveform detected by the transducer is distorted by frequency-dependent attenuation, dispersion, multipath propagation and mode conversion.

The sources and amplitudes of emissions are very diverse, ranging from crack growth in brittle materials (which may well be audible) to microvoid coalescence in mild steel which can be quiet SWE-wise. Some of the more common sources with examples of the materials involved are given in Table 4.1 (which is in no way exhaustive).

Table 4.1 Emission mechanism in metals

Emission source	Material
Twinning	Tin and titanium
Stress corrosion	Aluminium (7079 – T 651)
Martensitic transformation	Gold/cadmium alloy (47.5% Cd)
Hydrogen embrittlement	4340 steel
Plastic deformation (dislocation movement)	Magnesium, copper, brass, iron

Emissions due to the deformation of inhomogeneous materials such as concrete and composites are more complicated than those shown in Table 4.1; the failure of fibre composites may occur in either a brittle (low-work fracture) or in a fibrous (high-work fracture) manner; other aspects of fibrous failure involve debonding, matrix failure, fibre failure and pull-out which all contribute to the acoustic emission signature.

The effect of flaws is to intensify stress in their immediate vicinity, thus the material at the crack tip has used up more of its stress/strain resistance than has the remainder of the material. This generally means that it has also produced more emission. As the crack grows there are two main contributions to the crack tip: (a) plastic-zone growth in advance of the crack; (b) jumps of the crack front.

4.2.2 The Kaiser effect

Acoustic emission is not repeatable. The Kaiser effect is a basic law concerning the non-repeatability of acoustic emission. Since acoustic emission is generated by a material's reaction (deformation) to mechanical stress, it is useful to realise that most elastic bodies will yield no acoustic emission when the mechanical stress is re-applied on a second loading.

The Kaiser effect is noticeable in metals and other materials but some composite materials, such as fibre-reinforced plastic structures, are viscoelastic such that the load is distributed over a period of time [3]; they do not exhibit the classic Kaiser effect [4].

4.2.3 The felicity effect

Composites do not exhibit the irreversible Kaiser effect. If a composite is not held at load, but is unloaded immediately, acoustic emissions will occur at loads below the previous maximum. This is known as the felicity effect. The percentage of previous maximum load at which a material emits on a second loading is known as the felicity ratio.

The felicity effect and felicity ratio for the fibre breakage mechanism can both be good indicators of the residual strength (load carrying capacity) of a composite material.

4.3 Acoustic signal detection

In large metallic structures emissions arise from crack initiation and growth or by plastic deformation at the crack tip. In composite structures emissions resulting from sources such as fibre breakage, delamination, and matrix crazing are of concern.

To generate AE some form of stress (applied, thermal, residual, etc.) must be present to provide a release of energy from a potential defect. A dynamic situation must exist. Acoustic emission generated by many structures is irreversible.

A hypothetical AE waveform shown in Fig. 4.1 illustrates the characteristics generally used for analysis purposes. These include: number of threshold crossings (counts), peak amplitude, pulse duration, and rise time.

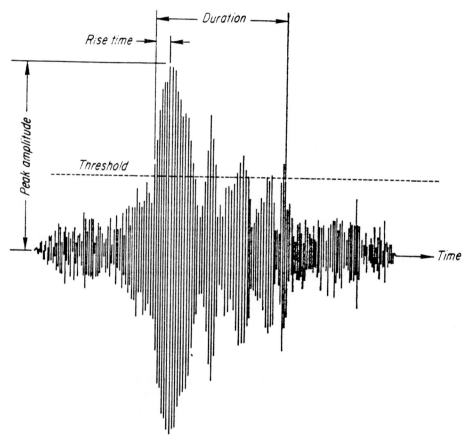

Figure 4.1 *Hypothetical waveform of an acoustic signal.*

The AE response of materials varies considerably (Table 4.2) and is also dependent on environmental and metallurgical conditions.

4.4 Energy release characteristics

When a crack grows energy is released as indicated in Fig. 4.2. Stress waves are a transitional form of energy. The amount of the energy change which is

Table 4.2 Factors that influence acoustic emission detectability

Factors resulting in higher amplitude signals	Factors resulting in lower amplitude signals
High strength	Low strength
High strain rate	Low strain rate
Anisotropy	Isotropy
Nonhomogeneity	Homogeneity
Thick section	Thin section
Twinning materials	Non-twinning materials
Cleavage fracture	Shear deformation
Low temperatures	High temperatures
Flawed material	Unflawed material
Martensitic phase transformations	Diffusion controlled transformations
Crack propagation	Plastic deformation
Cast structure	Wrought structure
Large grain size	Small grain size

translated into a stress wave depends upon the nature of the partition process and such factors as:

(1) The type of relaxation event, e.g. brittle (intrinsically noisy and of short relaxation time) or ductile (quiet and of longer duration).
(2) The rate of relaxation and repeated events.
(3) The microstructure (e.g. the effect of discontinuities, grain size and the presence of alloying agents).

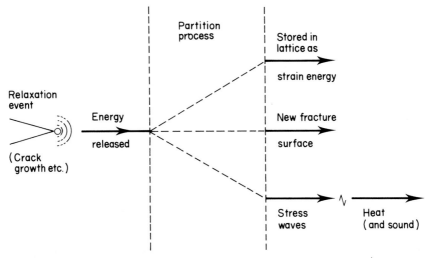

Figure 4.2 *Partition process: energy transformation and acoustic emission.*

The rise time of an acoustic emission signal is related to the lifetime of the deformation or fracture process from which it is generated. The amplitudes of the signal is also related to the magnitude of the deformation or fracture process. For example if the source of a wave is the growth of a crack then the rise time of the associated acoustic emission signal is proportional to the time taken

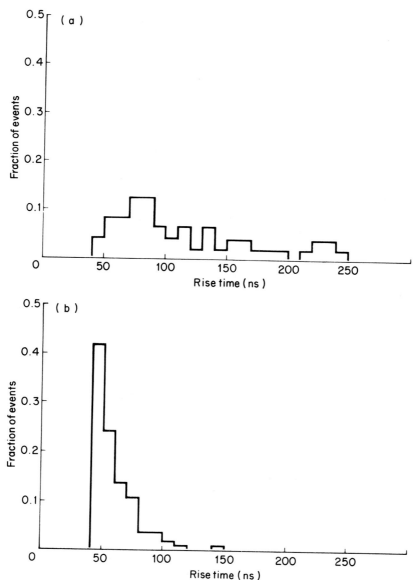

Figure 4.3 *Comparison of rise time distribution of an EN30A steel: (a) intergranular fracture; (b) ductile fracture.*

for the crack to grow, while the amplitude is proportional to the area over which the crack spreads.

It is possible to analyse acoustic emission data to obtain the velocities and areas of individual crack advance events and thus distinguish between different types of fracture processes (Fig. 4.3).

4.5 Stress wave propagation frequencies

Any abrupt incremental crack growth, plastic boundary zonal change, deformation or transformation generates a stress wave. This is initially a spherical wave front which propagates to the surface boundaries. Interaction with second phase particles, grain boundaries, lamination defects and reflections from the surface produces a complex field. Both compression and shear waves are generated internally and trains of surface waves on the external surfaces (Fig. 4.4). Three acoustic field regions are involved:

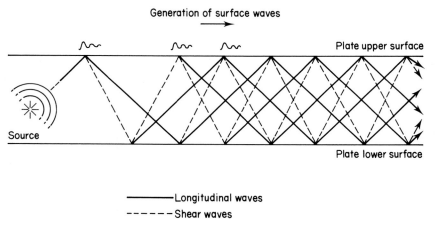

Figure 4.4 *Wave propagation within the wall thickness of a plate: single ray path from source.*

(1) Near field – A sensor on the surface responds to stress waves direct from the source (there will also be multitudinous reflections as well).
(2) Pseudo-near field – Depending upon the material geometry and attenuation factors, a sensor might respond to both direct line-of-sight stimulation by compression, shear or surface waves.
(3) Far field – Compression and shear waves are then mostly attenuated so that the less dispersive surface waves dominate.

It follows that for location studies it is better to work in the far field. Experience shows that a value of velocity close to that of the group velocity of the wave [5] is most suitable although inaccuracies can result owing to the velocities of different Fourier components of the wave train.

Longitudinal stress waves are less intense than are shear waves. Surface waves are rapidly attenuated by obstructions and other surface perturbations. These differences influence the spreading characteristics and accordingly the accuracy with which a source can be located.

Signals arrive at different sensors (as for example the four sensors in a quad array) at different times. The intervals are small but significant (thus differences of 100 μs might apply to spacing of 3 metres).

Small increments of crack growth can generate up to 1000 events per millimetre; the large size of this burst of emission depends upon: (i) microstructure of the material; (ii) failure mechanism.

Acoustic emissions have fairly broadband frequency spectra – usually covering a range from 0.1 to 1 MHz. Detection equipment operates between 0.06 and 0.65 MHz.

4.6 Instrumentation

4.6.1 Data acquisition in real time

By separating acquisition and detailed analysis into two separate yet compatible processing systems Data Acquisition in Real Time® (Dart) is able to operate independently or can be used as an advanced front end of the Dunegan/Endevco 1032 flaw detection system. Components of the DART include:

(a) Dual signal processor (DSP) boards – These provide logarithmic amplification, unique signal discrimination capabilities and circuitry to derive complete event descriptions for each of a pair of channels. Signal attributes include time of threshold crossing, time at peak amplitude, rise time, peak amplitude, duration, average signal level and counts. Thresholds may be set by channel or all channels can be adjusted to a variable signal-to-noise ratio and can be made automatically adaptive to a settable signal-to-noise ratio.

(b) Front end processor (FEP) board — This services interrupt signals from the DSP boards. In order to provide the user with the highest of data rates, the FEP board microprocessor has little to do beyond servicing channels. When the system is not busy acquiring data, the microprocessor capability is used to provide extensive diagnostics.

(c) Input/output processor (IOP) board – This formats data and provides communication with external devices. This second microprocessor also provides a 500-event buffer to store data while it is consolidating information and passing it to peripheral devices. Output formats include RS232 (300 or 9600 baud) serial data ports.

The combination of dual microprocessors and a unique conditioning circuitry truly marks a significant advance in acoustic emission instrumentation.

The logarithmic domain signal processing provides event features far in excess of the traditional amplitude, rise time, duration, counts and arrival time. The waveform shown in Fig. 4.5 represents a typical AE event in the logarithmic domain, where the exponential decay associated with material damping and

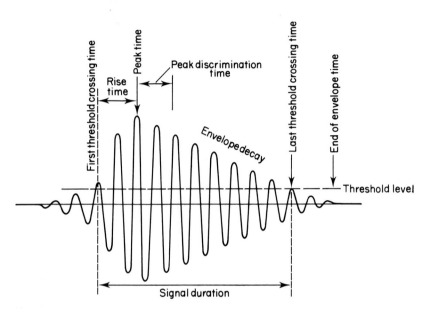

Figure 4.5 *Acoustic emission events as displayed in the logarithmic time domain.*

other factors appears as a linear decay on a logarithmic scale. With appropriate timing and envelope detection it is possible to:

● Provide an adaptive dead time based on event amplitude and optimized for the material under test.
● Allow the event discrimination to be set experimentally for the particular test situation.
● Maximize burst event rates by minimizing dead time.
● Discriminate and alert the operator against overlapping events.
● Eliminate electromagnetic interference early in the data chain.
● Avoid presentation of single events as dual events.
● Determine timing on the basis of signal peak.

Data logarithmic amplitude recording benefits from experience with linear amplifiers where gain can be set initially to what seems satisfactory but the next few bursts might go right off the scale for they may be too small even to see (Fig. 4.6 (a)).

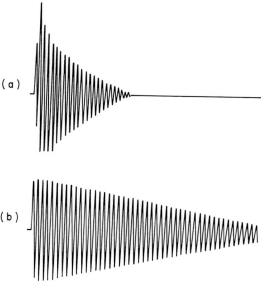

Figure 4.6 *Comparative recordings to show the effect of logarithmic amplification:
(a) broad-band AE – bursts which exceed linear scale range; (b) logarithmic amplification
used to contain AE bursts.*

With a logarithmic amplifier all signals are within a working range (Fig. 4.6
(b)). Big ones or small ones become measurable because the logarithmic amplifier tends to present the physical world in a comprehensible package (like any
dB scale). The logarithmic amplifier provides a new dimension to event descriptions with several advantages over the traditional linear amplifier.

Typically, a logarithmic amplifier dynamic range of 80 dB corresponds to a
linear dynamic range of 10 000: not an easy feat to achieve in a linear amplifier
without sacrificing stability and accuracy. For a logarithmic amplifier, however,
a 100 dB signal is only five times greater than a 20 dB signal. Thus the logarithmic amplifer captures AE signals of widely different amplitudes without loss of
stability or accuracy.

4.6.2 Dunegan/Endevco 8000 system®

When only a few channels with a sophisticated data analysis capability is
required, large-scale systems such as Dunegan/Endevco's 1032D® are too large
and the new 8000 has been introduced which has a high-speed data acquisition,
analysis and display in a self-contained table-top unit.

Incoming data can be processed from as many as eight sensors. High speed is
derived from distributed processing employing dedicated microprocessors for

data acquisition, characterization and display, thus allowing the built-in LSI 11® minicomputer to concentrate on data analysis and manipulation.

The built-in split-screen display maps the location of each active acoustic emission source and draws attention to areas of high activity. A box is drawn around each of these areas. All boxes are numbered and can be analysed simultaneously on the lower half of the screen. The lower half of the screen can also be used as a display area for graphs and tabulations based on the statistical analysis being performed on the data. Display can be changed during the test to enable users to identify and concentrate on 'significant' activity, which can often be masked by less significant activity.

Extended software is available to deal with the system for specific applications. Each material and structural geometry carries sound waves differently. For example, tests on thick complicated-geometry castings may require correction for deviation in the acoustic path, while tests on GRP structures may require very high data rates. The extended analysis software package enables the user to optimize the system to suit the specific application.

Data storage is via dual, double-sided, double-density 8-inch (200 mm) floppy disks providing one megabyte of storage for test data (approximately 40 000 events).

4.6.3 Pulse analysis: B & K type 4429 systems®

Acoustic signals are detected by transducers followed by possibly preamplification and conditioning before entering the analyser as shown in Fig. 4.7.

The B & K type 4429® acoustic emission analyser accepts up to four acoustic emission transducers connected to the four separate counting channels. These four channels are used in different ways according to which of the three operating modes is chosen. All three modes are based on the measurement of the same parameter, that is the time during which the AE signal level exceeds present

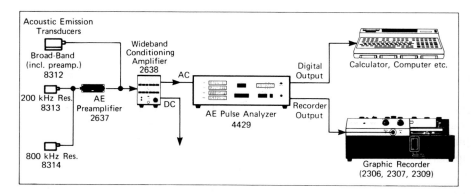

Figure 4.7. *B & K acoustic emission instrumentation system.*

limits. The input signal level is adjusted with respect to these limits and the system noise level by the wideband conditioning amplifier which has adjustable gain in steps of 1 dB up to 60 dB. The three operating modes are as follows.

(a) Weight mode

All four channels are used by measuring the time during which the signal level exceeds four preset trigger levels having an amplitude relationship 1, 2, 4 and 8. Multiplying (weighting) by the amplitude difference between these levels results in a measured value closely proportional to the area under the level versus time curve of the AE signal as shown in Fig. 4.8.

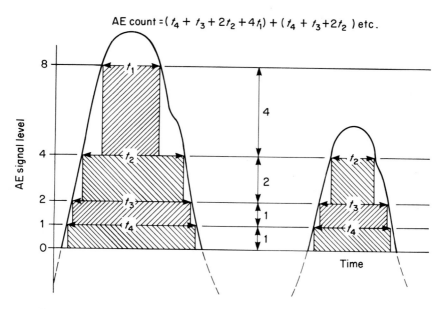

Figure 4.8 *Approximation to the measurement of the area under an AE curve.*

Multiplication is achieved by means of the components shown in Fig. 4.9. Comparators 4, 3, 2 and 1 are present to open pulse train gates during the period which the AE signal level exceeds 0.25, 0.5, 1 and 2 volts respectively. Pulse train frequencies of 1, 2 and 4 MHz are admitted to the respective pulse counter for each channel. Thus, if channel 3 or 4 gates are opened for 50 μs, a pulse count of $5 \times 10^{-5} \times 10^{6} = 50$ results; correspondingly if channel 1 gate is opened for the same period the pulse count will be $5 \times 10^{-5} \times 4 \times 10^{6} = 200$. The AE count accumulated in each channel is displayed on four 7-digit LED displays and the sum of counts in the individual channels, which represents the accumulated area under the AE curve, is shown in the Σ display.

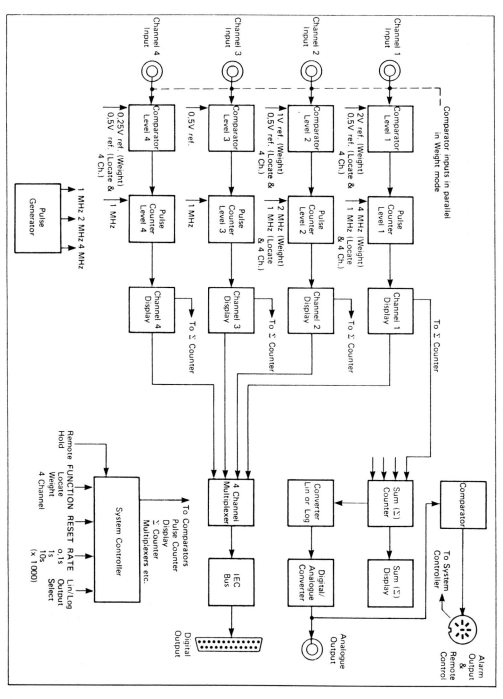

Figure 4.9 *Acoustic emission pulse analyser B&K type 4429: block diagram showing most significant components.*

A typical instrumentation system for evaluating AE activity by the weighting method is shown in Fig. 4.10.

Changes in the degree of AE activity with respect to time can be displayed and recorded by selecting the 'rate' function which reads out the total count accumulated dring a selected time period. Counts per 0.1 s, 1 s, 10 s, 100 s,

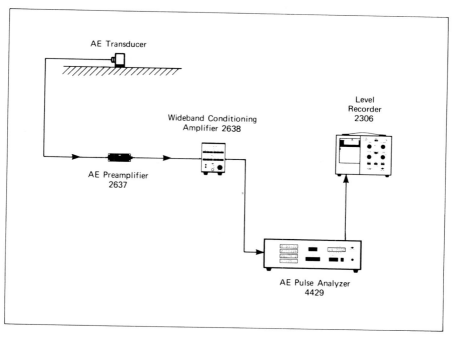

Figure 4.10 *Instrumentation set-up for evaluating AE activity using the B&K type 4429 in the weight mode.*

1000 s and 10 000 s are selectable. Readout and resetting with the selected time period is an automatic and continuous process. A typical rate plot is shown in Fig. 4.11.

(b) Four-channel mode

In this mode an AE count is made which is similar to the ring down count method. The four-channel mode registers the time (in μs) during which the AE signal level exceeds a common trigger level in each channel whereas the 'ring down' count method registers a count of 1 each time a preset threshold level is exceeded.

Because only one threshold level is used it is possible to count in up to four separate measuring channels simultaneously. The transducers can be positioned on four remote measuring points on a single proof-test object or on completely

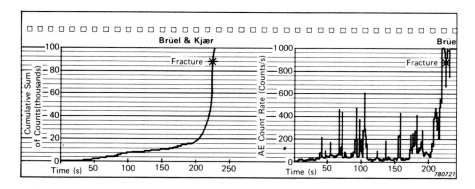

Figure 4.11 *Level recording during bending test on a notched specimen of glass fibre reinforced epoxy sheet.*

separate measuring objects. A typical instrumentation system is shown in Fig. 4.12. In the four-channel mode the sum (Σ) display indicates the total accumulated sum of counts in the four channels and, if desired, the count rate as described in the weight mode.

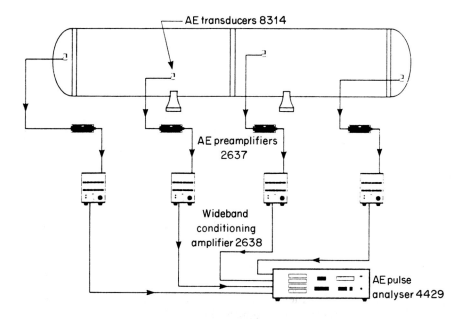

Figure 4.12 *Four separate transducers used in the four-channel mode to evaluate AE.*

(c) Locate mode

Operation in this mode enables the source of a burst of AE activity to be localized to a particular small area. An ultrasonic or X-ray examination of this area may then establish the nature and severity of the weakness.

Four AE transducers mounted on the test object are fed to the four input channels of the pulse analyser as shown in Fig. 4.13. After the system is reset, an AE pulse arriving at any one of the four transducers will admit a 1 MHz pulse train to the pulse counters associated with the remaining three channels. When the AE pulse reaches the remaining three transducers the associated counters are stopped, leaving a display of the relative times of arrival (in μs) of the AE burst at three transducers with reference to the fourth.

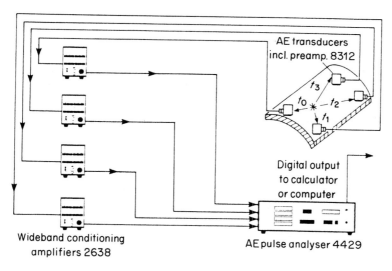

Figure 4.13 *Location of the source of AE.*

The relationship between the three times displayed defines the position on the surface of the specimen under which the burst apparently originated. For linear location systems, i.e. on pipelines, only two transducers are required.

Any number of pulse analysers may be coupled in parallel to form large-scale two- or three-dimensional location systems with the delay times being read out to a calculator or computer.

4.6.4 Sensor arrangement

A typical computer-based system uses several monitoring arrays. Each array has four piezoelectric sensors arranged in an independent unit. An array monitors one complete section of a structure, and a test system may have up to eight or more arrays. Multiple arrays may be placed anywhere on a structure because each array is independent of the others.

To locate a flaw, acoustic energy emitted from the flaw must reach all four sensors in an array within a certain time. The time allowed is based on the size of the array and wave speed of the material. To allow for wave speed, array patterns have been devised to test various shapes of pressure vessels, vessel heads, pipelines and other structures. These patterns can be mixed to monitor piping, vessel bodies and spherical or elliptical heads at the same time. In some cases, sensors can be left on the structure for future testing.

4.6.5 Computer processing

A block diagram of the computerized multichannel source-location system utilized in this work is shown in Fig. 4.14. The system accepts inputs from up to 32 transducers arranged in 4-transducer arrays for planar location and 2-transducer arrays for linear location. The system display oscilloscope (CRT) provides a real-time display of the location of each source in relation to the pos-

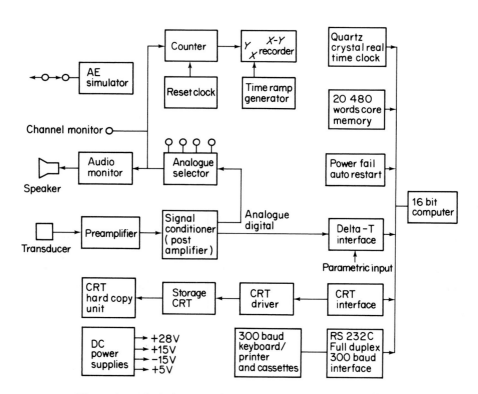

Figure 4.14 *Block diagram of computerized source location system.*

ition of the transducer array involved. Fig. 4.15 represents a hardcopy reproduction of a typical CRT display including the array transducers, active sources, and a histogram of events per array. In addition to the real-time display the important information associated with each event located by the system is printed out by a high speed teleprinter. A typical printout is shown in Fig. 4.16.

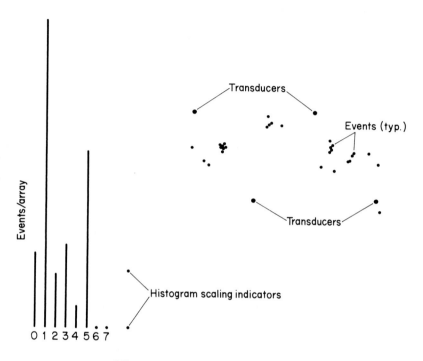

Figure 4.15 *Typical CRT presentation.*

The transducer used in each of these applications is a compression-sensitive, differential transducer with a resonance frequency of 220 kHz. The transducer measures 19.8 mm by 26.2 mm, weighs 26 grams and exhibits a peak sensitivity of -75 dB ref. 1 volt per microbar (0.1 Pa). The transducers were coupled to the structures through magnetically attached 300 mm long waveguides.

4.6.6 Dunegan/Endevco 1032 system®

In one application this system accepts input from up to 35 sensors, that is it can handle up to 8 separate arrays which can be distributed around a structure to provide overall coverage. An Interdata Model 7/16® computer is used to pro-

DATA RUN   2

EVENT	ARY	DT0	DT1	DT2	DT3	X	Y	COUNTS	TIME	PARAM
CAS ON										
1	3	384	0	20	340	2.0	-2.1	102	0	0899
73	3	379	0	19	335	2.0	-2.1	174	25	0948
699	0	1321	65	1160	0	49.9	-273.3	162	46	1185♦
781	3	1022	233	745	0	56.7	-80.0	8	50	1213
784	0	117	78	636	0	-115.9	-20.3	9	51	1215
826	0	1372	71	1211	0	50.4	-278.0	233	55	1238♦
854	0	113	78	636	0	-115.9	-19.9	16	56	1246
1862	0	1411	0	1228	70	25.1	-250.2	182	65	1292♦
1947	0	230	1256	0	750	105.7	44.7	30	70	1320
1974	0	456	726	410	0	104.7	-11.3	5	72	1333
2028	0	1462	0	1568	434	-65.9	-199.1	17	75	1367♦
2042	3	959	490	1098	0	-154.9	-93.2	37	76	1377
2043	0	162	92	578	0	-119.4	-25.6	16	76	1379
2046	3	890	0	1145	510	-78.4	-64.8	78	76	1380
2064	0	150	92	673	0	-116.5	-21.7	6	79	1406
2068	3	813	0	1431	540	-108.8	-27.6	2	79	1409
2072	0	1377	73	1289	0	62.6	-464.0	120	79	1412♦
2073	3	1250	0	1311	241	-58.9	-276.6	8	80	1412♦
2094	0	161	96	677	0	-116.9	-22.4	13	82	1433
2101	3	879	0	1492	601	-109.8	-28.0	1	82	1437
2143	0	122	65	642	0	-114.9	-21.2	5	87	1481
2152	3	1008	603	1057	0	-172.5	-97.2	5	88	1488
2164	0	122	65	643	0	-114.9	-21.2	7	90	1506
2189	3	923	39	1011	0	-128.4	-415.7	12	93	1533♦
2191	0	122	65	646	0	-114.8	-21.1	5	93	1536
2214	0	126	68	642	0	-115.1	-21.4	3	97	1559
2223	3	761	0	1080	568	-80.4	-42.7	8	97	1564
2226	3	0	1007	808	1101	-94.5	60.9	47	98	1575
2231	3	935	607	1074	0	-158.1	-72.7	18	99	1581
2259	0	1368	0	1152	49	29.7	-211.7	53	105	1620♦
2268	3	940	566	1076	0	-157.6	-79.4	9	106	1629
2291	3	0	942	812	1101	-86.4	48.3	78	108	1644
2336	0	1342	43	1156	0	44.4	-254.4	106	117	1700♦
2341	3	945	0	1180	662	-72.9	-57.1	2	118	1702
2355	0	141	91	669	0	-116.5	-21.1	15	122	1727
2361	3	942	615	1082	0	-158.1	-72.4	21	123	1732
2366	3	0	925	812	1101	-84.6	45.5	76	124	1738
2374	0	154	97	678	0	-116.8	-21.7	3	126	1752
2400	0	165	109	686	0	-117.6	-22.0	1	130	1775
2401	0	320	504	511	0	-159.4	-17.0	30	130	1777
2407	0	1418	82	1215	0	47.4	-217.1	85	132	1783♦
2415	3	918	590	1057	0	-158.0	-72.9	12	132	1789
2424	0	155	97	682	0	-116.7	-21.7	1	134	1798
2437	3	0	925	808	1001	-98.0	52.9	144	137	1814
2450	0	158	14	674	0	-110.7	-26.3	15	140	1829
2458	3	950	631	1082	0	-159.5	-72.3	3	142	1838

Figure 4.16 *Typical source location system teleprinter output.*

cess the data and presents results in the form of print-out and display on a storage CRT as well as storing data on floppy disc. A full historical record of a developing defect can be made and displayed on the CRT screen to provide direct visual indication of the severity of the defect in its position relative to the transducer array.

4.6.7 Data analysis

Two methods have been used to characterize acoustic emissions and are extensively used with composite materials; these are: amplitude distribution analyses, and counting threshold crossings.

(a) Amplitude distribution analyses

These are undoubtedly the key to detecting different mechanical failure mechanisms. All signals detected by the sensor are sorted according to amplitude, and then recorded in histogram fashion (Fig. 4.17).

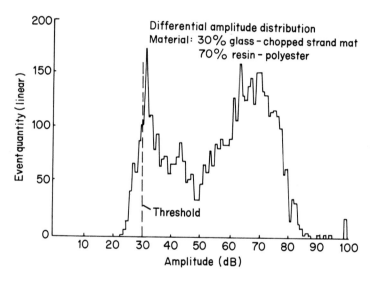

Figure 4.17 *Histogram amplitude distribution.*

Along the X-axis is a spectrum of amplitude values, 0–100 dB. Threshold of the entire system is variable but usually about 30 dB, so no data will be 'seen' below the threshold.

On the Y-axis is number of events. Each time an event of say 58 dB is sensed by the electronics, it is stored in the memory and simultaneously displayed as a displacement of the Y-value one unit in the 58 dB memory slot. A second event at 58 dB would displace the Y-value another unit, etc. The entire run of test data is accumulated in real time and then plotted post-test by the recorder for permanent documentation.

Data presented in the amplitude distribution were taken from a three-point--bend laboratory specimen. The material was hand lay-up randomly oriented chopped strand mat, 30% glass by volume. The double peak effect is indicative of two different failure mechanisms working within the material. Low amplitude signals, which are noticed at very low stress levels, are related to cracking of the matrix material. High amplitude signals have been associated with fibre failure.

Triple-peaked amplitude distributions have also been observed during laboratory experimentation (Fig. 4.18) with three-point-bending specimens.

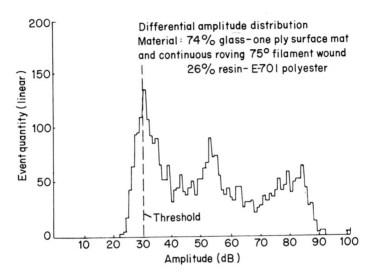

Figure 4.18 *Histogram showing matrix failure and fibre failure in a composite material.*

These data were from a specimen containing one ply of surface mat and several layers of crossing filament windings, oriented at 75° to the line of bending. Glass fibre in this specimen was 74% by volume. Peaks in the distribution correspond to matrix crazing, fibre-matrix delamination and fibre breakage, in ascending order.

(b) Counting threshold crossings

Such a data measurement is a very important method of quantifying AE data. Counts are the individual threshold crossings contained within each event.

Counting is acomplished using a variable gain system relative to a fixed threshold. Generally counts are plotted in real time versus a test parameter such as bending load (tensile load and pressure are other examples). A key to obtaining meaningful counts data rests largely in the proper selection of gain settings.

There are some very simple rules for setting counter gains to accept only data from fibre breakage signals; by noting the onset of fibre breakage relative to some parametric input it is possible to project the data and determine the residual strength of the structure. Fig. 4.19 depicts counts versus load for two separate loadings of a 0° fibreglass specimen.

A 60 dB total gain was used to record only the fibre breakage mechanism; this setting was determined through previous amplitude analysis experiments on similar specimens. The object of the first load excursion was to determine at what load fibre breakage was initially observed.

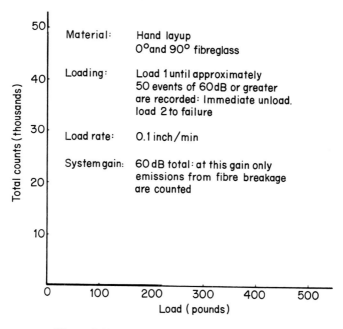

Figure 4.19 *Residual strength projection.*

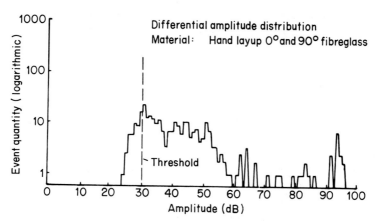

Figure 4.20 *Demonstration of the onset of fibre breakage.*

A mass of 240 lb (109 kg = 1.069 kN) produced 50 events which had peak amplitudes of 60 dB or greater (Fig. 4.20). After these fibre breakage events were observed and recorded, load was removed. The second loading was to failure; note the exponential increase in the number of counts past the last previous load.

This particular specimen had a residual strength of 2.29 kN. It is typical of this particular lay-up for fibre breakage to commence at approximately 50% of residual strength and this phenomenon is easily noticed when plotting counts from fibre breakage in real time.

4.6.8 Acoustic event magnitudes: two applications

A running average of the acoustic emission event magnitude in real time can be accomplished with Dunegan/Endevco instrumentation such as the Model 933® digitized interface which links 3000 Series counters and memories to the computational and plotting capability of the Hewlett-Packard 9825A Calculator® and 7225A Plotter.®

Application 1: tensile test

Two plots for the tensile test on a 7075—T6 aluminium tensile specimen are shown in Fig. 4.21. The average amplitude starts fairly high, falls to a low value during yield, then rises substantially before failure due to the formation of longitudinal cracks in the developing neck.

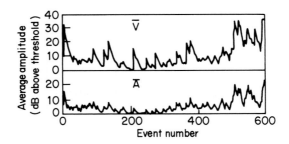

Figure 4.21 *Tensile test plot to failure.*

Note the figure shows plots of two different amplitude bar parameters. V is the average amplitude as measured in millivolts, whereas A is an average of the corresponding decimal numbers. V is more sensitive to individual large events, whereas A is more stable.

Both parameters are readily related to the more familier b-value (the slope of the cumulative amplitude distribution).

Record and playback

This technique is especially useful when an event occurs so quickly that it is not possible to record all the desirable plots in real time.

Application 2: cast iron

A test on a tensile test of a 3.5% carbon steel was completed in 90 seconds. The smoothed amplitude distributions for consecutive 10-second segments are shown in Fig. 4.22. The fine time resolution allowed us to recognize a single-peaked amplitude distribution. The amplitude of the peak rose steadily from 30 dB to 55 dB as stress and strain increased.

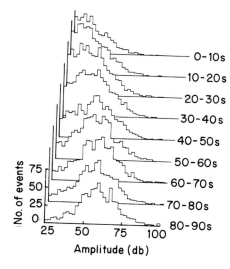

Figure 4.22 'Waterfall' event display.

Correlation plot

This is a plot in which each acoustic emission event is represented by a single point. The X and Y coordinates of this point correspond to two of the measured attributes of the event. A typical correlation of event attributes shown in Fig. 4.23 relates to the energy in the system and the emission counts – the solid line is a tentative theoretical curve based on the decay and the known parameters of the electronics.

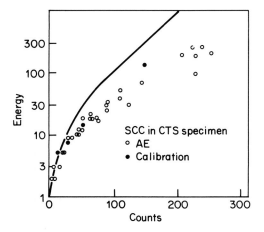

Figure 4.23 *Event/energy correlation.*

4.7 Piezoelectric polymer transducers

Polymer transducers based on piezoelectric polymer film are flexible with a wide range of sensing areas and in addition they have the following attributes:

● low density;
● excellent ability to follow severe contours;
● can be cemented to surfaces;
● because of the low mass a thin film transducer spread over a relatively large area has a minimal effect on the surface;
● high sensitivity;
● flat frequency response;
● are relatively easily and inexpensively fabricated.

Such properties compare favourably with crystalline and ceramic piezoelectric transducers.

A typical polymer is polyvinylidene fluoride PVF_2 which is manufactured in the US by Pennwalt Corporation under the trademark KYNAR®. The presence of fluorine in the fluorocarbon polymer PVF_2 renders this compound inert to almost every chemical. Properties exhibited by this material are:

● outstanding weather and heat resistance;
● highest achievable piezoelectric activity among all polymeric materials;
● unaffected by long-term (20 years or more) exposure to sunlight and other sources of ultraviolet radiation;
● low moisture absorption;
● high chemical resistance;
● high tensile and impact strengths;

- high resistance to fatigue and creep;
- flexible in thin sections for film, tubing and coated wire;
- has a broad useful temperature range from below $-62\,°C$ to $149\,°C$;
- melting point ($171\,°C$);
- high dielectric strength in thin sections.

4.7.1 Applications

Feasibility studies and evaluations have been carried out as follows:

(a) National Bureau of Standards, Gaithersburg, Maryland, USA

A programme to develop a non-destructive evaluation method using piezoelectric polymer transducers and Fourier transform vibrational spectroscopy found that the mass and modulus of the PVF_2 transducer was so small that the principal vibrational modes of an acrylate bar tested were not disturbed. Spectra obtained from vibration, excited by dropping steel balls or impacting the object with a pendulum bob, were interpretable in terms of normal vibrations. Introducing a defect into the bar caused changes at the dominant frequency and resulted in greater coupling of energy into modes that were only weakly excited in the absence of the defect.

(b) Pennwalt Technological Center, King of Prussia, Pennsylvania, USA

Examined the frequency spectrum produced when a steel ball struck a stressed riveted aluminium tensile bar (it had been stressed at the Naval Air Development Center, Warminster, Pa, USA approximately two million times until a crack developed under one of the rivets). The electrical signal from an attached PVF_2 thin film transducer was captured by a commercial multichannel digital recorder and fast Fourier transform analyser. Qualitative differences in the spectra before and after the crack had been enlarged (by further stressing) were apparent.

(c) Naval Air Development Center, Warminster, Pennsylvania, USA

Acoustic emissions from stressed metal specimens have been detected by the use of PVF_2 sensors. Single highly damped pulses with almost negligible ringing produced: (1) in a ceramic transducer a ringing waveform of considerable duration; and (2) in a PVF_2 transducer a single pulse with a rapid ring-down.

Such excitation pulses produce a signal from the detector for which the frequency spectrum will be proportional to the frequency response curve of the transducer. Thus the conventional narrow-band ceramic transducer has a prolonged response to a given excitation, whereas the PVF_2 transducer has a response that itself approximates to the excitation pulse. Monitoring tests using PVF_2 transducers indicated the existence of an emission-rate-related failure precursor in graphite/epoxy and a frequency-shift-related failure precursor in graphite/epoxy and a frequency-shift-related failure precursor in boron/ aluminium. Acoustic emission signature analysis testing has identified PVF_2 as a promising transducer material for use in acoustic emission sensors.

(d) General Electric Corporation, Schenectady, New York, USA

Acoustic imaging devices are being developed for detecting defects in metal parts of aircraft engines. There are plans for incorporating PVF_2 into the imaging device.

(e) Stanford Research Institute, Menlo Park, California, USA

An ultrasonic camera has been developed which will be used for acoustic imaging in conjunction with PVF_2 transducers.

(f) Stanford University, Palo Alto, California, USA

Ceramic turbine blades are being tested with PVF_2 transducers bonded to the curved surfaces. Such transducers can be installed at numerous junctions and diffusion bondings in aircraft and large areas tested in fractions of a second. Thus the testing time can be shortened, and the accuracy can be increased.

Interdigital surface wave transducers are under development. Surface waves can be excited (generated) and used to detect surface imperfections and interfacial structures on various materials. Interdigitated transducers can be fabricated easily from PVF_2; they would be efficient, tuned, variable frequency generators of acoustic signals. Microspaced photolithographic arrays can be deposited with uniform spacings 8–15 µm wide. Then sound waves from 40 to 200 MHz can be produced (the higher the frequency, the greater the resolution of microstructures). At these ultra-high frequencies microcracks could be detected.

At the higher frequencies steerable bulk shear waves can be produced. Such waves are important in imaging work in testing the integrity of bonds and interfacial laminations.

(g) Naval Air Rework Facility (Code 340), Jacksonville, Florida, USA

Inspection techniques which are believed to be able to lend themselves to improvement by the use of PVF_2 transducer configuration include ultrasonic pulse-echo techniques, in both the longitudinal and shear wave modes in the MHz range, are used to evaluate structural parts of aircraft for crack-like defects and to inspect engine components, airplaine wing skins, and supporting machined stiffeners. Polymer transducer elements can be prepared with selectively activated areas; hence shear waves can be generated, controlled, and detected utilizing sensitive PVF_2 material.

(h) Naval Air Engineering Center, Lakehurst, New Jersey, USA

Piezo-polymer transducers could be highly advantageous for inspecting airplane wing surfaces. A PVF_2 thin film transducer with an area 100 mm square would facilitate inspection procedures that currently utilize a 12 mm square ceramic transducer.

(i) Naval Air Propulsion Test Center, Trenton, New Jersey, USA

Attention has been given to the use of polymer transducers in the evaluation and monitoring of helicopter structures showing that their high sensitivity and flexibility makes them particularly suited for:

(1) monitoring the fatigue life of the helicopter rotor hub and vertical shaft;
(2) monitoring the bending of flexible shaft single-rotor aircraft;
(3) detecting the delaminations in rotor blades;
(4) detecting impending structural failures in hinges and landing gears.

(j) Boeing-Vertol Corporation

High frequency sensors and peripheral data processors have been developed which make use of the resonance of a piezoelectric transducer to detect periodically occurring sharp rise-time signals emanating from cracks or spalls in the bearing elements or bearing race or other defects in helicopter transmissions.

The Boeing system is based on transducer resonance between 20 kHz and 300 kHz. It would be useful to replace the ceramic with a polymer transducer but the resonance frequency of PVF_2 in its thickness mode of vibration is higher than 10 MHz. Other than increasing its thickness by a factor of 30, another mode of vibration (the lengthwise vibratory mode) can be utilized. This mode has already been successfully employed in an electret condenser phonograph cartridge. It has a flat and highly responsive sensitivity from d.c. up to 40 kHz. Note that the actual (high frequency) resonance can be controlled by tailoring the lateral dimensions of the polymer thin film transducer.

Another conceivable method for detecting the fast rise time signals from cracks or similar defects is to utilize the flat frequency response of PVF_2 piezofilm. In the form of a thin film, the PVF_2 transducer (in its thickness mode) has a frequency independent transfer function ranging from d.c. up to 50 MHz.

Proper design of supporting system electronics then permits a real-time defect signature to be detected and analyzed.

4.8 Stress corrosion cracking

A practical investigation of acoustic emission was developed by P. N. Peapell and E. C. Walkley of the Royal Military College of Science, Shrivenham, UK.

Noise is usually emitted within the range 0.1 to 2.0 MHz, which has the advantages of: (1) freedom from cultural environmental noise; and (2) low signal source attenuation. Stress corrosion cracking develops so rapidly that these emissions can be readily recorded over a 30 minute test.

Thus a cold-drawn alpha brass immersed in an aqueous mercurous nitrate solution quickly produces a response. With a signal from a Dunegan/Endevco D92011® piezoelectric transducer amplifier 108 dB and filtered to pass frequencies between 0.1 and 2 MHz the output was recorded as standard ring-down counts at a 0.5 volt trigger level.

A Bryans *X–Y* recorder® recorded output as acoustic emission rate (counts/second) against time to produce a graph as in Fig. 4.24. The crack propagation can be made audible by the incorporation of an audio amplifier/speaker system in the equipment.

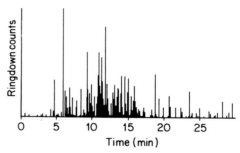

Figure 4.24 *Acoustic emission rate during stress corrosion cracking: brass.*

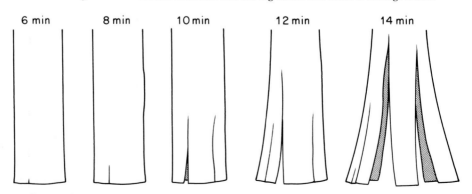

Figure 4.25 *Visual changes in a cold drawn alpha brass during stress corrosion cracking.*

A series of diagrams, Fig. 4.25, illustrate the progress of visible cracks. Little is apparent until after 6 minutes – yet acoustic emission results showed significant bursts of emission initiating at about $4\frac{1}{2}$ minutes (Fig. 4.24). Subsequent acoustic emission activity concentrated between 6 and 18 minutes from the start provided advance warning of crack initiation and growth.

4.9 Case study: riveted component

Crack growth from rivet holes under fatigue loading was monitored through acoustic emission in a laboratory evaluation [6].

Rivet fretting was detectable at frequencies below 150 kHz but to overcome this noise transducers were used with a relatively high natural frequency

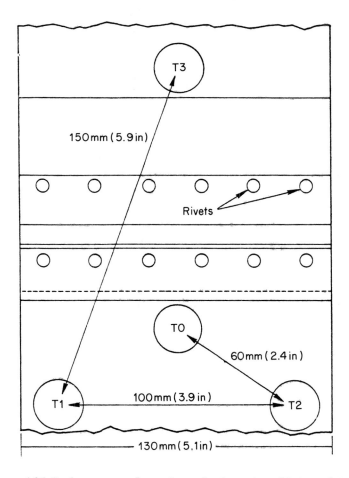

Figure 4.26 *Coplanar array of transducers for the testing of fatigue of rivets.*

(450 kHz) which is sensitive to crack growth, and were arranged in a centred equilateral triangle configuration to provide planar location of AE sources (Fig. 4.26).

Two sources of emission labelled A and B were identified in the locations shown in Fig. 4.27. A data window analysed the emission characteristics from the two sources. The amplitude distributions of the events from the sources provided valuable information concerning the operative mechanism – a narrow band of amplitudes (Fig. 4.28) observed from both sources suggested a single mechanism could be responsible. Analysis of rise-time distribution showed the events to exhibit fast rise times (a feature often observed with crack growth).

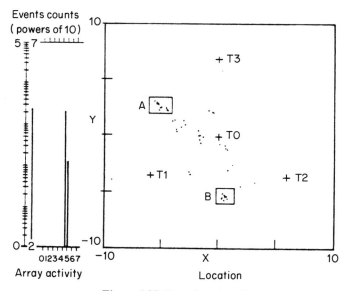

Figure 4.27 *Event location plots.*

A plot of the number of AE events as a function of the number of fatigue cycles showed that significant activity occurred after 45 000 cycles, followed by a quiet period and a further burst of activity at 65 000 cycles. Further use of the data windowing established that source A produced the initial burst of activity at 45 000 cycles and source B the activity at 65 000 cycles.

A pulser superposition technique precisely located sources A and B and identified two rivet regions. Pulser superposition is generally used to confirm the location of sources of emission but was essential in this case where a distorted transducer configuration was employed.

To confirm that sources A and B were produced by fatigue crack growth the sample was fractured in tension to produce failure in the riveted region. Fracture surfaces characteristic of fatigue were observed in the regions specified by

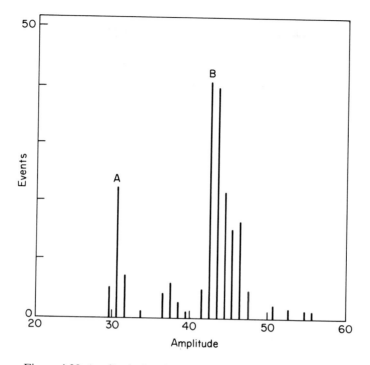

Figure 4.28 *Amplitude distribution: fatigue cracks in fifth array.*

AE, with the crack lengths being between 1 and 2 mm (0.04 and 0.08 in) long. It is significant that ultrasonic inspection under static conditions failed to detect cracks in regions A and B.

4.10 Proof testing: pressure vessels

Although a vessel may suffer subcritical crack growth in service by mechanisms such as fatigue or stress corrosion cracking, it may still pass a routine proof test. AE inspection is now regularly employed during proof tests on pressure vessels to provide 100% volumetric coverage of the vessel and detect the presence of such mechanisms (Fig. 4.29).

Repeat proof tests separated by a period of in-service operation should be acoustically quiet – providing no structural degradation has occurred. Conversely, the presence of significant activity during a subsequent proof load indicates that some degradation mechanism has occurred.

Since emissions only occur while a crack is actually growing, this will be only while the vessel is being pressurized the first time, unless on subsequent pressurization the previous maximum pressure is exceeded.

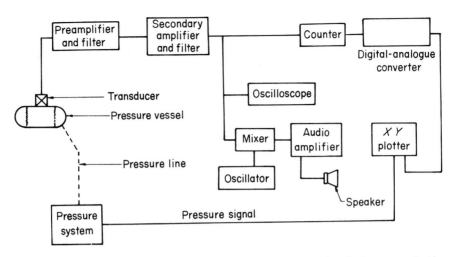

Figure 4.29 *Pressure vessel: block diagram of AE instrumentation during pressurization.*

A problem with using AE during proof testing is that a leak, however small, will mask any emissions from growing cracks – the emission resulting from even a few tiny drops of liquid per minute escaping will be very much stronger than emissions from developing cracks. Thus if any leak does develop, the test must be stopped immediately and the leak eliminated since the proof test only allows one chance to detect any defect.

4.11 Cracked pressure vessel

The British Royal Naval Scientific Service pressure tested a cracked pressure vessel. A circumferential crack around a thread root in one of the screwed end caps of the vessel was found during ultrasonic examination. Simple stress wave interrogation (acoustic emission) during pressurization indicated that this crack grew by a small amount and then arrested at a pressure of about 22.5 M Pa Support for this view was provided by subsequent ultrasonic examinations, which indicated 1 mm over a length of about 18 cm. No other crack increments were detected in this area, either by AE or ultrasonics, during further application of pressure and the bottle burst elsewhere at a pressure of about 52 M Pa. Excessive noise from water leakage in the apparatus prevented AE being used above a pressure of 30 M Pa and swamped any possibility of anticipation of the final failure mode.

Examination of the area of the original defect indicated that a crack about 5 mm deep × 180 mm long had been present before pressure testing and that at some time it had propagated about 2.5 mm. This gave confidence since it confirmed the indication of arrested cracking detected by AE, and showed that AE could be conducted with remarkable simplicity (despite noisy pumps in an exposed test site). The large defect observed in the bottle and responsible for its rejection from service did not cause early failure nor influence the final failure mode.

4.12 Deep diving chamber

Conventional proof testing of the pressure chamber complex used at the British Royal Naval Physiological Laboratory was combined with an acoustic emission analysis. Four interlocking pressure chambers (the largest, 8 m deep by 5 m diameter) form an all-welded structure operating at 3.4 MN/m^2. Acoustic emission examination during the proof pressurizing did not detect any subcritical defect extension and thereby gave further confidence in the overall integrity of the vessel.

4.13 Case study: pipelines and piping systems

Above-ground and buried piping has been successfully monitored by AE both during hydrostatic and in-service inspections. Pipelines and piping systems can be monitored prior to service, during hydrostatic requalifications proof tests or, in certain instances, during on-stream overpressurizations.

Sensor spacing depends on pipe thickness, diameter, type of paint or coating, soil conditions if buried, and the product in the line. Spacings of up to about 250 m between transducers on buried piping and 60 m in refineries and chemical plants are not uncommon.

Some specific case studies include the following.

4.13.1 Pre-service hydraulic tests

Approximately 1200 feet (367 m) of 3 inch (76 mm) to 16 inch (406 mm) diameter 304 stainless steel piping was monitored during pre-service hydrotests. Eleven defective welds were detected and later confirmed by ultrasonics.

4.13.2 Hydrostatic requalification

More than 2500 feet (762 m) of 12 inch (30.5 cm) and 24 inch (61 cm) diameter above-ground piping was inspected and located two AE sources ranked number one and two in severity. These were confirmed as rejectable weld defects, then repaired. The source ranked third in severity was also confirmed, but the size was within acceptable limits.

In another study, various sections of 6061-T6 aluminium piping in an ethylene facility were monitored during hydrostatic testing. Three of the five reportable and recordable AE sources were confirmed through visual examination as cracks.

4.13.3 On-stream testing

Overpressurization of 5% provides the basis for an AE inspection which with a 100 ft (30.5 m) refinery amines piping revealed several areas of caustic cracking.

In another application, 550 feet (168 m) of refinery steam line was monitored during an on-stream overpressurization. Cracks were detected in three expansion loop elbows and in a longitudinal seam weld. On-line monitoring of the steam line under static conditions revealed that the emission was cyclical in nature and associated with fatigue.

A third study covered monitoring of 1200 feet (367 m) of buried natural gas transmission pipeline at each of ten compressor stations. The only excavation required at each site consisted of four narrow access holes. Each test was performed on-stream with natural gas as the pressurizing medium. The AE inspections revealed two rejectable welds which were subsequently repaired.

4.14 Case study: overpressurization

Start-up and cool-down periods of petroleum and chemical reactors may lead to hydrostatic and on-stream overpressurization. Under such conditions acoustic emission provides complete volumetric integrity. Specific cases quoted by Dunegan/Endevco include:

4.14.1 Catellitic cracker (hydrostatic testing)

Instrumented with 24 transducers the computerized source location system pinpointed five sources of acoustic emission which were all confirmed as cracks and weld defects, of which the most severe was a 10 inch (25.4 cm) long crack located in an area not normally inspected by traditional NDT techniques.

4.14.2 Catellitic cracker (start-up/cool down)

After 24 hours surveillance during cool-down and removal from service, two sources were identified and later confirmed, one as a crack and the other a slag inclusion in a weld.

4.14.3 Hydrocracker (see also Section 4.20)

For many structures the thermal stresses associated with start-up and cool-down may exceed the stresses imposed during operation or applied during proof testing. High temperature transducer/waveguide assemblies make monitoring possible at shell temperatures of 530 °C. An inservice test of two hydrocrackers carried out two months prior to a scheduled shutdown itemized several areas requiring additional inspection and their severity. The second highest emission came from a crack in the heat-affected zone of a head closure weld; the fact that this had 'talked' or propagated during a 5% overpressurization indicated that the crack had been growing in service.

4.15 Blast furnaces

Stress corrosion cracking is an international problem in blast furnace stoves resulting in costly maintenance.

The problem arises from the fact that in operation the environment contains a small proportion of nitrous and nitric oxide which is thought to condense into a weak acidic solution. Although an aluminium vapour barrier is bonded to the inside face of the shell, further protection and monitoring of the structure is thought desirable.

Research has shown such stress corrosion cracking produces a relatively large number of high amplitude emissions which permit the detection of any incipient failure in service, even against high levels of background noise.

British Steel at their Redcar plant introduced a large AE monitoring system at the transition girder region of each of four stoves serving the blast furnace. A configuration is also planned with possibly a multiplexed 128-channel system to monitor comprehensively the critical welds on the blast furnace stove domes.

4.16 Case studies: paper mill digesters

Inspection tests of a Kamyr continuous digester involved two phases:

(1) locating AE sources and indicating their relative severity followed by confirmation of defects by inspection; and then
(2) checking the structure after the repair of the cracks that AE discovered.

A single weld in the transition area near the top of the digester was instrumented in the first phase with four sensors and monitored during a 10% overpressurization. The digester was heating but was isolated from the rest of the system. The test took only 45 minutes.

This first inspection resulted in detection of over 4000 acoustic emission events, which were reported to the owner as four grade A (severe) and several grade B (considered to be recordable and reportable) sources.

One week later, during a regular shutdown, the AE sources were confirmed as hundreds of transverse cracks in the weld metal. Although the cracks were all relatively short and shallow (within corrosion allowance), their activity during that 10% overpressurization indicated that the cracks were also growing during normal operation.

Following the repair of the cracks, the entire structure [which is 190 ft (55 m) in height] was instrumented with 28 sensors and subjected to an even higher pressure during a hydrostatic test.

During this second test, a total of just 55 widely scattered AE events were detected by the computerized source location system. This dramatic decrease in AE activity confirmed the success of the weld repairs and the current overall integrity of that structure.

Several other digesters have been monitored with acoustic emission in a similar manner following the repair of defects. In each case, the AE inspection confirmed that all of the defects had been eliminated.

4.17 Case study: glass reinforced plastics

Materials made from glass reinforced plastics (GRP) are copious emitters of acoustic emissions. Degrading mechanisms such as matrix crazing and fibre breakage can result from process-introduced defects, maltreatment or corrosion. Discrimination between the various mechanisms is possible using amplitude distribution analyses. Tests are generally performed using a number of strategically placed individual transducers [7].

High frequency transducers are placed near critical regions such as nozzles, whereas low frequency transducers are used to provide general coverage. The preferential attenuation associated with the high frequency transducers provides a zonal location.

4.17.1 Minesweeper hulls

An exploratory programme at the Admiralty Materials Laboratory (UK) to determine the extent to which AE analysis could and should be used for structural validation within MoD(N) was introduced at a time when the building of a GRP minesweeper hull was being considered, and problems arose of inspecting the hull. This was incorporated into a proposal for a degradation index approach. Preliminary work indicated that the degradation index or some other critical parameter such as strain might provide the basis for a rational approach to validating composite and monolithic structures.

Experiments showed that in composite materials the nature of the emissions characterized their source since it is fairly easy to distinguish between resin crazing, fibre/resin interface debonding and fibre breakage [8]. Fatigue tests on a 2/3-scale model of joint details confirmed that as defects extended, their emissions increased; they also revealed that material and joint configurations used were insensitive to the presence of defects such as gross delaminations for the particular test conditions. While it was easy to find an 'acoustic window' through which to listen to specimens under test in Instron and other testing machines, emission source could not be characterized by analysing the frequency content of single pulses (since each burst of energy excited the whole specimen into resonance).

4.18 Case study: cryogenic ammonia tank

During filling to a higher-than-normal level this tank was monitored at each of 64 transducer locations. A 4 inch (100 mm) diameter plug of insulation was removed, sensors attached magnetically and the holes re-insulated to prevent icing-up. Sensors were only attached to the cyclindrical portion of the tank.

Icing was detected in two locations where insulation was defective and stress emissions were located in the dome section at some distance from the sensors.

4.19 Case study: spheroids

Monitoring in conjunction with hydrostatic proof testing has been reported by Dunegan/Endevco in relation to several 66 ft (20 m) and 81 ft (25 m) major diameter Horton spheroids. Such monitoring uses only from 32 to 64 channels of instrumentation and a complete operation of set-up, testing and dismantling is accomplished typically within five working days.

Experiences with such spheroids include:

(1) Considerable AE in the bottom quadrant of a 66 ft (20 m) Horton spheroid. Visual examination attributed this to settlement.
(2) A section incorrectly welded for several metres was detected by the considerable AE from the overstressed weld.

4.20 Case study: hydrocrackers

Two 22.5 m high hydrocracker reactors, fabricated from A-387-61T grade E steel and clad on the inner surface, were simultaneously monitored during an on-stream overpressure test. Each hydrocracker was instrumented with 16 high-temperature transducer/waveguide assemblies. The acoustic emission source location system monitored the reactors while they were being pressurized from 80% to 106% of the normal operating pressure.

Emission commenced at well below the working pressure and continued throughout the inspection. This pattern has been observed on several insulated reactors and is associated with bolt noise or in some cases actual cracked tack welds. It is important, therefore, that the location of insulation support rings and other vessel anomalies be known prior to an acoustic emission test in order that activity of this nature be correctly interpreted.

In accordance with the ASME proposed standard for AE monitoring, all acoustically active sources detected during the test were graded using analytical techniques such as event rate, count rate, and source activity as a function of location and/or pressure. The classification of AE sources by grade requires the following actions:

Grade A – If it occurs during pressure build-up, pressurization should be halted. Confirm results as relevant by other non-destructive examination methods.

Grade B – Recordable and reportable for future comparison.

Grade C – Further evaluation or correlation not required

A total of nine insignificant sources were detected by the source location instrumentation on the two vessels. Eight of the nine were confirmed as code-acceptable shell discontinuities by a post-test ultrasonic inspection. Two of these sources were due to shell laminations. The ninth source detected was the insulation support ring attachment point.

This test was conducted immediately prior to a scheduled shutdown, in order to assess the general integrity of the vessels as well as to pinpoint areas requiring additional inspection during the shutdown.

The owner of these vessels anticipates future monitoring of these structures during similar on-stream proof tests. The magnetically attached transducer/waveguide assemblies have been left in position on these reactors, for future monitoring purposes.

4.21 Case study: hydrotreaters

Two hydrotreaters, fabricated from SA-387-D steel, were instrumented with twelve transducer/waveguides and simultaneously monitored during an on-stream proof test. Six grade C and B sources were detected.

A third hydrotreater, fabricated from A-204-B carbon steel and containing an internal stainless steel liner, was monitored in a similar on-line proof test. The stainless steel liner was plug welded to the inside surface of the shell on 300 mm centres.

The insulation support rings on the vessel run parallel to the circumferential and longitudinal welds forming weld inspection channels. Since the rings are offset from the welds by only 75 mm and run along their entire length, it is very difficult to determine whether emission received in these areas comes from the insulation support rings or from the shell weld seams. For example, this hydrotreater contributed seven discernible sources, with longitudinal and circumferential welds superimposed. The source designated S6 was classified as a grade B source and ranked number one in terms of overall activity. A visual inspection only revealed cracked insulation support ring tack welds, but an ultrasonic inspection of the same area revealed a 76 mm long, crack-like indication in the weld metal near the intersection of the circumferential and longitudinal welds. The remaining sources were associated either with weld inspection channels or with the vessel internal liner.

Emission from insulation support rings can be expected from most insulated vessels. The emission is sometimes extraneous in nature and due to relative growth of the shell with respect to the rings, as in the case of the hydrocrackers discussed earlier. Even when the emission is real as in the case of cracked tack welds, it is normally not important since the welds are not pressure-containing components, and there is very little possibility that a crack could propagate into the vessel shell. When these rings are located in parent metal regions, it is fairly easy to interpret the emission based on the activity as a function of pressure and location. When the insulation support rings are located at welds, nozzles or structural support points within the expected locational accuracy of the source location system, the real-time interpretation of defects becomes very difficult. In order that shell defects be not overlooked, all emission from questionable areas must be assumed to be real and shell related until proven otherwise by alternative techniques.

4.22 Case study: absorber tower

A 50 m tall absorber tower, fabricated from A-212-B carbon steel, was monitored with acoustic emission during a pneumatic requalification proof test. Six planar arrays of four transducers were attached to the tower using magnetic holddowns. The absorber tower had been taken out of service to facilitate the repair of several hydrogen embrittled areas. After the repairs were performed, including the replacement of several nozzles, the vessel was monitored during pneumatic proof testing and operational start-up. The absence of AE from the previously cracked areas confirmed the successful repair of those regions. However emission did occur near the top of the vessel (an area supposedly not susceptible to hydrogen embrittlement cracking). This area was one of several that were active during proof test and start-up.

As a result of the acoustic emission observed from the vessel during these inspections, and the history of hydrogen embrittlement cracking, the transducers, preamplifiers and associated cables were left on the vessel for future monitoring. Two areas of considerable emission were observed during a subsequent on-line surveillance period. The first area was the region near the top of the vessel, that had been active during the proof test and operational start-up.

The second source was associated with a previously quiet area near the centre of the vessel.

Although an ultrasonic shell inspection of both areas indicated no defects, the vessel was temporarily taken out of service for an internal inspection. This was done because of: (a) the large amount of emission received on-line from these two areas; and (b) internal process problems affecting the operation of the absorber.

Personnel inspecting the inside of the tower found a tray down-comer or trough that was almost completely eroded away at the exact elevation of the emission. In addition, several improperly fastened tray manways were observed at the top of the tower, the other area of activity. One of the tray manways had fallen through to a lower tray while others had been cast aside, effectively negating the function of the trays. From a process standpoint these manway and trough problems resulted in gas coming through the tower much faster than normal, resulting in the process problems experienced.

Once the manways were correctly fastened to the trays at the top of the vessel and the eroded tray trough replaced, the process problems were eliminated and emission from these areas ceased.

It is interesting to note that while the detection of shell anomalies was the real purpose of this AE inspection, two unexpected problems associated with the vessel internals were detected and proved to be highly pertinent.

4.23 Case studies: quality assurance

Examples of acoustic emission testing used to assure the quality of materials include the following.

4.23.1 Swiss watches

The stainless steel or titanium used for watch springs is produced in small billets (approx. 1 m × 1 cm × 0.4 cm). These billets are given a controlled stretch to straighten them; AE is used to grade the material as acceptable (or not acceptable) for spring manufacture. (This technique is now being extended to the assessment of material to be used in surgical implants.)

4.23.2 Foodstuffs

Acoustic emission has been used in food research for assessing characteristics such as the crispness of crisps and the ease of scooping soft ice cream from the freezer.

4.23.3 Surface coatings

Cracking of brittle surface coatings, such as the surface layer of aluminium diffused into titanium turbine blades, is tested by AE methods. The technique has also been used to survey surface defects in a braided glass surface using a rolling hertzion test to interrogate the surface.

4.23.4 Mechanical processes

The use of AE to monitor machining is less advanced, but recent work in Japan has shown that the technique can be used to detect tool wear. The application of AE to deformation processes is perhaps the least developed of the three areas to date. One example concerns a drawing process. The cases for small arms ammunition are produced by drawing; preliminary work has shown that AE could differentiate between good and defective products.

4.23.5 Heat treatment

Thermally activated processes such as martensitic transformations can easily be detected using AE, for example clinking in large castings or delayed cracking in the brittle heat affected zone adjacent to a weld during tempering. Recent work has shown the potential of using AE monitoring for controlling the rate of chromium plating. As the current density increases in an electrolytic cell, the deposition rate increases; at a critical current density the plating starts to crack Using AE, in conjunction with some feedback, it should prove possible to plate rapidly with a minimum of cracking.

4.23.6 Honeycomb structures

The aluminium skin/phenolic honeycomb interface of the F111 tailplane is susceptible to water ingress and subsequent corrosion and debonding. If the skin is heated, using a heat gun, the rate of reaction increases so much that the corrosion activity can be detected. On the basis of the AE data the corroded area is marked out and the skin cut off, replaced and repaired.

4.23.7 Geological faults

For large civil engineering and small-scale geological structures such as mines and dams, we have to use lower frequencies to obtain the longer range required. Using low frequencies this AE instrumentation can monitor slope stability, detect the onset of rock falls in mines and assess foundation efficiency. It is even being used to monitor the San Andreas fault in Southern California.

References

1 Kaiser, J. (1950), 'Untersuchungen über das Auftreten von Geräuschen beim Zugversuch', *PhD Thesis*, Technische Hochschule, Munich.
2 Arrington, M. (1975), 'Acoustic emission; an introduction for engineers', *Chartered Mechanical Engineer*, April, 53–5.
3 Rotem, A. and Baruch, J. (1974), 'Determining the load–time history of fiber composite materials by acoustic emission', Technion-Israel Inst. of Technology, Haifa, Israel, MED Report No. 44, March.
4 Jessen, E. C., Spanheimer, H. and Deherrera, A. J. (1975), 'Prediction of composite pressure vessel performance by application of the Kaiser effect in acoustic emission', ASME Paper H300-12-2-037, June.
5 Kolsky, H. (1963), *Stress Waves in Solids*, Dover, New York.
6 Davis, R., Hickman, J. and Peacock, M. (1981), 'Acoustic emission as an NDT tool for the process industry', *Metal Progress*, February, 68–73.
7 Fowler, T. J. (1979), *Proc. ASME Conf. on AE Testing of FRP Equipment*, Chicago, May.
8 Birchon, D. (1971), 'The philosophy and practice of structural validation', *Proc. ASTM Conf. on Predictive Testing*, Anaheim, California, April.

Further reading

Acoustic Emission Monitoring and Ultrasonic Examination Correlation on a Reactor Vessel', ERPI NP 921—Project 449—Final Report, October 1978.
Dunegan, H. L. (1975), 'Quantitative capabilities of acoustic emission for predicting structural failure,' in *Prevention of Structural Failure*, Am. Soc. For Metals.
(1975), 'Monitoring structural integrity by acoustic emission', ASTM STP 571.
Bentley, P. G., Dawson, D. J. and Parker, J. A. (1974), 'Instrumentation for acoustic emission testing of steel pressure vessels,' *Proc. 2nd AE Symp.*, Tokyo.
Pollock, A. A. and Wadin, J. R. (1979), 'Application of acoustic emission as an on-line monitoring system for nuclear reactors, NUREG/CR-0605, January.
Tatro, C. A., Brown, A. E., Freedman, T. H. and Yanes, G. (1979), 'On-line safety monitoring of a large high-pressure high-temperature autoclave,' *ASTM Symp. on Acoustic Emission Monitoring of Pressurized Systems*. Fort Lauderdale, Fla., January.
Sinclair, A. C. E., Formby, C. L. and Connors, D. C. (1976), 'Fatigue crack assessment from proof testing and continuous monitoring,' in *Acoustic Emission*, Applied Science, Barking.
Egle, D. M., Mitchell, J. R., Bergey, F. H. and Appl, F. J. (1973), 'Acoustic emission for monitoring fatigue crack growth,' *ISA Trans.*, **12**, no. 4, 368–74.
Fisher, V. T., (1977), 'Anwendung verschiedener Verfahren der Schallemissionanalyse beim Ermüdungsrißforschritt in Schweißverbindungen,' *Materialprüfung*, **19**, 11.
Norita, K., Kawahara, M., Ishihara, K. and Fuji, T. (1977), 'Fatigue crack growth analysis and its non-destructive monitoring in a spherical vessel containing weld defects, *Proc. Third Int. Conf. on Pressure Vessel Technology*, Tokyo, April.
Lucia, A. C., Marozzi, C. A., Terranova, A., Franchi, M., Galli, M., Pagella, M., Ubaldi, M. and Vergani, L. (1979), 'Final report on contract 184-77-PIPGI on detection of fatigue crack formation in reactor pressure vessels.'

Lucia, A. C., Testa, M., Brunnhuber, R., Franchi, M. and Galli, M. (1977), 'Real-time monitoring of pressure vessels by acoustic emission', *Proc. ANS Topical Meeting on Thermal Reactor Safety*, Sun Valley, USA, August.

Lucia, A. C., Marozzi, C. A. and Terranova, A. (1979), 'Detection of fatigue crack formation in nozzle weldings of pressure vessels', ASME Paper 79-PVP-101, June.

Lucia, A. C., Marozzi, C. A. and Terranova, A. (1979), Study Contract No. 1025/12-SISP.I 'Relationship between fracture mechanics and acoustic emission parameters,' Final Report.

'Proposed standard for acoustic emission examination during application of pressure', available from the American Society of Mechanical Engineers, United Engineering Center, 345 East 47th Street, New York, NY 10017, USA.

Silk, M. G. (1984). *Ultrasonic Transducers for Nondestructive Testing*, Adam Hilger Bristol/Boston.

Williams, R. V. (1980), *Acoustic Emission*, Adam Hilger, Bristol.

5
Visual inspection

5.1 Introduction

To the capabilities of the human eye (with its attendant weaknesses) and the human memory have been added a range of techniques which make visual inspection even under the most exacting of circumstances a rational possibility.

Fibre optics are used in industry for inspection of cavities where access is difficult and where viewing is only otherwise possible through time consuming dismantling and uneconomic downtime. The first endoscopes (a term covering all instruments which give views of the interior of enclosed spaces and cavities) were developed for medical purposes. The word endoscope is of Greek origin and freely translated means 'seeing inside'. This development began with a cytoscope, signalling the beginning of modern urology. Technical instruments known as borescopes followed; these are rigid instruments using a lens system and equipped with miniature lamps for illumination. Modern glass fibre illumination replaced these lamps in the next generation of instruments, giving a considerably increased amount of light and making photography possible.

Extensive use of electronics has made it possible to extend considerably the scope of visual inspection and video recording. Even model aircraft have an important role as vehicles for photographic inspection.

5.2 Inspection: eyes and illumination

The limitations of the human eye mean that, with insufficient illumination, the finest optical designs are virtually worthless. In most examinations, we are dealing with relatively low light level situations compared to normal daylight conditions.

Considering the mechanism of the human eye (Fig. 5.1), for sharp vision the image formed by the lens must fall upon a small central area at the rear of the retina called the fovea. This tiny area is packed with approximately 150 000 cone cells and ganglion nerve endings. It is the cone cells which account for our keen perception of fine details, shape, size and colour. Altogether, there are about 7 million of them in the eye. Cone cells are not sensitive at low light levels, therefore the eye also contains some 100 to 150 million rod cells for twilight vision. These are scattered throughout the periphery of the retina and are primarily for detecting motion, not fine detail.

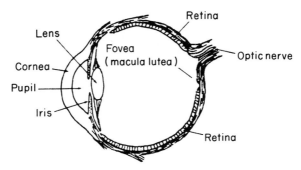

Figure 5.1 *The human eye.*

The human eye is amazingly efficient: the pupil can stop down so that one can see in bright noonday sun or it can open wide permitting vision in an almost dark room (a light level change of about a millionfold). It takes the eye about 10 to 15 minutes to make the change (as when going into a movie theatre) and can take half an hour to become fully dark adapted. Fighter pilots flying night missions wore red goggles before take-off to prepare their eyes for low light level viewing.

Comparative light intensities and requirements are given in Table 5.1.

Table 5.1

Outdoor sunlight	55 000–99 000 lux
Outdoor shade	11 000 lux
Indoor by a window	1100 lux
Recommended light for:	
ordinary reading	275 lux
watch repairing	
fine machine work	550 lux
needlework	
inspection	1100(+) lux

A fibrescope must be used when light has to be projected a distance of about 0.5 m, for example into a carbon-coated cavity as in the combustion chamber of a large jet aircraft engine or industrial turbine, and where small entry ports limit the total scope diameter the size of the light guide bundle within is greatly reduced. If the total outside diameter of a scope is only 2 mm or 3 mm (which includes the wall thickness of the probe) there is not much room inside to fit both the lens system and light guide.

One way to keep the illumination system as efficient as possible is to make the light guide fibres continuous from light source to scope tip (50% of the light is lost wherever there is an interface or connection). Another way to increase brightness is to use more powerful lamp sources, but this is not quite so simple as a lamp with a larger filament size cannot be focused to a fine spot at the end of the fibre bundle, so any gain in output is minimal. A larger lamp produces so much heat that it melts the end of the fibre bundle and burns it, forming a carbon layer which reduces the light output to almost nothing. So above 150 watt, it is necessary to use either a mercury-arc or xenon lamp. These can produce significantly more light but require a high voltage starter and intricate electronics.

5.3 Magnifiers and optical aids

Visual examination can be assisted by low-power magnification for which a range of devices is available [1];

- measuring magnifier
- fixed focus magnifiers
- torchlight magnifiers
- pocket microscope
- measuring microscope ($\times 20$, $\times 40$, $\times 80$, $\times 100$)
- depth measuring microscope

5.4 Modern borescopes

Rigid borescopes have an optical viewing system of lenses, and not of fibres. Borescope images are perfectly clear, not made up of many thousands of tiny dots of light. A coherent fibre bundle is not needed and therefore a borescope costs a fraction of the price of a fibrescope. Ease of use, robust construction and its availability in different lengths, diameter, angles of view and fields of view make borescopes a popular inspection investment (Fig. 5.2).

5.4.1 Slenderscope

This instrument has a diameter of only 1.7 mm. It is used to inspect small-bore cross-drilled holes, also for spray nozzles on diesel fuel injectors [2].

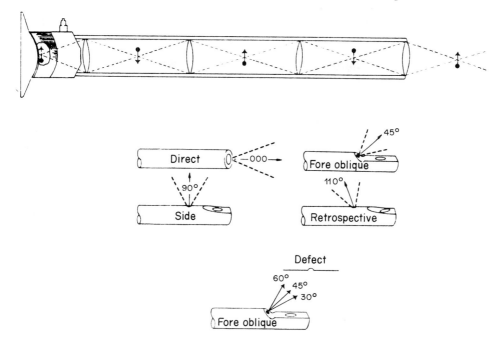

Figure 5.2 *Rigid borescope lens assembly (upper) and oblique viewing heads (lower).*

The slenderscope has replaced X-ray inspection techniques, with considerable cost savings, for the inspection of internal profiles of two welds on small-bore hydraulic tubing (of which one particular aircraft utilizes 600 such components).

On an aeroplane the slenderscope was used to inspect cross struts inside the tail plane structure. Rivets were removed and the instrument inserted through the rivet holes; afterwards it was made good by rivetting with the next larger size rivet. The alternative was a very expensive, time consuming stripdown of the whole tailplane.

5.4.2 Rigid extendible endoscopes

These instruments are suitable for the inspection of large hollow bodies, long pipes and all otherwise inaccessible areas. The basic equipment comprises an ocular tube with focusing control, electrical connection and monocular attachment, an objective tube, transformer connection lead, cleaning set and instrument case. The length of a basic system can be increased by the addition of extension tubes such that the model TEW-24-10 is 24 mm diameter and has a length of 16.5 m (54.2 ft).

Interchangeable objective heads which are available include:

- Wide-angle fish-eye head
- 90° lateral head
- 0° lighting ring
- 110° retroviewing objective head
- Fore-oblique viewing objective
- 60° or normal angle
- 35° narrow angle
- 90° wide angle

The eyepiece tube fits into the rear end of the focusing tube. Three types of eyepiece are available:

- A standard, straight, single ocular tube
- A binocular attachment
- A 90° circular tube

Extensive use of these rigid extendible endoscopes includes the detailed inspection of gun barrels, boiler and condenser tubes, wave guides, helicopter blades, boreholes, gas cylinders (the interchangeable head allowing every part of the cylinder to be examined), pressure vessels, headers, cavity walls, paper drying drums, gas mains, air receivers and torpedo bodies.

5.4.3 Borescope application: ducting inspection

The build-up of dust, bacteria, etc. in air conditioning plant tends to be gradual and therefore not necessarily noticed by the occupants of the building where the plant is installed.

Satisfactory maintenance of such plant sometimes involves specialist contractors who carry out a full survey of the plant and ducting as a preliminary to providing a quotation for their services to clean up the ducting, etc.

The conventional way of inspecting this ducting, etc., has meant that inspections could only be carried out in the areas where access panels, filter boxes, inlet and outlet grilles, etc., existed. Thus many metres of ducting and pipework interconnecting these plants could not be inspected. One of KeyMed's [2] customers, an experienced company in this field, investigated the use of borescopes for this application. They found that the access holes which are already needed for the introduction of cleaning chemicals, etc., could also be used as access points for the borescope to view the inside of the pipework and ducting that previously had not been accessible.

The borescope gave them access to view the areas that previously had eluded them but they still required documentary evidence to present to their clients. As a matter of course with their standard surveys, they had produced photographic as well as laboratory reports as supporting evidence concerning the debris, etc. found in the ducting. The use of the Olympus OM2 camera allowed them to extend that photographic evidence to cover the views that were obtained through the Olympus industrial borescopes.

Similar applications in the civil engineering field have involved the use of borescopes for the checking of tie bars in cavity walls, flue ducting, beam ends, box girder bridge sections, etc.

5.5 Fibrescopes

The industrial fibrescope (Fig. 5.3) is a flexible instrument. A multilayered sheath protects two fibre optic bundles each composed of tens of thousands of glass fibres. Each individual fibre consists of a central core of high quality optical glass coated with a thin layer or cladding of another glass with a different refractive index. This cladding prevents the light which enters the end of the fibre from escaping or passing through the sides to an adjacent fibre in the bundle. The bundle used to carry light from the external high intensity source to illuminate the object is non-coherent (in the sense described in the next paragraph) and is called the light guide bundle. These fibres are generally about 30 μm each in diameter. The size of the entire bundle is determined by the diameter of the scope.

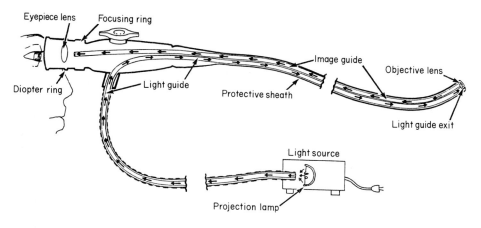

Figure 5.3 *Industrial fibrescope.*

A fibrescope works on the principle of light being reflected internally along the inside of a rod of glass (Fig. 5.4). Each fibre consists of two different types of glass, the core with a refractive index of 1.62 and a coating glass with an index of 1.55. The lower refractive index of the coating glass maintains total internal reflection. To be flexible, the 'rods' in practical terms are very fine fibres of glass, each one approximately one third the width of a human hair; a fibrescope contains many, many thousands of such fibres. In random positions light would be transmitted through the bundle and not as an image. An image can only be passed if every individual fibre is in exactly the same position at each end, i.e. the top left-hand corner fibre at one end must be the top left-hand corner at the other, second from the left at one end must be second from the left at the other, etc. The position of the fibres between the ends is of no consequence.

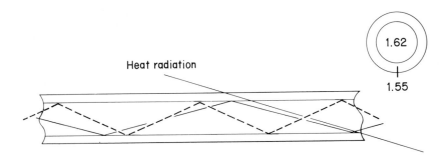

Figure 5.4 *Internal reflection within a glass fibre. All visible wavelengths are transmitted together with near UV light, but IR wavelengths are absorbed. An image guide bundle consists of as many as 120 000 geometrically arranged fibres of diameter 10 μm each. Refractive indices of the central fibre and its cladding are shown on the cross-sectional view.*

Another bundle called the image guide is used to carry the image formed by the objective lens at the tip of the scope back to the eyepiece. This is a coherent bundle which means that the individual fibres must be precisely aligned so that they are in an identical relative position to each other at their terminations. Image guide fibres are in the 9 to 17 μm range as their size is one of the determining factors of resolution, although the precision of alignment is more important.

A real image is formed on both highly polished faces of the image guide. Therefore to focus a fibrescope for different distances, the objective lens at the scope tip must be moved in or out, usually controlled remotely at the eyepiece section. A separate diopter adjustment at the eyepiece is necessary to compensate for differences in eyesight.

The quality of the image is largely governed by the precision with which the matrix at the ends is formed. The 'Olympus' bundle aligns the fibres almost perfectly in the three axes. This can only be achieved by firstly ensuring a very high consistency in the diameters of the fibres – if supplied with 6 footballs, 6 tennis balls, 6 table-tennis balls and 6 marbles, and asked to form these in a perfect square on the floor, it would be almost impossible.

Fibrescopes usually have a controllable bending section near the tip so the observer can direct the scope during insertion and be able to scan an area once inside. Fibrescopes are made in a variety of diameters, some as small as 3.7 mm, and in different lengths up to about 2 m or more, with a choice of direction of view at the tip.

The industrial fibrescope is basically similar to the medical instrument. The basic features of the industrial fibrescope are as follows:

(1) The light guide for illumination purposes only and consisting of fibres in a random format, runs from the distal end of the fibrescope (which is inserted into the area for inspection), up to the control handle and continues without a break from the control handle to the light guide connector which plugs into the light source.
(2) The image bundle with its precision matrix, otherwise known as the coherent bundle, runs again from the distal end up to the control handle and eyepiece.
(3) From the distal end fine cables run through tubes to the control handle and terminate at an angulation control which, through a push–pull process, enables the operator to angle the last few centimetres of the fibrescope at the distal end for the purposes of steering the fibrescope through awkward internal galleries or for scanning internal areas. In the case of the Olympus industrial fibrescope this angulation is to 120° to each side.
(4) The two bundles and cables are then encased in a four layer construction sheathing including a neoprene sheath which makes the instrument water-proof from the distal end to the control handle.
(5) The control handle then incorporates a focusing control or controls to ensure that the image is of the highest possible quality and to cater for dif-ferences that occur in each person's eyesight.

Accessories which can be used with fibrescopes range from television to 35 mm cameras, and light sources to angled eyepiece adaptors. Closed-circuit TV cameras can be fitted to all instruments via an adaptor; this enables groups of engineers to study and discuss the same view simultaneously. Closed-circuit TV can then, of course, also incorporate video tape recordings and these can be par-ticularly useful for training puposes. Video recorders can be used for re-interro-gation and as a visual record for comparison.

Use of 35 mm cameras to provide factual recording is growing in industry. Most single lens reflex 35 mm cameras can be fitted to the instruments by using adaptors. A permanent record of the view obtained can be invaluable for reference purposes, reports, training purposes, and even in publicity. Olympus OM2 is particularly suitable, being a small lightweight camera. it offers auto-matic exposure so that even if the lighting conditions change in the middle of an exposure, the camera will adjust; consequently an engineer unversed in photography can obtain photographs.

A powerful light source is a most important accessory. Typically, the KeyMed KLS2 incorporates two 150 watt quartz halogen lamps and in the event of one lamp blowing, it is a simple matter to switch to the other, thus eliminating lamp change at what may be a particularly inconvenient moment. Each lamp has an iris control so that the light intensity can be varied according to the conditions and so that the colour temperature of the light remains constant ensuring it is white at all light levels. Other light sources, mostly in the medical instrument range, offer extremely high intensity illumination and, in some cases, have facilities for pumping and sucking of liquids, air, etc.

A steerable catheter is another valuable accessory derived from medical applications. The steerable catheter has proved very useful in industry for the retrieval of foreign bodies and objects.

5.5.1 Fibrescope applications

Combinations of fibrescope design are available to meet the wide range of needs of industry. Fibrescopes are currently used widely in the oil, aerospace, power generation, shipping, military, chemical and nuclear industries.

Applications vary enormously, but include for example, inspections of turbines, pipework, pressure vessels, compressors, heat exchangers, airframe, diesel engines, and boilers.

In some industries the use of fibrescopes is mandatory. For example, in the case of gas turbines, the manufacturer of the turbine will actually recommend in the maintenance manual the use of a particular fibrescope specification to inspect certain critical areas and to define the access holes to be used.

In the aerospace industry both engine manufacturers and airlines are particularly interested in the adoption of any system which could check the internal parts of the engine without the necessity for it to be removed from the aircraft and dismantled. The use of the fibrescope has made this possible. This inspection technique has now been accepted to the extent that access ports are actually designed into the engine prior to manufacture. This is now the case on many more recent engines, including the Rolls-Royce RB211 (aero industrial versions), Olympus 593 which powers *Concorde,* and RB199 which powers the new *Tornado,* as well as engines manufactured by Pratt and Whitney and General Electric.

5.6 Optical fibres: direct fracture sensors

Normally inaccessible parts of structures can be effectively and cheaply monitored for cracks by means of optical fibres attached by appropriate adhesives and observed remotely and automatically. The technique [3] was developed by the National Maritime Institute and Cranfield Institute of Technology for an underwater structure, but it is evident that the principle could be applied widely.

Optical fibres 125 μm in diameter, suitably degraded to fracture under minimal stress, are firmly attached to steel and concrete surfaces. Light from a light-emitting diode at one end of the fibre assembly is detected by means of a photodiode detector at the other end. Fracture causes a loss of the light signal at a control station.

The fibres can be produced cheaply and are suitable for retrofitting on existing structures. The advantages of an optical signal include its freedom from electromagnetic interference and its ability to be used in a range of hostile and potentially flammable environments.

5.7 Sewer monitoring

In the UK a specialist organization [4] surveys some 500 000 metres of sewer per year. Contracts vary in size between those taking less than one working day and others in excess of 70 000 metres. Surveys are carried out with high resolution closed circuit TV cameras on most drains and sewers. However, in larger sewers and culverts specialist teams carry out surveys by man-entry methods and colour photographs are taken of salient features. The photographs are referenced by distance measurement.

Detailed records, both written and photographic, are kept on file of each contract and the organization is often called upon to retrieve and copy information from this store.

Standard equipment can be used to survey pipes from 100 mm diameter upwards and consists of a closed circuit TV camera and cable, winch and winch wire, control unit and monitor. Photographs are taken of the monitor screen to enable the log of the survey to be illustrated. Various additional items of equipment are available to enhance the quality of the survey and to tackle difficult situations like single access point sewers:

(1) Video tape recordings are taken of part of all of the survey and can include a digital display of camera position, pipe diameter, location, date, time and weather conditions, etc.
(2) In pipes of 225 mm internal diameter and larger, *in situ* colour photographs may be taken giving very high definition prints which are invaluable in assessing the more obscure features within sewers such as degradation due to sulphate attack.
(3) Knowledge of the exact location of features in a sewer can often be essential. Accordingly a locating probe may be mounted inside the camera so that buried manholes, junctions, faults, etc. can be located on the surface, while being viewed by the camera.

A need exists for an efficient analysis of the data produced; this may range from a discussion giving a résumé of the condition of the sewers surveyed to a full computer analysis of the survey. This highlights the number of cracked pipes, leaking joints, condition of manholes, etc. in individual sewer lengths and in the area as a whole.

5.8 Miniature closed circuit television

Specially designed camera systems allow engineers to investigate and report on conditions quickly, safely and efficiently. Ability to 'assess the inaccessible', whether it be underground, underwater or in areas of high radiation, offers a major advantage in that inspections can normally be undertaken without committing plant and services to long periods of costly 'down time'. The range of applications extends from helping to maintain both nuclear and fossil fuelled power stations to investigating geological boreholes and examining offshore structures.

The R20 sub-miniature camera [5] was originally developed to facilitate the inspection of small diameter tubes in nuclear power stations. The small size and impressive picture quality enables engineers to inspect other small cavities previously inaccessible to closed circuit TV cameras. With an outside diameter of 17.3 mm (0.68 in) the sub-miniature camera has the following features:

(1) Resolution: 400 TV lines;
(2) Cable length: 50 metres for maximum resolution;
(3) Angle of view: 60° approximately;
(4) Voltages: 110/240v, 50/60 Hz;
(5) Forwarding and right-angle viewing.

The R93 [5] is a small portable camera with superior resolution and only 40.5 mm (1.6 in) outside diameter. The model R253 system [5] is designed for underwater use and incorporates alternative light sources as well as remote focusing and iris facility.

5.9 Electronic camera units

Inspection tasks for which electronic camera units (ECU) have formed an integral system unit include the on-line control of:

(1) strip width monitoring;
(2) diameter gauging;
(3) flaw detection;
(4) length measurement.

These are based on a Linescan camera [6] which uses as its image sensor a linear photodiode array (LPA). These arrays consist of a shift register and a single line of photodiodes. The number of photodiodes in the array can be up to 1024, but the particular array used in the camera depends on the application. A block diagram of a typical array is depicted in Fig. 5.5.

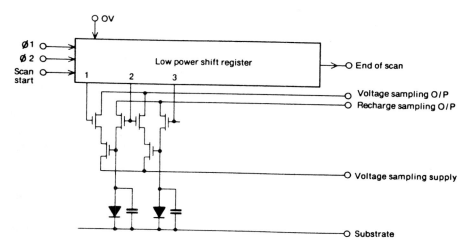

Figure 5.5 *Block diagram: photodiode array.*

The LPA scans a single line and produces an electrical signal which is proportional to the light intensity across the scanned line. Thus any scene which is imaged onto the array will be reproduced as a series of electrical pulses, the amplitude of each pulse being directly proportional to the light intensity.

The concept of using LPA in remote measurement is based on semiconductor design and manufacturing techniques. These enable the photodiodes within the array to be very accurately dimensioned. Since the dimensions of the diodes are effectively fixed, the array itself can be regarded as a precision scale.

In practice the array always uses a photographic lens to image the scene onto the photodiodes. Any optical system can be very easily and accurately calibrated so that its magnification is known.

If an object such as a bar is imaged onto the array, the image of the bar will fall onto the array as a shadow. By counting the number of diodes which are in this shadow and knowing the magnification of the lens, the bar's diameter can be determined, without contacting it. The converse applies if the object is a slit, i.e. the image would be in the form of white bar. By counting the number of diodes illuminated the slit's size would be determined.

It is on this basis that linear arrays are used in dimension gauging.

5.9.1 Linescan camera

Linear self-scanning photodiode arrays containing from 64 to 1024 elements are incorporated in the 4000 series and the 'M' series of linescan cameras [6].

(a) 4000 camera

This camera contains the array and one printed circuit (PC) card. The array is mounted on a PC card directly behind the lens, such that the main axis of the array is on the centre line of the lens. The array card mates with the main camera PC board by means of Varelco stakes.

The main PC board contains TTL to MOS clock drivers and video processing circuitry together with a digital comparator. The array is operated in the voltage sampling mode which enables a very simple video circuit to be employed.

The video signal is fed to a digital comparator which has an adjustable threshold level control. This enables the video waveform to be digitized in the camera. The main inputs to the camera are $+5$ V and -30 V voltages together with three TTL clock pulses. The outputs from the camera are the video output and the digitized video. All the inputs and outputs are carried by a single multi-core cable which connects the camera to the ECU processor.

(b) 'M' camera

This camera contains five small printed circuit cards in addition to the array PC card. The five PC cards are arranged on a mother board which connects with the PC boards and the array card by Varelco stakes. The array carrier board is bolted to the front of the camera directly behind the lens thread.

The reason there are more PC cards in the M-series camera is that the array has dual shift registers and the photodiodes can only be operated in the recharge mode. The basic design philosophy behind the M camera is different from the 4000 camera in one fundamental aspect, that of operating distance from the processor. In many industrial gauging applications the actual measurement area may be hazardous either in terms of temperature or atmosphere, or both. In these circumstances the processing electronics is usually required to be remote from the measurement area.

The M camera has the capability to operate up to 100 m remote from the ECU processor. This is achieved by using line drivers and receivers. Synchronizing clocks are generated by the processor and control the camera. The rest of the array clocks are generated in the camera itself.

The input signals to the M camera consist of two clock signals and an unregulated d.c. power line. The output from the camera is the preamplified charge pulses from the photodiode array.

All cameras used with the ECU systems have 42×1 mm lens threads. This is the thread size used on all screw-in Pentax/Praktica lenses, thus a very large selection of lenses is available.

5.10 Television tube: miniature cameras

Entirely new constructional techniques used in the 80XQ Plumbicon TV camera tube [7] result in miniaturization for the new one-man ENG and EFP camera units.

The performance of the 12.5 mm ($\frac{1}{2}$ in) 80XQ shown in Fig. 5.6 compares with that of the earlier larger 17 mm ($\frac{2}{3}$ in) tubes yet its useful target area is only 4.8 × 6.4 mm (8 mm scan diagonal). The total volume occupied by the 80XQ and its deflection coil assembly (DT1120) is only 22 × 78 mm³: the tube plus assembly weighs 65 g.

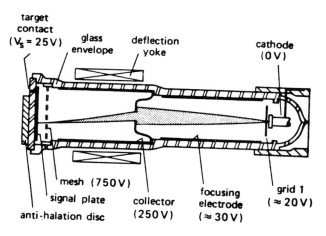

Figure 5.6 *Miniature camera 80XQ in which the glass envelope supports and locates the principal electrodes.*

Not only are dimensions scaled down but the tube has also been tailored to meet the special requirements of integrated circuits. This makes it possible to get so much into a single camera pack. Power economy has been achieved with this tube; a 9 V heater (to suit camera batteries) only takes 50 mA. Further economy is obtained by using electrostatic focusing and highly efficient deflection coils with impedances tailored to suit the drive circuitry.

Particular emphasis has been placed on achieving low geometric distortion and stable registration. The glass envelope of the 80XQ supports the principal electrodes: the target, focusing electrode, collector and grid.

5.11 Television pipe inspection

A remotely operated motorized carriage carrying a television camera has been designed for visual inspection of lengths of pipework by traversing the camera along 50 m of pipe and around bends the equivalent of two revolutions. Facilities are also incorporated to allow retrieval of any foreign bodies discovered by the television inspection [8].

The carriage is suitable for operation in pipes with diameters of 203 mm (8 in) or more, can climb slopes steeper than 45° and can negotiate bends of only 380 mm (15 in) radius in 254 mm (10 in) bore pipes.

The vehicle consists of a body on which are three spines, each carrying two motors driving pairs of reversible wheels. The motor assemblies can swivel on their axes to enable eight wheels to be in contact with the pipe at all times and thus provide good traction under most conditions. Each pair of motor assemblies can be expanded by a turnbuckle, adapting the vehicle for operation in pipes of larger diameter.

5.12 Television surveillance: Swedish nuclear plant

At Oskarshamn I a TV surveillance system has been installed inside the containment to detect and identify any abnormal conditions as early as possible [9]. It consists of twelve TV cameras and spotlights. Seven are permanently focused on some critical components known to cause trouble, such as the reactor circulation pumps and valve gaskets. Other cameras are manoeuvred by remote control from the control room. It is possible with this arrangement to inspect visually almost the whole space inside the containment and, with zoom lenses, to study most of the components and pipings in great detail.

To make it easier to locate a faulty component or a leakage with a TV camera without too much scanning, three cameras are equipped with microphones which can be used together to pinpoint a fault that produces enough noise.

At this plant, a small rupture occurred in the pipe between the main relief valve and the pilot valve. An increased leakage could have led to the opening of the main valve and a partial blow down of the primary coolant. The ordinary containment leakage system recorded an increase in the leakage rate but it was still well under the permitted level. Nevertheless steam-leakage from the crack was easily visible using the TV system and the station was quickly shut down. The repair took only one day. The risk of a blow down to the containment and a long outage was consequently reduced. The TV surveillance system thus not only enhanced the safety of operation, but also increased plant availability.

5.13 Television monitoring of water turbine installations

Water turbine installations include a dam, often far removed from a long machine hall, and screening plant that stretches along the whole length of the hall. The headwater level and the tailwater are important for the safety of a hydraulic power station and also for the safety of the environs. Central television monitoring stations with remotely controlled panning and vertically tilting cameras are installed for improving safety and as a measure of rationalization. Each television camera is in circuit with a monitor unit in the control room and if necessary, can be switched over to a large television receiver. To permit one operator to exploit this effective visual observation of the whole hydraulic power station in the headwater and tailwater areas, at intervals or continuously (e.g. in the event of a high water level or debris-laden water), not only by day but also by night, each television camera unit is fitted with a remote-controlled searchlight.

Before starting control manoeuvres, every opening in the dam is checked for freedom from floating objects with the television installation. The television cameras then supervise raising or lowering procedures for all sluice gates.

The headwater is observed by television cameras a long way ahead of the dam or the screen in front of the machine hall, to detect swimmers, pleasure boats and water skiers quickly and warn them of the risk they run by means of loudspeakers. Particularly successful is supervision of the headwater for approaching oil slicks, whose removal before entry into the hydraulic power plant can be effected satisfactorily. Especially useful has been the supervision of the headwater, the dam, and also the machine hall screen in the event of a substantial amount of drift ice and debris.

The machine hall screen is also monitored by a television camera for fouling. In addition, the degree of fouling is determined by measurements of differential pressure, and cleaning at intervals is carried out accordingly.

The television camera for the tailwater takes over supervision of boat locks; also, shore control in the region of the dam is carried out by television cameras, particularly in the case of high water. Before actuation of all sluice gates in the dam installation, a check is made with television cameras in the vicinity of the tailwater to ascertain whether people are still present in the danger zone, despite the loudspeaker warnings.

A favourable disposition of cameras in the machine hall makes it possible to carry out an overall visual observation from the control room in the hall of all water turbine installations, even in areas that are difficult of access. Since, with appropriate adjustment of the sensitivity of television cameras, even oil vapour, oil spray, and particularly the slightest signs of smoke can be detected. A television monitoring system for machine installations is an extremely effective precaution against fire.

5.14 Low light level television camera tube

High sensitivity is obtained from a range of tubes produced under the Mullard trademark NEWVICON℠. Two tubes are 16.9 mm ($\frac{2}{3}$ in) diameter, 108 mm ($4\frac{1}{4}$ in) long both with magnetic deflection but respectively magnetically and electrostatically focused. A third tube is 25.4 mm (1 in) diameter, 159 mm ($6\frac{1}{4}$) long and is magnetically deflected and focused.

High sensitivity under the low light leve. conditions as when the scene viewed by the camera is illuminated at only a 'dim twilight' level (1 lux) is achieved by the NEWVICON tubes from the adoption of a heterojunction photoconductive target composed of cadmium and zinc. They are more sensitive than silicon vidicons and twenty times more sensitive than a standard vidicon. Resolution is excellent, i.e. 650, 600 and 800 TV lines respectively. The tubes' spectral response is maximum at 750 nm and cut-off is at 900 nm (1 nanometre $= 10^{-9}$m). Newvicon tubes' heater consumption is low at 95 mA at 6.5 V.

5.15 Image intensifiers

Image intensifiers enable man to see in levels of illumination that would normally be prohibitive. Modern imaging devices are so developed that image perception is limited only by statistical fluctuations in the photon flux incident on the input optical system.

In an image intensifier, an image of the scene being viewed is intensified electro-optically and viewed directly on a luminescent screen. The initial image is focused onto the input face of a transparent photocathode; this then emits a corresponding 'image' of photoelectrons which is accelerated onto the screen. As each photon from the scene is made to produce several tens of photons at the screen, the final image is greatly intensified.

There are two main categories of intensifier [10]:

(1) Active image converters

They normally require the scene to be artificially illuminated, for example, by infrared radiation. The image intensifier then converts reflected infrared radiation to an intensified directly-viewable image.

(2) Passive image intensifiers

Visible and near-infrared radiation is intensified. They operate in available light, require no additional illumination and provide clear vision in conditions of near total darkness.

Electrostatically- and magnetically-focused devices have been developed. Only electrostatically-focused passive image intensifiers are considered in this description. Four main types apply:

first generation – single-stage types; cascade types;
second generation – inverting microchannel plate types;
                      double proximity-focused microchannel plate types.

The average number of photons emitted per second by an object in a night scene is linearly proportional to the brightness of the object. The fraction of these photons that falls on the retina of the eye of an observer depends upon the size of the pupil, that is upon the state of dark adaptation, and upon the distance of the observer from the object. The number of photons received per second from any one element will vary with time as a result of the random emission process.

The process of seeing 'detail' in a scene depends on the ability of the eye to detect a difference in illuminance at adjoining areas on the retina. This can be enhanced by an instrument which:

(1) Captures a larger fraction of the photons emitted by the scene.
(2) Has a higher quantum efficiency than that of the unaided eye;
(3) Has a spectral response which is matched more closely than the eye to the night sky illumination (for example, the S25 photocathode used in image intensifiers).

Table 5.2 shows scene illuminances for a range of lighting conditions.

Table 5.2 Range of scene illuminance (all values in lux)

Direct sunlight	10^5
Bright daylight	10^4
Overcast day	10^3
Very dark day	10^2
Twilight	10
Deep twilight	1
Full moon	10^{-1}
Quarter moon	10^{-2}
Starlight	10^{-3}
Overcast starlight	10^{-4}

5.15.1 First generation image intensifiers

(a) Electrostatically-focused single-stage image intensifiers

As shown schematically in Figure 5.7 light from the object is focused on to the input to the tube by means of an objective lens and transmitted through the fibre optic window on to the semitransparent multi-alkali photocathode. A corresponding electron pattern is emitted. These electrons are then accelerated by a voltage of up to 16 kV on to an aluminized phosphor screen.

An inverted intensified photon image is generated and transmitted through the output window to appear as a directly-viewable image on the screen of the intensifier. The level of intensification depends mainly on the photocathode sensitivity, the accelerating voltage, and the tube magnification. The typical

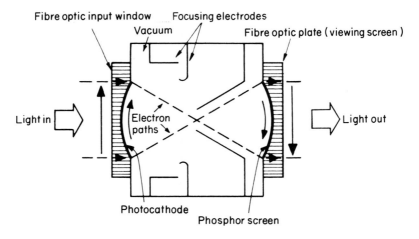

Figure 5.7 *Electrostatically focused single-stage image intensifiers.*

luminance gain of an intensifier of this type would be between 100 and 2000 (a gain of 2000 gives an extremely bright image of a moonlit scene; a gain of 5000 to 10 000 gives a useful image for scenes illuminated by overcast starlight).

(b) Electrostatically-focused cascade image intensifiers

For a gain of 50 000 or greater, single-stage electrostatically-focused intensifiers can be coupled in cascade, Fig. 5.8. High-resolution vacuum-tight fibre optic windows link individual stages.

A three stage cascade image intensifier requires an EHT of about 45 kV devived from an oscillator. The integral-oscillator cascade image intensifier has a well-defined automatic brightness control characteristic and a fast response to sudden changes in scene brightness.

5.15.2 Second-generation image intensifiers

Signal-to-noise ratio at the retina of the eye for detail in the scene is limited by fluctuations in the impinging photons (more exactly, of photoelectrons in the first stage). Subsequent links in the chain should not degrade this ratio.

The noise contribution of subsequent stages and of the eye should not exceed the amplified photon noise. In a first-generation intensifier this requires the gain be at least 50 000. Frequently this level of intensification can only be achieved by employing three amplifying stages which makes the first-generation intensifiers prohibitively massive.

With a three-stage cascade intensifer, the persistence of the three phosphor screens in series can be disturbing when observing scenes with moving

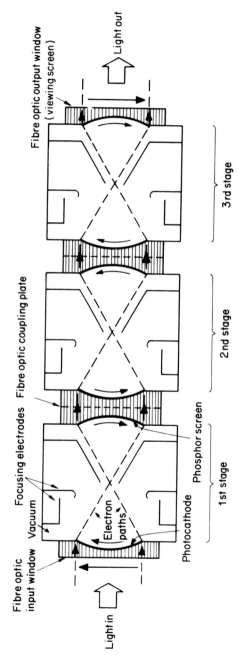

Figure 5.8 *Electrostatically focused cascade image intensifier.*

Light in

Fibre optic input window

Vacuum

Focusing electrodes Fibre optic coupling plate

Electron paths

Photocathode

Phosphor screen

1st stage

2nd stage

3rd stage

Fibre optic output window (viewing screen)

Light out

highlights: bright tracks streak the image. Although the persistence of a type P20 phosphor is sufficiently short (a few milliseconds) for a single-stage tube, in multistage tubes the afterglow remains visible for several seconds as a result of the amplification. Phosphors with a much shorter persistence are available, but they are unsuitable because of their poor luminous efficiency. Thus image persistence is also a limitation of first-generation image intensifiers.

Tubes which use microchannel plates for electron multiplication are termed 'second-generation image intensifiers'.

(a) Microchannel plates

A microchannel plate is an array of minute channel electron multipliers, which amplify an electron beam containing spatial information. Each cylindrical channel (see Fig. 5.9) in the array combines the function of the diode structure of a conventional photomultiplier with that of the resistor chain that distributes supply voltage to separate dynodes.

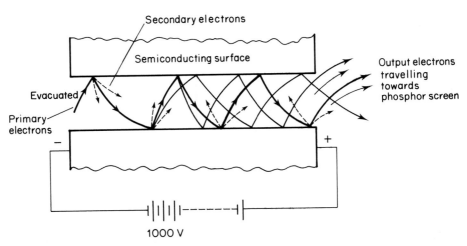

Figure 5.9 *Electron multiplication in one channel of a multichannel plate.*

The inner surface is made slightly conducting by a reduction process, and will emit secondary electrons when bombarded with primary electrons that have been accelerated by an electric field. Electrons entering a channel strike the wall and produce secondary electrons which are accelerated along the channel by an axial electric field applied between the electrodes. Transverse energy of emission causes the electrons to traverse the channel so that they again collide with the channel wall and produce still more secondary electrons. This process is repeated many times so that a large number of electrons emerge from the channel output and are accelerated across a narrow gap to a phosphor screen.

(b) Inverting microchannel plate image intensifiers

The microchannel plate is mounted just in front of the luminescent screen (Fig. 5.10). The photoelectrons impinge on the microchannel plate and, after multiplication in the channels, are accelerated towards the screen which is mounted close to the plate. A field strength of about 5 kV/mm across the gap between microchannel plate and screen gives a satisfactory modulation transfer function.

Electrons reaching the plate parallel to its axis would travel through the channels without impinging on the channel walls, giving rise to an area of reduced gain in the centre of the screen. This is avoided by setting the channels at a small angle to the optical axis of the system.

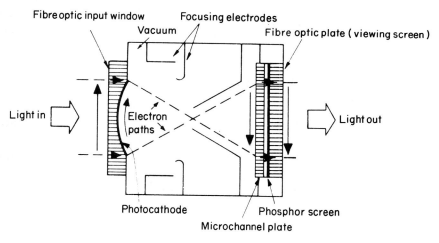

Figure 5.10 *Inverting microchannel plate image intensifier.*

Normally, the electron lenses of intensifiers of this type are designed to achieve unity magnification of the image from input to output. However, to meet certain operational requirements, both magnifying and demagnifying intensifiers have been developed.

It is possible to design the electron lens configuration so that positive ions produced at the microchannel plate are prevented from reaching the photocathode. The aluminium coating of the microchannel plate could be dispensed with, allowing the primary electron energy to be reduced to about 1 kV. This improves the secondary emission. Image distortion is also reduced. In this tube, secondary electrons produced in the closed area between the channels are not recaptured, as in the case of the double-proximity tube (see later), but are collected by the positively-biased anode. They are thus prevented from reducing the low-frequency contrast.

The noise factor (F_A) of the tube shown in Fig. 5.11 ranges from 2.8 to 3.0, and thus lies between the values for a first-generation inverting type and a second-generation-inverting type. The tube is well protected against adverse effects of highlights. This has been achieved by choosing the power supply so that the cathode voltage exceeds the input voltage of the microchannel plate as soon as the cathode current becomes excessive.

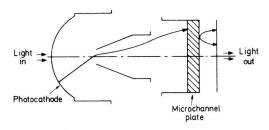

Figure 5.11 *Magnifying microchannel plate image intensifier.*

(c) *Double-proximity microchannel plate image intensifiers*

In the double-proximity tube, the microchannel plate is mounted a few tenths of a millimetre from the photocathode. A voltage of 100 V across a gap of 0.2 mm provides focusing sufficient for good reproduction. A flat image intensifier tube of small size and low weight is obtained (Fig. 5.12).

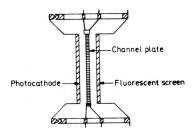

Figure 5.12 *Double proximity microchannel plate image intensifier.*

5.16 Aerial surveys: model aircraft

Costs of aerial surveys are expected to be reduced by the use of radio-controlled model aircraft (Fig. 5.13).

Figure 5.13 *Model aircraft used for industrial aerial inspection.*

5.16.1 Water flow study

Water flow on Dartmoor, Exmoor and Bodmin Moor, England, has been aerially photographed by remote control. An extreme estimate of cost saving gives the cost of observing an estuary's full tide cycle at about £2000 using helicopters, compard with about £50 for a radio-controlled model. By using cameras with infrared sensitive film it has been possible to detect pollution.

The British South-West Water Authority has developed a special model aircraft for the purpose, designed by Mr Jeff Hoare, Assistant Engineer, who also assisted in its construction [11]. The plane is built mainly of plastic covered balsa wood, has a wingspan of 2.6 m ($8\frac{1}{2}$ ft) and carries a camera. A problem of 'camera shake' caused by engine vibrations was eliminated by mounting the single engine on top of the wing and placing four layers of foam blocks at various places between it and the camera mounted in the nose.

The model aircraft weighs less than 5 kg (11 lb) to satisfy the British Civil Aviation Authority's regulations. It also had to be designed to be almost stall-free, easy to control and capable of very slow flying, although it could reach about 20 m/s (45 mph).

It normally works at a height of between 122 m (400 ft) and 457 m (1500 ft). The range of the ground-based radio control is about 1.2 km ($\frac{3}{4}$ mile) to left and right, but the length of flight can be extended considerably by following the aircraft from ground in an open-top vehicle.

5.16.2 Geological and geographical study

A radio-controlled model aircraft was used for geological and geographical photography, built by Professor P. Fouche and Mr. P. Schoeman of the University of the North in South Africa [12].

5.17 Underwater inspection

5.17.1 Direct visual inspection

Divers undertaking the inspection of underwater structures such as those used for offshore rigs can report their observations in three ways:

(1) Handwritten notes made by the diver. Using these and his memory, he makes a full report after the dive (a most unreliable method).

Under cumbersome underwater conditions a diver can only make simple notes and crude sketches. Physiological and psychological restrictions of working in a hostile environment can affect his ability to think clearly and to write. His full attention should be directed to the task at hand; mental recall at the surface of what he has seen underwater only a short while before is no easy task.

(2) 'Running commentary', where the diver gives a continuous description of what he sees to the people at the surface and this is tape recorded.

Diver-to-surface communication systems are standard equipment in the North and Celtic Seas. During a dive, the diver can report what he sees, and this can be recorded, together with surface instructions to the diver. Commentary alone is not particularly satisfactory for the following reasons:

   (a) the diver may not be suitably knowledgeable in structural defects to observe correctly;
   (b) his description may be inaccurate or inadequate;
   (c) divers' voices in a helium atmosphere become distorted even when a helium speech unscrambler is used.

Commentary is usually employed in conjunction with video techniques when it is very useful.

(3) Photographic or video methods, i.e. television and still cine photography.

Television recordings of inspection which can be monitored from the surface at the time of recording are not general practice. There are three methods:

   (a) diver-operated systems;
   (b) systems mounted on manned submersibles;

(c) systems mounted on unmanned submersibles.

With diver-operated systems persons on the surface watching the TV monitor can direct a diver where to point his camera, and a commentary from the
diver usually backs-up his camera work. He may also take stills or cine film
of areas of particular interest, though the use of cine is not yet common. On
submersibles, the TV camera is remotely controlled, either from within the
submersible by the observer (watching on the monitor) or, with an unmanned submersible, from the observer/operator at the surface. Normal photography can be undertaken with both types of submersible.

Video systems do not provide very fine detail (less than 0.5 mm) nor do
they give good colour quality. Still photography can give very high resolution through the use of 70 mm film, and both still and cine give accurate colour reproduction. Continual improvements in TV cameras and recording
techniques are being made.

Detectable crack sizes

In any situation the visual detection of defects is limited to the order of a hair-
line crack. A diver can usually see such a crack *if he knows it is there* but he is
unlikely to be able to locate such a fine crack without prior knowledge.

On a close visual inspection there are a variety of physical factors which can
affect detection of a defect:

(a) Weld surface – If cleaned off, properly illuminated and the weld face and the
    finishing suitably prepared, there is some likelihood of detecting a crack.
    Detection is more likely if the weld is dressed off, presenting a smooth back-
    ground. A very fine crack may be detected if a shadow is cast on it.
(b) Concrete – The surface is almost certainly likely to be rough. A defect is
    therefore harder to detect than for steel. Complete cleaning of concrete
    areas cannot be considered and therefore cracks may be hidden by marine
    growth. Many cracks in concrete are unimportant; it is not the detection of
    cracks that is important, but their size, nature and growth rate. Chemical
    changes in concrete may not be evident visually because the appearance of
    concrete, its colour and texture vary with each batch.

For general inspection, or inspection where the camera has to look at the subject from a distance, underwater visibility becomes an important factor. Typical
visibilities encountered in the North Sea vary from about 4 m in the southern
area to 18 m in the northern area.

5.17.2 Television inspection

A typical underwater TV camera for use to continental shelf depth is about
500 mm long and 100 mm diameter. The type and amount of lighting required

depends on the particular situation, though for typical visual inspection, a single lighting unit attached to the camera is normally satisfactory, occupying about 300 mm³. Although slightly smaller than TV cameras, still cameras are also enclosed in cylindrical housings and require separate lighting units. In general, cameras are already quite compact. With appropriate research and development, they are expected to be reduced in size.

All types of camera can do close-up work, but there is one available which is specifically designed for this task. The diver requires no skill to operate it, he simply places the frame over the target and pulls the trigger.

Although this new camera cannot be mounted on submersibles there is no restriction on the others which provide a useful back-up to the TV system. However, it is more difficult to remotely point the camera in the right direction as there is no direct monitoring of its position and direction.

5.17.3 Stereophotography

High quality stereophotographs of ships' hulls can be taken by divers without previous experience of underwater photography. A low cost system developed by the US Navy uses a single camera mounted in a tray which takes photographs from two positions. Visibility is enhanced by a Plexiglass box containing clear water placed between the camera and the hull. The system provides detailed pictures which will show such features as corrosion pits, paint blisters, cracks and fouling organisms [13].

5.18 Photogrammetry

Defined as the 'science of measurement from photographs', photogrammetry has been adopted for the accurate acquisition of positional data and adopted by Imperial Chemical Industries plc (ICI) as a suitable tool for information retrieval. An interesting report [14] discusses one application: its use during the 'revamping' of a process plant piping installation.

Photogrammetry involves the processes of recording, measuring and interpreting photographic images and as such includes:

(1) Metric photogrammetry – involving accurate measurements and computations to determine precise positions in space.
(2) Interpretive photogrammetry – dealing with the recognition and identification of objects.

Three-dimensional effects may be involved. The basic principles of obtaining three-dimensional information from photographs are the same for all photogrammetric applications – it is necessary to combine two-dimensional positional information from a minimum of two photographs.

This can be understood from the human experience of stereoscopic depth perception. In Fig. 5.14 both eyes are viewing a number of points on a terrain. When eyes function normally each acts in a way similar to that of a camera at the time of exposure, and light rays converge to focus on the retina. The brain combines the information from the two eyes and qualitatively assesses the relative positions of the viewed points. If eyes are replaced by a pair of cameras, light rays from a point will impinge on the photographic plates in the two cameras at distinct positions. Knowing the camera positions, their focal lengths, and recording the positions on the photographic plates where corresponding

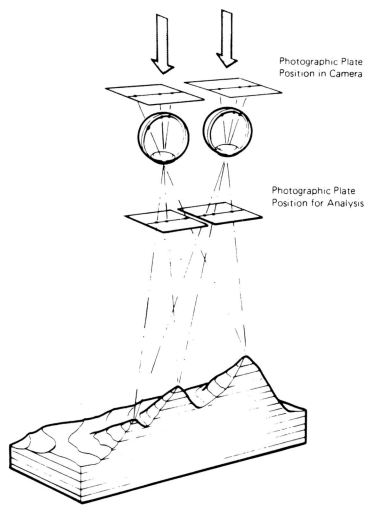

Photographic Plate
Position in Camera

Photographic Plate
Position for Analysis

Figure 5.14 *Principles of photogrammetry.*

rays impinge, enables the mathematical determination of three-dimensional positions of points in the terrain. These can be programmed into a computer from the system of coordinates so obtained.

The process of abstracting and interpreting the geometric information for pipes, structures and equipment is illustrated in Figs 5.15 and 5.16. Suitable groups of points are measured from photographs. The positional information about each point is processed on-line and then for the whole group of points the

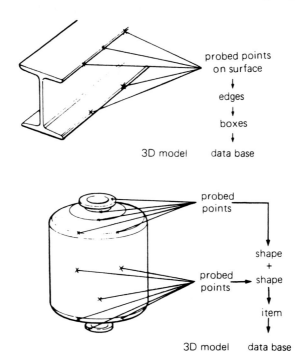

Figure 5.15 *Measurement of structures and equipment.*

corresponding engineering parameters are computed on-line and stored in a data base. The data base can be accessed at any time to examine and display the information.

The facility developed by Bracewell and Klement [14] includes:

(1) An ergonomic work station incorporating audio output to prompt the operator.
(2) A comprehensive software package including three-dimensional graphics, a data base and a large number of sophisticated interpretation routines enabling totally interactive on-line data measurement and interpretation.
(3) Interfaces to other three-dimensional computer-aided design modelling systems, e.g. plant design management systems (PDMS).

One application of this technique was undertaken at a Shell UK Oil (SUKO) refinery where a cost effective method was required of producing accurate as-built piping drawings of an extremely congested intersection of two pipe trenches, to enable additional pipes to be routed through the area. ICI provided the photogrammetric survey service and generated the necessary as-built drawings as a subcontractor to Babcock Woodall-Duckham. The main benefits which were sought by SUKO were: improved accuracy, comparable cost, improved confidence in the data and reduced elapsed time for completion of the work.

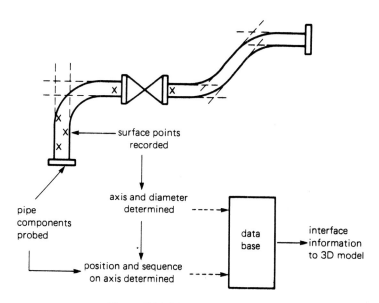

Figure 5.16 *Measurement of pipes.*

A minimum of 70% of the pipes in the surveyed area were in ICI's system and the information interfaced to PDMS. The information included:

(1) For unlagged pipes – the true pipe diameter, flange, tee and bend positions and valve positions showing the true overall length of each valve and the obstruction volume of the headgear;
(2) For lagged pipes – the overall obstruction volume, including in-line components;
(3) For other items, e.g. main pipe supports and access platform support steel – the overall obstruction volume.

The main information was presented as scale drawings, e.g. plans, elevations, sections and isometric views, up to a maximum of twenty. They were selected by SUKO, and produced within two calendar months of completion of site photographic work.

5.18.1 Underwater photogrammetry

In 1978 divers discovered serious damage to the main support legs of an oil production platform in the North Sea at a depth of 100 m. This happened during piling operations, when a shackle broke and one of the main foundation piles weighing 240 tonnes fell about 40 metres before hitting the jacket. The damage, consisting of severe buckling and tears, covered an area of 6 m by 2 m at the change in section from conical to cylindrical.

To regain structural strength and obtain recertification by Lloyds it was necessary to make repair patches to cover the holes and buckled areas. These needed to match the distorted area very closely.

A topographical map of the damaged area was required and for this purpose the National Engineering Laboratory (NEL), East Kilbride, Glasgow was approached and their Applied Photography Section employed [15]. Underwater photogrammetry was considered a most likely method of templating and NEL undertook responsibility for coordinating the development and interpolation of the photogrammetric survey.

Photogrammetry has long been used in terrestial surveying, and occasionally in engineering topographical studies. However, to the best of our knowledge, this was the first occasion on which it would be used for a survey in the North Sea. The stereo photogrammetric survey was made with a pair of 35 mm underwater cameras mounted in a frame which was specially designed for use with a locating grid, which acted as a reference plane and also as a means of achieving the required stereo overlap between adjacent horizontal and vertical stereo models. The cameras were calibrated and fiducial marks fitted in their focal planes.

A detailed programme for the survey was then drawn up, including instructions for the diving team to carry out the photography. Once the photographs were obtained they were analysed to give the raw data necessary for the next step. This produced a detailed photographic plot of the area of interest. By devising a special series of curve-fitting programs, the Systems Software Division at NEL obtained the type of information shown in graphic form in Fig. 5.17.

This was then passed to the manufacturers of the repair patches, where it was compared with mechanical pinboard template data. A series of templets were made, as shown in Fig. 5.18, such that it was then possible to produce patches to give a satisfactory fit. These patches were lowered into place in April 1978 and fitted satisfactorily with a minimum of modification. The time and expense saved was very considerable.

5.18.2 Deformation monitoring: mining equipment

Periodic monitoring of vertical deformations of structures in lignite mines

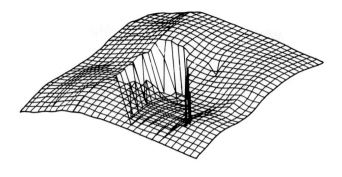

Figure 5.17 *Computer graphic representation of damaged leg.*

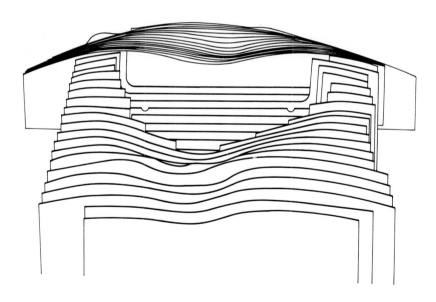

Figure 5.18 *Model of damaged leg derived from photogrammetric data.*

worked by VEB Braunkohlewerk, Welzow in the German Democractic Republic is reported to use photogrammetric methods developed by VEB Carl Zeiss. Measurements on pit equipment and conveyor bridges carried out so far were based on the method of time-base parallax photogrammetry. This measures movements of an object parallel to the image plane of a camera. The movements are determined by stereoscopic measurement of coordinate parallaxes in pictures recorded successively (time basis) from the same station and with the same orientation. This technique can be used to great advantage for stationary measurements of vertical deformation in overburden conveyor bridges. The method requires relatively few instruments and can be adapted to a diversity of measuring conditions.

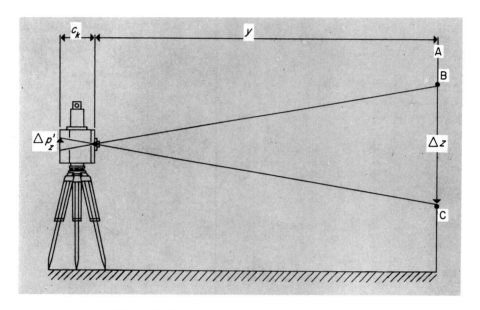

Figure 5.19 *Parallax photogrammetry technique.*

The principle is shown in Fig. 5.19 as applied to measurements of vertical deformations. The deformation parallaxes P_z' are a measure of the vertical movements z which occurred at the object. These are measured as accurately as possible on a stereocomparator. It is in the nature of the measuring principle that it allows only relative measurements. This means that for the present purpose only different static states, resulting from a change in loading conditions for example, can be compared in relation to one another.

A Stecometer, which is a precision stereocomparator made by VEB Carl Zeiss, Jena, records data automatically. The zero-load picture (initial state, for instance, an empty overburden conveyor bridge) is compared with that in the

Table 5.3 Compilation of photogrammetric deformation measurements carried out at overburden conveyor bridges

Object to be measured	Camera	Camera distance (m)	No. of measuring points	measuring accuracy attained (mm)	Job
Strip mine: Spreetal Overburden conveyor bridge (II) F42	Photheo 19/1318 UMK 10/1318	330–420 145–265	25 24	± 9 ±12	Measuring the effect of different loading conditions of the bridge flat-belt conveyors on the vertical bending behaviour of the overburden conveyor bridge
Strip mine: Welzow-Süd Overburden conveyor bridge F60 (without infeeding conveyor)	UMK 10/1318	340	38	± 10	
Strip mine: Welzow-Süd Overburden conveyor bridge F60 (with feeding conveyor)	UMK 10/1318	425	40	± 10	
Strip mine: Nochten Overburden conveyor bridge F60	UMK 10/1318	324	38	± 15	Measuring vertical deformations during backing up of F60 on to main supports after completion of assembly
Strip mine: Welzow-Süd Infeeding conveyor for F60	Photheo 19/1318	280	52	± 5	Measuring vertical deformations occurring on connection of F60 infeeding conveyor after completion of assembly

photogram (a given state of loading, for instance). Image scale and hence camera-to-object distance determine the accuracy of measurement. It also depends to a very great extent on the success or failure of eliminating during plotting such image and orientation errors as occur inevitably while taking the picture.

Conspicuous image points are selected to serve as control points from image regions of the stereopairs consisting of equi-oriented pictures which are not affected by the movement. By this means the image points should show no parallaxes. They are used to improve the measuring points in between. Most of the plotting is done by machine; at the same time the mismatches occurring at the pass points are corrected by means of a computer program which has been used successfully for many years.

Table 5.3 summarizes typical photogrammetric measurements carried out on overburden conveyor bridges.

The measurement result is plotted in Fig. 5.20. The deformation values can be read off directly. The photogrammetric measurement does not yield the deflection curve for the infeeding conveyor immediately because the latter has been moved in the vertical at the connecting end while being backed up on to the main bridge. The actual amounts of deformation are transformed numerically to the supporting points of the infeeding conveyor later.

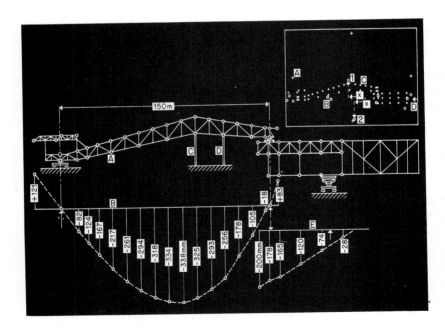

Figure 5.20 *Outcast coal dredger: deformations determined by parallax photogrammetry.*

5.19 The helium–neon gas laser

In the gas laser a helium–neon gas mixture is contained inside a long quartz tube with optically plane mirrors at each end. A powerful radio-frequency generator is used to produce a discharge in the gas so that the helium atoms are excited or 'pumped' to a higher energy level. In turn the neon atoms are excited to a higher energy level, by collision with the excited helium atoms, so that a population inversion is obtained, i.e. most of the neon atoms are in a higher energy state than the normal ground state. Stimulated emission the occurs as the neon atoms decay to a lower energy level. This process is called laser action.

The light beam which emerges from the laser is: (a) monochromatic; (b) coherent; and (c) intense. Moreover, a gas laser produces a continuous rather than a pulsed beam with only slight beam divergence. The beam can be easily manipulated, controlled and detected. Only simple optical components are required to expand or focus the beam to the desired spot size at any distance. This makes the laser a valuable tool for many applications, including:

(1) Alignment, inspection and measurement;
(2) Direct writing onto photographic emulsions and films;
(3) Non-destructive testing;
(4) Holography;
(5) Audio-visual data links;
(6) Character recognition;
(7) Non-contact printing;
(8) Audio and optical disc systems.

Many companies manufacture solid-state gas lasers for a wide range of applications. As an example some details are given of the range of helium–neon lasers available from Hughes Aircraft Company (Table 5.4).

Table 5.4 Typical outline specifications from the Hughes range of gas lasers

Model number	Minimum output power at 632.8 nm TEM$_{00}$ (mW)	Beam diameter at $1/e^2$ points (mm)	Beam divergence (full range) (mrad)	Starting voltage (kV dc)
3121 H	1.0	0.63	1.3	10.0
3122 H	2.0	0.63	1.3	10.0
3124 H–P	4.0	0.81	1.0	12.0
3170 H	10.0	1.37	0.6	12.5

Further details are available in literature from the Electron Dynamics Division, Hughes Aircraft Company, 3100 West Lomita Boulevard, Torrance, California

90509, USA. The models produced include external mirror lasers small enough to fit into a 'hip pocket'. They have Brewster windows and factory adjusted mirrors, resulting in high power output at a given distance.

5.20 Holographic monitoring methods

5.20.1 Pulsed laser holography

Stress corrosion cracks on an aircraft wing have been detected by means of this technique using a TRW System [16] with particular emphasis on the integrity of the wing plank slice. A three-step procedure was adopted:

(1) Preload the specimen in a way that will tend to open up the cracks;
(2) Suitably strain the structure so that the locally weakened areas will respond in a discontinuous fashion;
(3) Make a double-exposure hologram of the specimen, capturing a relative displacement of its surface between exposures.

Internal pressurization was used to effect the preload. The wing was purged and pressurized with dry nitrogen, since the wing contained residual fuel. Maximum pressure reached during holographic inspection was 16 kPa (2.3 lbf/in^2) above atmospheric. Notwithstanding the complexity of the wing structure, it was hoped that even this low pressure would effect a small preload in the right direction and increase the probability of finding any cracks present. Unfortunately, the effect of this type of preload was imperceptible.

In situ straining was caused by using an electromagnetic impulser so as to approximate the type of disturbance created by an impacting pendulum. A solenoid delivers an impact to the test structure; electrical energy stored in a capacitor is discharged on command into the solenoid coil. This accelerates the plunger at a rate dependent upon the voltage level to which the capacitor is charged. The electronic circuitry is packaged in a chassis box and mounted on the impulser base plate. The impulser clings to the test surface by means of three suction cups on which a continuous vacuum is drawn. The plunger does not strike the test surface directly, but transmits through a load distribution block to prevent damage to the structure. A timing signal is produced by a light beam path mechanism built into the upper housing of the impulser; it synchronizes the mechanical impulse to the firing of the laser.

To make a hologram the following procedure was employed:

(1) The optical subsystem was positioned over the selected area and brought to a predetermined height, in this case about 1.5 m (60 in);
(2) A film plate (in cassette) was inserted into its nest and the photographic shutter was cocked;

(3) The laser storage capacitor was charged to the required voltage level;
(4) The 'fire' button was depressed, initiating the following sequence of events:

    (a) the photographic shutter opened;
    (b) the impulse introduced a stress wave to the wing surface;
    (c) the laser emitted two pulses which were timed to record the differential displacement of the surface;
    (d) the shutter closed.

(5) The exposed film plate was removed and developed;
(6) The hologram was reconstructed and the fringe pattern analysed for indications of cracks.

5.20.2 Other possible applications

Following feasibility studies [17] recommendations were made to apply the TRW mobile holographic inspection system to other problems such as:

(1) Detection of loose fasteners in aircraft structures.
(2) In-flight evaluation of deformations and stresses in aircraft structures.
(3) A system study to determine the optimum size of area to be inspected at one time.
(4) The TRW concept of 'structural signature' to assess fatigue damage in complex aircraft structures. (The concept is based upon comparing the structure against itself periodically. As the structure experiences progressive duty cycles, changes in the 'structural signature' of a highly redundant structure should result from fatigue or reduction in cross-section of a structural component.)
(6) Non-destructive testing of helicopter blades, the CH53 helicopter in particular.
(7) Measurement of residual stresses in aircraft components.
(8) Aircraft engine components. TRW is currently under contract with the Naval Air Engineering Center to develop pulsed laser holographic inspection techniques which will eventually permit inspection of jet engine turbine blades without requiring disassembly of the blades from the turbine wheel.
(9) An automatic film processing technique.
(10) Recently developing photopolymer recording media for making high efficiency phase holograms, which need no chemical processing and which now can be viewed immediately.
(11) Automated data reduction to permit use of the holographic inspection technique on a broader basis which will not require highly trained personnel to interpret the holographic results. (TRW has conceived an automated data reduction technique and has performed preliminary investigations to verify that the technique is indeed feasible and that it offers significant cost

savings in the long run. By use of viewing optics and a TV camera, the fringes from a reconstructed holographic interferogram can be read directly, while simultaneously monitoring the areas under investigation, into a digitizer the output of which can be fed to a minicomputer or a sophisticated digital computer to obtain location and extent of defects automatically.)

5.20.3 Case study

A P-3 aircraft wing with a service history in the US Air Force known to involve considerable corrosion cracking was examined under pulsed laser holography. Several cracks were found.

One was a 50 mm (2 in) crack on the underside of the right wing, another about 25.4 mm (1 in) long was located on the upper surface of the right wing, near the inboard fuel tank access.

Another strong holographic signal indicated a probable crack in the unmarked area on the left wing. It was located on the upper surface near the inboard engine nacelle. No crack had been found in that area by the ultrasonic probe and since cracks were found in the laboratory which were missed by ultrasonics, the interferogram probably indicated a crack.

5.21 Moiré fringe: linear measurement

This technique could be used more extensively for structural monitoring.

A Moiré fringe is produced through the relative movement of two linear rulings. Thus if parallel lines are contact printed onto a transparent base they provide with the original parallel rulings the appropriate action pair for a Moiré fringe displacement monitor.

If such a print is placed in exact register with the ruling from which it was printed, no light will come through. If the print is rotated at a very small angle to the original, the crossover will appear as black lines, substantially straight and at right angles to the lines on the ruling (Fig. 5.21).

The Moiré fringes produced by the superposition of two gratings which are imperfect (either because the lines are not quite straight or the interval not quite constant) are distorted. A grating may be tested by superposing it on a perfect grating and viewing the Moiré fringes.

If two gratings are superposed so as to show the Moiré fringes and one is moved a distance x (perpendicular to its rulings), the Moiré fringes move a distance $y = x/\psi$ (see Fig. 5.22). The movement of the fringes can be made 100–1000 times as fast as that of the grating by adjusting ψ. It is possible to attach one grating to the base and the other to the carriage of a travelling microscope and to count the passage of the Moiré fringes across the field of an auxili-

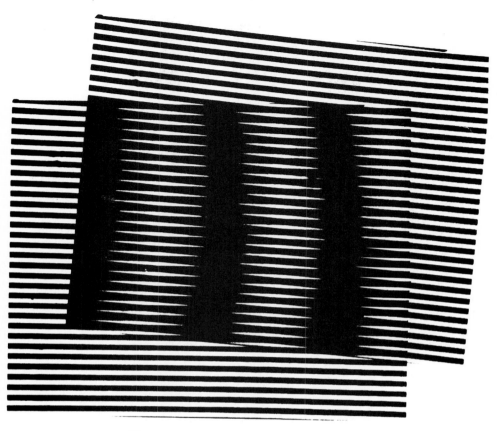

Figure 5.21 *Enlarged view of Moiré effect.*

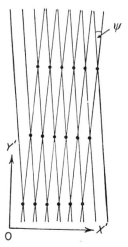

Figure 5.22 *Crossed grating at an angle ψ showing Moiré effect.*

ary microscope. This count may be made automatically by photoelectric cells. Since one Moiré fringe passes every time the carriage moves through one grating interval, the distance moved (expressed as a multiple of a grating interval) can be read on a counter. The fringes thus form a measuring scale.

5.22 Radiological examination

Radiological examination is based on the fact that X-rays and γ-rays can pass through materials which are optically opaque. The absorption of the initial X-ray depends on: (1) thickness of material; (2) nature of the material; (3) intensity of the initial radiation; factors which affect the transmitted radiation. The primary methods for detecting the transmitted radiation make use of: (a) a fluorescent screen (fluoroscopy); (b) a photographic image (radiography).

With fluoroscopy the transmitted radiation produces a fluorescence of varying intensity on the coated screen. The image is positive so that the brightness of the image is proportional to the intensity of the transmitted radiation; it is low

Table 5.5 British Standards for radiological practice

BS2600/1962	General recommendations for the radiographic examination of fusion welded joints in thickness of steel up to 50·8 mm (2 in.)
BS2737/1956	Terminology of internal defects in castings as revealed by radiography
BS2910/1962	General recommendations for the radiographic examination of fusion welded circumferential butt joints in steel pipes
BS3451/1962	Testing fusion welds in Al and Al alloys
BS3971/1966	Image quality indicators (IQI) and recommendations for their use.
BS4080/1966	Methods for NDT of steel castings
BS4097/1966	γ-ray exposure containers for industrial purposes
BS499/1965 Part 3	Terminology for fusion welding imperfections as revealed by radiography
BS1500/1958	Fusion welded pressure vessels for use in chemical, petroleum and allied industries
BS2633/1966	Class 1 metal arc welding of steel pipelines for carrying fluids
BS3351/1961 BS4206/1967	Piping systems for the petroleum industry Methods of testing fusion welds in Cu and Cu alloys

in cost, produces speedy results and has a scanning capability. Disadvantages are: (1) lack of a record: (2) generally inferior image quality.

Radiography provides a permanent record when sensitized film is exposed to X-rays. Flaws and voids, being hollow, show up as dark areas, whereas refractory inclusions appear as light areas since they absorb more radiation.

Specifications relevant to radiological practice are listed in Table 5.5.

5.23 Proton scattering radiography

During studies with intense beams of energetic protons at UKAEA Harwell (first used in nuclear physics) employed in combination with photographic films, it was noticed that sharp shadows were cast on the films by the edges of the massive steel bending and focusing magnets when the beams were incorrectly adjusted. Even very thin objects placed in the beam often showed up quite sharply defined. The phenomenon was subsequently recalled and led to a new type of radiography using protons [18] from the 160 MeV proton synchrocyclotron at Harwell.

Radiograph production depends on the multiple Coulomb scattering of energetic charged particles as they pass through the electron clouds round the atoms. Repeated small deflections produce a small net deflection of an individual particle emerging from any material. Particles are not absorbed by this process hence the selective absorption involved with X-rays or neutrons does not apply. Intensity changes at an edge in the material. High intensity proton radiographs show the object as an outline whenever there is a rapid change of thickness. This is in contrast to an X-radiograph (recorded on film) where the different thicknesses of the object produce a different shading of the areas. These two representations of an object are accepted with equal plausibility by the mind. Xero-radiography, in which a charged selenium plate uses the modified charge distribution rendered visible by dusting with a fine powder, has many of the general features of proton radiography; whereas the edge pattern is fundamental to proton scattering, it is entirely a property of the recording device with xero-radiography. Details of proton radiography based on the theory of multiple scattering are described by West and Sherwood [19, 20].

5.24 Radioisotope techniques

It is usual to consider radioisotope techniques as falling into two distinct groups: 'sealed-source' techniques and 'unsealed-source' or 'radiotracer' techniques.

In a sealed-source application the radioactive material remains encapsulated throughout the course of the measurement. Radiation from the source is directed at the medium of interest and changes in the intensity of the transmitted or scattered radiation can be analysed and related to the properties of the medium.

Sealed-source techniques constitute an important class of investigational techniques. Apart from their use in radiography – surely the most widely used of all radioisotope techniques – they find extensive use in thickness and corrosion monitoring, determination of the density profiles of process vessels, level detection, the location of blockages and various types of elemental analysis.

In a radioactive tracer technique the radioactive material is introduced into the medium of interest. By appropriate selection of the chemical and physical form of the injected radioactive material, so that it behaves as a true tracer, detailed information on the movements of the medium can be obtained by measurement of the emitted radiation.

The uses of radiotracers for the diagnosis of plant faults are perhaps less well known than sealed-source applications, but are no less valuable. This is amply illustrated by the work of the Physics and Radioisotope Services Group of Imperial Chemical Industries plc (ICI) who carry out each year several hundreds of radioactive tracer studies on process plant.

5.24.1 Radiotracer techniques of flow measurement

Among the most frequent applications of radiotracers are measurements of the flow rates of process material. Such measurements are often undertaken to provide on-line calibration of installed flowmeters but, no less importantly, they are also carried out to provide much-needed flow information in systems which are without flow indication.

In this latter context, the information provided is often unattainable by any other means and can be particularly relevant since an accurate knowledge of the flow in a specific section of process plant can often provide clear (and early) indication of the onset of plant malfunction.

Although there are several ways in which radioactive tracers can be applied to the measurement of flow rate, two techniques have been found to be particularly applicable to studies on process plant. These are known as the 'pulse-velocity' and 'dilution' techniques.

Plant fault diagnosis

Calibration of flowmeters by radiotracer methods is particularly useful in verifying overall plant mass balance, particularly by measuring flow rates to drains to assess total effluent flows. The following are typical examples.

(a) Feedwater distribution – waste-heat boiler. Tube failure in the system shown in Fig. 5.23 caused several unscheduled shutdowns. Feedwater distribution to the various boiler passes was measured using a radiotracer pulse-velocity technique involving an aqueous ^{24}Na sodium carbonate solution injected in the feedwater pump and velocity detectors. This showed the flow was asymmetric

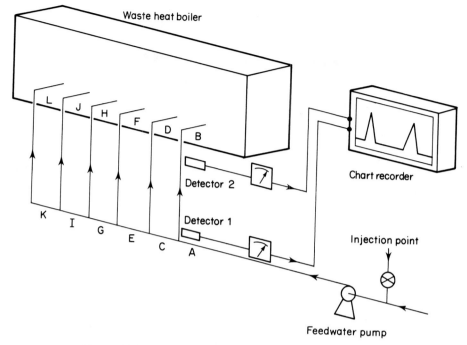

Figure 5.23 *Distribution of feedwater to a waste-heat boiler.*

with the legs IJ and KL taking excessive quantities of water while the other legs were starved – and these were the tubes which regularly failed.

(b) Investigation of internal damage in a regeneration column. Radiotracer flowrate measurements were used to investigate the inefficient operation of a regeneration column associated with a gas absorption system (Fig. 5.24).

Under ideal conditions, spent liquor from an absorption tower, after passing through the top bed of the regeneration column, is taken off from the draw-tray through the heater. The heated liquor is then returned to the lower half of the column and finally passes from the column bottom.

The abnormally low temperature of the regenerated liquor coming from the bottom of the column suggested the possibility that the heater was being by-passed.

Under normal conditions, the flow rate of liquor through the heater should be approximately equal to that from the column bottom. This was checked by using radiotracer pulse-velocity techniques to measure the two flows using the injection points and detector sitings shown in Fig. 5.24. Two interesting facts were revealed.

Firstly, no tracer movement between detectors D_1 and D_2 was observed. This indicated that there was zero flow through the heater and suggested that the

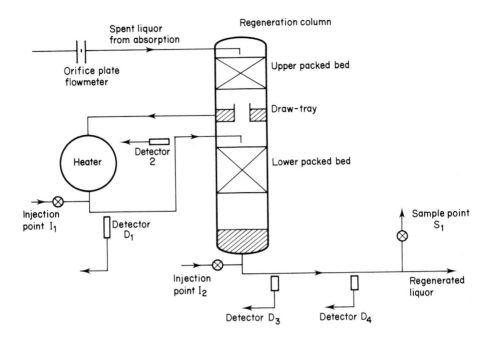

Figure 5.24 *Regeneration column: investigation of internal damage.*

draw-tray inside the column was severely damaged. Secondly, the measured velocity between detectors D_3 and D_4 was abnormally high and implied a flow rate four times greater than that indicated by an orifice plate on the liquor inlet to the column. This discrepancy was so great that it was decided to remeasure the flow using the radiotracer dilution technique to give an absolute measurement of volume flow rate. This was done by injecting tracer at a constant rate at injection point I_2 and sampling from the sample point S_1.

The results of this measurement were in agreement with the orifice plate readings thus suggesting that the result of the pulse-velocity measurement was anomalously high. The only possible explanation was that the cross-sectional area used to calculate the volume flow rate from the measured velocity was erroneously large. This implied that the bore of the exit pipe from the bottom of the reactor was restricted, possibly by deposits on the walls.

It was therefore recommended that at the earliest opportunity the plant should be shut down so that the structural integrity of the draw-tray could be examined and the liquor exit line checked for deposits. On carrying out this examination, it was revealed that the draw-tray had collapsed onto the lower packed bed and that the exit line was partially full of ceramic debris from the packing rings.

This application of radiotracer flow measurement techniques benefited production in two ways. Firstly, the identification of the draw-tray failure while the plant was still on-line enabled maintenance effort to be directed immediately at the source of the problem, thereby reducing downtime. In addition, the detection of the deposits in the liquor exit line initiated a thorough cleaning operation of the pipework and thus forestalled blockages which ultimately would have led to an unscheduled shutdown.

(c) Measurement of laydown rates in a tubular reactor. Substantial deposition of solids on the walls of a liquid-phase tubular reactor on an organic chemical production unit made it necessary to shut down the plant every six months so that the blockages could be removed. Since the duration of each shutdown was two weeks, some 8% of the yearly production was lost in this way.

In an effort to avoid these shutdowns, it was decided to attempt to control the deposition rate by periodically purging the reactor with a solvent and to monitor the effectiveness of this treatment using a radiotracer residence-time measurement. The first measurement was carried out with the reactor in a semi-fouled condition.

A pulse of radiotracer was injected into the organic feed and two radiation detectors mounted respectively at the inlet and exit monitored the passage of tracer through the reactor. From the response curves, the mean residence time was computed. The observed mean residence time, together with the known liquid feed rate to the reactor, facilitated the calculations of the free volume of the reactor. The reactor was then purged with solvent and the residence-time measurement repeated. The free volume of the reactor was again calculated and was found to be unchanged, thus implying that the purge had not been effective.

Subsequently, several alternative types of purging operation were tried until the measured mean residence time indicated an increase in the free volume of the reactor, implying that the deposits had been removed.

On the basis of these results the most effective purging technique was selected and was thereafter carried out on a regular basis. Adopting this procedure, it was found to be possible to prolong the operating life of the reactor between shutdown to well over a year with a consequent increase in production.

(d) Measurement of by-passing in a feed/effluent exchanger. A progressive drop in the conversion efficiency of a gas-phase reactor was attributable either to loss of catalyst efficiency or to internal by-passing in the associated feed/effluent exchanger (Fig. 5.25).

Radiotracer residence time measurements were carried out to investigate the latter possibility. A pulse of gaseous radiotracer (^{41}Ar) was injected into the gas feed to the exchanger and a radiation detector was positioned on the effluent gas exit line, as shown, to record the residence time distribution of the gas.

In the absence of by-passing, the residence time distribution would have been a single peak corresponding to the plug-flow of the tracer through the system. However, examination of the detector response curve revealed the presence of a subsidiary peak prior to the main peak. This peak corresponded to the fraction of the gas which had taken a shortened route through the system. The presence

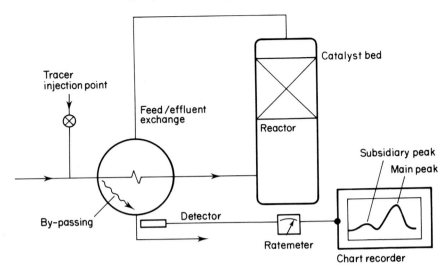

Figure 5.25 *Feed/effluent exchanger: bypass testing.*

of internal by-passing was thus established. From the ratio of the areas of the subsidiary and main peaks it was possible to calculate the fraction of the feed gas which was by-passing the reactor and to show that this was sufficient to account for the observed loss of conversion efficiency.

5.24.2 Leak detection

The high sensitivity of radiotracer techniques makes them extremely useful for the location of leaks on all types of process plant. Large savings can result from the ability to rapidly localize the source of product loss – no trivial problem on large-scale production units. Perhaps the most difficult type of leakage to detect is internal leakage – for example from the tube to the shell side of a heat exchanger. This type of malfunction can also be extremely costly, affecting as it does the quality of the product.

The principles of radiotracer leak testing are best illustrated with reference to a frequently encountered problem – the leakage of water into distillation columns. A typical distillation system is shown in Fig. 5.26. Water in product can originate either from a steam leakage in the reboiler or from a water leakage in one of the product condensers.

The reboiler is tested by injecting a gaseous tracer such as ^{133}Xe into the reboiler steam supply and sampling the vapour from the return line to the column. Any steam leak will be revealed by the presence of ^{133}Xe in the samples and the size of any leak detected can be inferred from the measured concentration of tracer. In a similar manner, water leakage from the product condensers

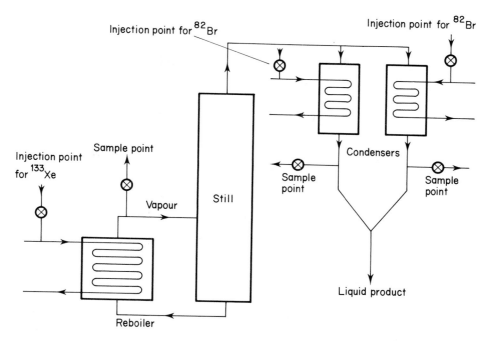

Figure 5.26 *Distillation system: water leak testing.*

can be investigated by injecting an aqueous solution of KBr (incorporating ^{82}Br) into the cooling water, sampling and assaying the condensed product. Again, the presence of tracer facilitates the localization of the leak.

All of these tests are, of course, carried out while the column is on-line so that the faulty vessel is identified prior to shutdown. Shutdown time is thereby minimized with resulting savings.

5.24.3 North Sea grout monitoring techniques

Techniques for monitoring the way North Sea installations are fixed to the sea bed use radioactive tracers and gauges. These monitor the position and the density of the large quantities of cement grout that are used to fix the piles which support the structures above the sea bed. For density measurement the gauges are fixed to the platform during construction. For positional monitoring, tracers are added to the grout when it is being mixed and are detected by means of separate logging equipment which is run down the pile.

Research and development work into these techniques was carried out at UKAEA Harwell by members of the Hydrological Tracers and Coastal Sediments Studies Project in the Nuclear Physics Division. Wimpey Laboratories Ltd, to whom rights have been licensed, have developed a complete package to enable the techniques to be applied quickly and efficiently in the often adverse offshore conditions.

5.25 Neutron radiography

This technique is capable of revealing differences between materials with similar X-ray absorption characteristics and can be used to examine complete products, particularly to assess whether they have been correctly assembled.

Neutrons are uncharged particles of great penetrating power. A metal foil placed in contact with photographic film and exposed behind the subject to a neutron beam becomes radioactive and emits β- or γ-rays. This secondary radiation exposes the film and produces an image which includes both metallic and non-metallic components.

Although X-ray inspection can be performed without dismantling a machine the technique is of limited use when a component of interest is encased, hidden behind a thick piece of metal or is organic (e.g. 'O' rings in hydraulic valves). An investigation into the potentialities of neutron radiography (X-ray) was undertaken as part of its Analytical Rework Program by the Naval Air Development Center (NAVAIRDEVCEN), Warminster, Pennsylvania, USA for investigating new and improved techniques of non-destructive inspection of Naval aircraft and aircraft components (aimed at minimizing disassembly and reducing inspection time). In support of this project, Intelcom Rad Tech conducted a program for the development of Californium-252 based neutron radiography (n-ray) as a non-destructive inspection technique.

Initial results show that n-ray is able to penetrate material and disclose hidden discontinuities and defects. Direct or visual access to the component is not required for a successful inspection. Surface corrosion, even when hidden behind metallic structures several centimetres thick, can be detected.

The feasibility study showed that the n-ray technique is capable of non-destructive inspection of a number of aircraft components, some of which are not amenable to inspection by any other non-destructive inspection technique. Examples are:

(a) Examination of composites.
(b) Detection of hidden surface corrosion.
(c) Examination of adhesive-bonded honeycomb assemblies.
(d) Detection of water in honeycomb.
(e) Inspection of printed circuit boards.
(f) Inspection of explosive devices encased in metal.
(g) Detection of blocked cooling passages in turbine blades.
(h) Coking of fuel manifolds and ignitors.
(i) Examination of metal-brazed honeycomb.
(j) Inspection of oil filters for foreign substances.
(k) Inspection of critical bearings for internal lubricants.

A penetrating radiation of neutrons is applied in a similar manner as that of X-rays to obtain a visual image of the internal form of an object.

Visualization is effectively the result of the transmitted neutron beam being detected such that the intervening body produces a shadow. The lower density of the image corresponds to portions of the object that are more effective in attenuating the direct neutron beam.

N-ray and X-ray techniques complement each other in that the relative absorption characteristics of most elements are essentially reversed. Thermal neutrons are highly attenuated by light elements (principally hydrogen). X-rays are more attenuated by heavy elements. With n-rays it is possible to inspect light elements or certain defects encased in or behind heavy elements. It can detect imperfections in thick samples through which X-rays cannot penetrate, or in samples containing hydrogenous matter or contaminants whose neutron-absorbing and scattering properties differ significantly from the base material.

5.25.1 Neutron radiography system

A large thermal neutron flux at the object and imaging system is important and forms the basic element of the n-ray system. Gamma rays must be at a minimum. A sharp, clear picture of the object is required. This means that the basic components consist of:

(1) A neutron source;
(2) A moderator and collimator assembly;
(3) An imaging system.

These are shown diagrammatically in Fig. 5.27. Californium-252 is a suitable neutron material with a very intense point source and relatively few γ-rays. Because the fission is of high energy (average neutron energy exceeding 1 MeV) it is necessary to reduce the average energy by placing the ^{252}Cf source in a material that has a high density of hydrogen. Hydrogen has the unique ability of

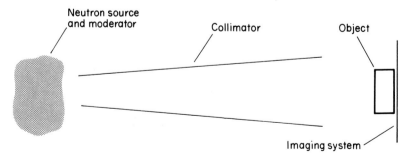

Neutron source and moderator

Collimator

Object

Imaging system

Figure 5.27 *Basic components of an n-ray system.*

slowing down the neutrons in the fewest number of scattering collisions, and thus in the smallest volume.

Photographic film provides a permanent record of a neutron radiograph but the film is relatively insensitive to neutrons. Direct imaging by neutrons is not feasible. Sensitivity is improved by using a converter screen which captures the neutrons and ultimately emits a radiation to activate the film. Converters fall into two general classes:

(1) Metallic screens that emit low energy γ-rays, X-rays and electrons;
(2) Scintillators (loaded with neutron absorbers) that emit visible light.

Metal converters are used with standard X-ray film and scintillators are employed with light-sensitive film.

Exposure may be performed by either a direct or a transfer method. The direct exposure method as illustrated in Fig. 5.28 is the most common technique.

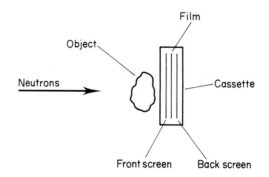

Figure 5.28 *Imaging for n-ray by direct exposure method.*

Gadolinium is the most common converter screen used in neutron radiography. Such screens can be used only in the direct-exposure method as gadolinium emits soft γ-rays and low-energy electrons promptly upon capturing neutrons. Extremely high-resolution radiographs are possible by using a very thin gadolinium converter, less than 0.05 mm (0.002 in) thick.

Photographic films are used to record images in all the imaging systems described above and can be processed in the usual manner.

5.25.2 Radiographic inspection: concrete road bridge

An 8 MeV linear accelerator (Radiation Dynamics Ltd, Super-X radiographic linac) was used to examine concrete road bridge components of up to 1.6 m (5 ft 3 in) thickness [21].

After precautionary tests in the relative safety of a laboratory, arrangements were made to gather crack information from an examination of the road bridge over the Swaythling Brook on the A27 trunk road at Southampton, UK. The interested parties were:

Bridge Engineers	Hampshire County Council
Bridge Engineers	Southampton City Council
Police	Southampton City
Health and Safety Executive	London
Linac Operations	Rolls-Royce Ltd, Advanced Projects Department, Bristol
Operational Safety	EMSc Division, Harwell
Radiologists	NDT Centre, Harwell

It was agreed that five minutes of X-ray exposure time would be permitted, subject to a safety survey with the linac at one-tenth power and then at full power, proving the predicted radiation levels in the surrounding environment. Five minutes' exposure time was considerd to be suitable for radiographing the key points on the structure as required by the bridge engineers. It was decided to close the A27 fully from midnight on Saturday 14 October 1978 to 6 am on Sunday 15 October, with partial reopening at 6 am and full reopening at 8 am. A schematic diagram of the radiographic exposures for the linac and for γ radiography using an iridium-192 source is given in Fig. 5.29.

Radiographs revealed the following:

(1) Linac radiograph of 1118 mm concrete. Reinforcing 6 mm rods and a crack 250 mm deep from the front face. Film-Industrex C; 0.5 mm lead screens; 3 min exposure.
(2) Linac radiograph of 1575 mm concrete, 6 mm rods and crack. Film Test X-H; 0.5 mm lead screens; 16 min exposure.
(3) Radiograph of bridge through deck and beam underneath; 870 mm concrete. Showing reinforcements, metal clips and loose compaction. Film NDT 75; 0.5 mm lead screens; 28 s exposure.
(4) Low density radiograph showing cracks extending from deck and under the beam. Film NDT 65; 0.5 mm lead screens; 28 s exposure.

5.26 Polarization image detector

An image display unit has been developed by Professor G. F. J. Garlick, Dr G. A. Steigmann and Mr W. E. Lamb of the Department of Physics, University of Hull, UK which provides a picture of the degree of optical polarization over a scene or surface. The polarization image detector is sensitive to surface structure rather than reflection intensity or colour, and so provides a new dimension in

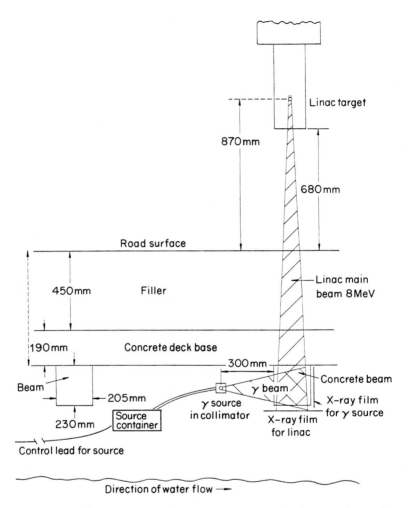

Figure 5.29 *Transverse section of Swaythling road bridge during radiography.*

remote sensing. The picture produced is independent of normal scene intensities.

A television camera unit is used to scan a scene or surface through polarization analysers, producing output signals representing light intensities. These signals are fed to an electronic arithmetic unit, which produces a composite signal representing the degree of polarization. This signal is then used to modulate the brightness of a display tube, thus giving a 'picture' of 'degree of polarization'.

The polarization image detector can be made in several forms, according to specific needs. In its most sensitive form it can give good image contrast for 0.1% polarization. It has the resolution of a conventional 625-line television picture, is easily transportable, and can be used in real-time or recording modes.

5.27 Photography

5.27.1 Photographic records

Monitoring is based on the comparison of measurements. Of the various techniques a comparison of visual records over a period of time can be very effective. Photographs must represent a true record of the image being received and therefore photographs must be faithful salient reproductions. Professional technical photography is a skilled profession in its own standing. With this in mind the following notes have been prepared from information published in standard text books [22–23].

5.27.2 Photographic concepts

Photography is a two-stage process. First we form an image of a subject by physical means, and then we make a permanent copy of this image by chemical means. The image may be the shadows cast by X-rays or γ-rays, or the trace of a moving beam of electrons or invisible images formed by infrared or ultraviolet rays, but normally it is the visible light image formed by the glass lens of a camera.

5.27.3 Light-sensitive emulsions

Photographic emulsion consists of discrete crystals or 'grains' of silver bromide whose size depends upon the presence of silver halide solvents when the emulsion is made. When the silver bromide grains are developed metallic silver grains are produced – their coarseness may depend upon 'speed'.

Silver grains of a developed emulsion are distributed in a random fashion, one of the characteristics of which is that there are bound to be areas of relatively high concentration and areas of lower concentration, i.e. areas of higher and lower optical density. On magnification to an extent considerably lower than that necessary to reveal the individual grains, therefore, one sees the variation in density as a mealy or granular effect, known as 'graininess'.

To avoid graininess it is desirable to use negative material of the finest grain practicable and an exposure which will keep the negative densities as low as possible. The density of the print will clearly have an effect, as a highlight of zero print density will have no graininess, irrespective of the corresponding higher graininess of the negative. Prints usually cover a density range from near zero to near maximum density, so that normally little control of print density is possible, but a high-key print is usually less liable to show graininess than a low-key print.

5.27.4 Photographic film

A high-speed photographic material requires less exposure to produce a usable negative. Negatives have ASA, DIN and BS speed ratings. Small grains of silver halide have the least sensitivity, accordingly 'slow' films are fine grained but are of low sensitivity with little tone and extremely high contrast. The principal types of film are:

(1) Process films – these are high-contrast slow films used for copying diagrams, graphs and other subjects where immediate tones are not required. Most of these films are not colour sensitized, because exact tone reproduction is not the main aim, a non-colour-sensitive film can be used in a bright orange safe-light. They can be obtained in orthochromatic or panchromatic form if desired.
(2) General purpose 'ordinary' films – these are intended for a continuous tone reproduction and lie between high-contrast and high-speed films.
(3) Fast panchromatic films – can be used with both artificial light and daylight, where medium to soft contrast is required with high speed and balanced colour sensitivity. A correction filter may be needed as these are not so sensitive to green light.
(4) Reversal films – a reversed image is obtained by removing the negative silver image and exposing the remainder of the silver halide which when developed forms a second silver image. This second image is then a direct or positive copy of the original subject; for example, the parts of the film that were originally unexposed are now completely black and vice versa.
(5) Special films:

   (a) X-ray film – made such that the action of X-rays on an emulsion is to produce a latent image, as some of the energy is absorbed by the silver halide crystals. Because the emulsion is almost transparent to X-rays, the grains must be highly concentrated within the layer. X-ray emulsions are difficult to produce, it is often more convenient to record the rays indirectly. The image is projected on to an intensifying screen placed in contact with the film, which under the action of X-rays emits light, usually in the blue and ultraviolet regions. As emulsions are naturally sensitive to these parts of the spectrum, it is possible to use a conventional blue sensitive film.
   (b) Polaroid-Land film – produces a finished print within 10 seconds of the exposure. It consists of a fast panchromatic emulsion which, after exposure, is brought into contact with a receiving layer containing physical development centres. Small pods of developer are then made to burst and develop the latent image in the negative. The chemicals diffuse through to the receiving layer, where the unused silver halide from the negative emulsion is converted into silver, so giving a positive image. Very high speeds – up to about 30 000 ASA – can be achieved with this type of material.

(c) Nuclear emulsions – use smaller silver halide grains than for ordinary emulsions. They are less sensitive to light. A track is produced in the emulsion by the interaction of charged particles, such as protons and γ particles, with the silver halide crystals. The emulsion is usually very concentrated to ensure that the particles are intercepted, and it is sometimes so thick that it is self-supporting and is not coated on a base. It is possible to build up a stack of layers, so that a three-dimensional record of the particle tracks is obtained.

(d) Autoradiographic emulsions – basically similar to nuclear emulsions they are designed to be very sensitive to the action of nuclear particles. The basic technique in autoradiography is to place the radioactive sample immediately in contact with the emulsion. In this way the precise location of naturally radioactive sources or radioactive tracers can be recorded.

(e) Infrared emulsions – may be sensitized to either the red or green regions by the addition of a sensitizing dye. Infrared plates must be kept very cool (about 0 °C), because ordinary heat radiation from objects at room temperature is sufficient to cause fogging over a period of time.

(f) Colour films – these consist of three emulsion layers: the top one is sensitive to blue light only, the middle layer is sensitive to blue and green light, and the lowest layer is sensitive to blue and red light. In order to prevent blue light from exposing all three layers a yellow filter consisting of finely divided silver is coated just below the blue sensitive layer. Incorporated with each emulsion there is usually a colourless substance known as the colour former or colour coupler. During development this forms a dye which has a colour complementary to the colour to which the emulsion is sensitive.

5.27.5 Underwater photography

Photographic records are considerably affected (Fig. 5.30) by the conditions. The amount of direct sunlight will depend upon the elevation of the sun, geographical latitude, season and time of day. Water absorbs light of different wavelengths preferentially at different depths, therefore colourings change when photographed underwater.

Table 5.6 Visibility according to type of water

Rhein	0.1–0.5 m
Lake Constance	1.3 m
Lake Geneva	2.4 m
Bavarian mountain lakes	5–10 m
Riviera	10–25 m
Corsica and Eastern Mediterranean	20–40 m

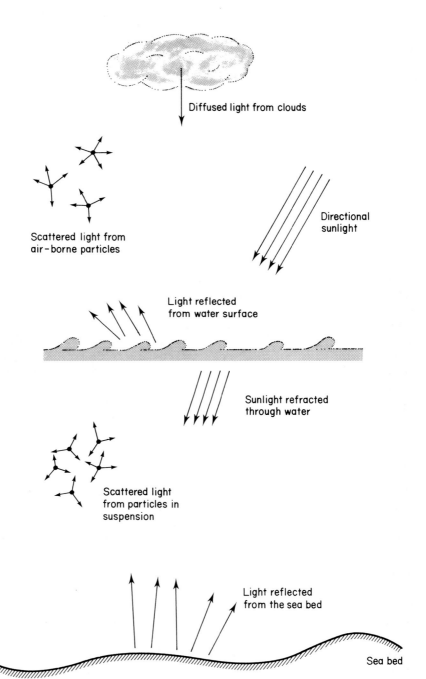

Figure 5.30 *Factors affecting underwater photography.*

Visibility reduces underwater. This is due to the absorption of light energy and to turbidity. Typical values are given in Table 5.6.
Visibility is greatly restricted in rivers and harbours due to industrial and shipping effluence.

5.27.6 Underwater photographic equipment

Six basic requirements should be included in the specification for underwater equipment. It should:

(1) Be watertight.
(2) Be able to withstand hydrostatic pressure;
(3) Maintain hydrostatic balance;
(4) Have hydrodynamic streamlines;
(5) Be corrosion resistant;
(6) Have mechanism controls functionable from underwater.

Normal cameras may be placed into underwater covers which still allow access to the controls, they also need a parallax-free view finder, flashlight illumination and interchangeable filters.

For underwater use the lens should be of short focal length with consequent wide angle of view and good depth of field. The lens should have a wide maximum aperture and be equipped with an automatic preset iris diaphragm. The angle of view of a lens is reduced by $\frac{1}{4}$ when submerged in water, and the focal length is effectively increased by $\frac{1}{3}$. A lens of normal focal length is thus converted into a telephoto lens, a wide angle lens almost into one of normal focal length and a telephoto lens into one of even greater focal length.

5.27.7 High-speed motion analysis

Events which take place at very high speeds can be photographed in colour by a new Polaroid high-speed recording system to be subsequently analysed in slow motion or frame by frame minutes later [24, 25].

This system consists of three key components: the Polaroid Polavision Phototape Cassette, the Mekel 300 Instant Analysis Camera, and the Polaroid Polavision Analyzer.

The Polavision Cassette delivers the results in full colour, instantly. It is a light-tight, completely self-contained film cartridge that requires no threading or manipulation and packages all the colour chemistry needed to process the entire 11.8 m (39 ft) film length instantly.

The camera is a precision recording instrument. Its intermittent pin-registered film transport, operated by microelectronic circuitry, allows one to document high-speed events with needle-sharp accuracy at rates from 4 frames per second up to 300 frames per second.

The Polavision Analyzer does the rest. Electronically sequenced and controlled by a rotating logic disc, this variable speed, portable display device automatically winds, processes and projects the film onto its built-in 30.48 cm (12 in) diagonal screen in just 90 seconds.

Uses which have already been reported include:

(1) Increased productivity of an automatic screw machine through use of the Polaroid High Speed Recording System to fine-tune machine alignment

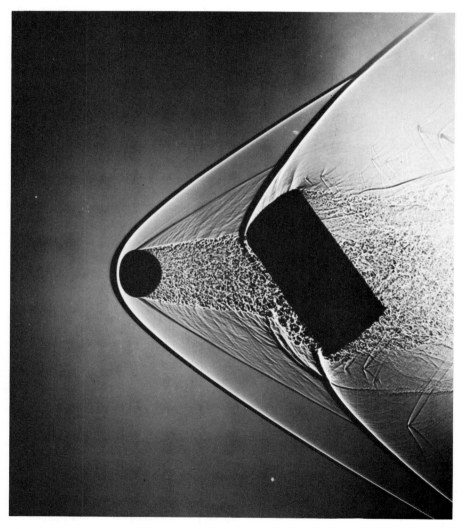

Figure 5.31 *Steel ball and sabot fired from a 37 mm cannon shown by a shadowgraph exposed for slightly less than $10^{-6}s$ and revealing the shock front, back-turbulence and rotation of sabot.*

with instant colour movies. The incidence of the driver knocking screws out of the jaws has been virtually eliminated.

(2) Aberration in a metal-forming tool identified when the colour movie, viewed in slow motion, clearly showed a worn cam to be the cause of the problem, permitting immediate correction with little downtime.

(3) Critical part on an electrical switch seen to be subject to severe strain because of flexing of its mounting bracket.

(4) Web-cutting process showed that the web was fully stopped and in place prior to cutting, this knowledge allowed the time of the pause to be shortened which led to an increase in production.

(5) A movie of a gel coating operation showed that uneven coating was caused by the gel being applied while the web was still in motion. Immediate minor adjustment of the flow greatly reduced rejected material.

(6) Trouble with high-speed plastic injection mould – found distortion in parts was caused by a momentary sticking as it was being ejected.

Some typical high-speed motion photographs using special cine cameras are shown in Figs 5.31 and 5.32.

Figure 5.32 *35 mm camera used to measure truck movement in conjunction with other cameras photographing the truck interior in an investigation of transit damage as part of an insurance claim investigation.*

References and list of companies

1 AMT Sales, 7 Oaklands Close, Shalford, Guildford, Surrey, UK (Tel. 0483-60314)
2 Hanwell, A. E., Slenderscope, Key Med, Stock Road, Southend-on-Sea, SS2 5QH, UK. (Tel. 0702-616333)
3 Hockenhull, B. S., *et al.* (1980), 'Optical fibre witness device for monitoring the integrity of off-shore structures' Report T80.4032(c), *Tech. Alert Report*, Technology Reports Centre, Department of Industry, Orpington, Kent, BR5 3RF, UK. (Tel. 0689-32111)
4 Harley, R. I., Insight Inspections Ltd, Hanley Road, Malvern, Worcs. WR13 6NP, UK. *also* Centreline/Raymond International, 365 W. Passaic Street, Rochelle Park, N. J. 07662, USA.
5 Rees Instruments Ltd, Old Woking, Surrey, GU22 9LF, UK.
6 'Solid state cameras and image information handling', Fairchild Camera and Instrument (UK) Ltd, 230 High Street, Potters Bar, Herts. EN6 5BU, UK.
7 '80XQ TV Camera Tube' Mullard Ltd, Technical Publications Dept, New Road, Mitcham, Surrey, CR4 4XY, UK.
8 Telespection Techniques in Industry *Process Engineering,* March 1979, p. 15.
9 Reisch, F. and Forsberg, S. (1977), 'Inspectors should be an operator's friend', *Nuclear Engineering,* May, 63–4.
10 Etherington, K. F. (1977), 'Image Intensifiers – an introduction', Mullard Technical Co. Rep. No. 136, October, pp 227–39, Mullard Ltd, Torrington Place, London, WC1E 7HD, UK.
11 Hoare, J. (1980) 'Model aircraft for aerial surveys' *Water Research News,* December, WRC Medmenham Laboratory, Henley-on-Thames RG9 1EW. *Also* 'Surveys from model planes' (1981), *Sunday Telegraph,* 8 February, p. 5.
12 *South African Digest,* 5 August 1983, p. 1, Private Bag X152, Pretoria 0001, South Africa.
13 Mittleman, J. 'Underwater stereophotography for hull inspection' Tech. Alert Report M140/AD-A081 322 (Paper £3.50) Dept. 1, Technology Reports Centre, DOI, Orpington, Kent, BR5 3RF, UK. (Tel. 0689-32111).
14 Bracewell, P. A. and Klement, U. R. (1983), 'The use of photogrammetry in piping design', *Proc. I. Mech. E.,* **197A,** No. 30, 1–14.
15 Welsh, N. (1979), 'Underwater photogrammetry', *NEL Newsletter,* No. 6.
16 TRW Corporation, 1501 Morse Avenue, Elk Grove Village, IL. 60007, USA. *also* 23555 Euclid Avenue, Cleveland, Ohio 44117, USA.
17 Koury, A. J. Maintenance Technology Dept, Naval Air Systems Command, Washington, DC 20361, USA. Paper MFPG-33, April 21-23, 1981, AIR-4114C.
18 West, D. (1974), 'A general description of proton scattering radiography' AERE Harwell Report R7757, June.
19 West, D. and Sherwood, A. C. (1972), *Nature,* **239,** 157–9.
20 West, D. and Sherwood, A. C. (1973), *Non-destructive Testing,* **6,** 249–57.
21 Pullen, D. A. W. and Clatton, R. (1981), 'The radiography of Swaythling Bridge', *Atom* 301, November, 283–8.
22 Engel, C. E. (1968), *Photography for the Scientist,* Academia Press.
23 Schenk, H. and Kendal, H. (1954), *Underwater Photography,* Cornell Maritime Press, Cambridge, Maryland, USA.
24 John Hadland (PI) Ltd, Newhouse Laboratories, Bovingdon, Hemel Hempstead, Herts, HP3 0EL, UK. (Tel. 0442-832525)
25 'Polavision instant high speed motion analysis system' John Hadland (PI) Ltd, Newhouse Laboratories, Newhouse Road, Bovingdon, Hemel Hempstead, Herts, HP3 0EL, UK.

Further reading

Charlton, J. S., Heslop, J. A. and Johnson, P. (1975), *Physics in Technology,* **6,** 67.

Ditchburn, R. W. (1963), *Light,* Blackie, Glasgow.

Kingslake, R. (1965), *Applied Optics and Optical Engineering,* Academic Press, New York.

Levenspiel, O. (1962), *Chemical Reaction Engineering,* Wiley, New York.

(1973), *Nuclear Techniques in the Basic Metal Industries: Proceedings of the 1972 Helsinki Symposium,* International Atomic Energy Agency, Vienna.

(1971), *Nuclear Techniques in Environmental Pollution: Proceedings of the 1970 Salzburg Symposium,* International Atomic Energy Agency, Vienna.

(1962), *Radioisotopes in the Physical Sciences and Industry: Proceedings of the 1960 Copenhagen Conference,* International Atomic Energy Agency, Vienna.

(1966), *Radioisotope Instruments in Industry and Geophysics: Proceedings of the 1965 Warsaw Symposium,* International Atomic Energy Agency, Vienna.

(1967), *Radioisotope Tracers in Industry and Geophysics: Proceedings of the 1966 Prague Symposium,* International Atomic Energy Agency, Vienna.

6
Leak monitoring

6.1 Introduction

The integrity of containment vessels such as air receivers and pipelines can be established by determining their leakage characteristics. On a gross-leakage basis this provides an indication as to the size of the defect. On a leak-source location basis this provides the means for then applying crack monitoring techniques.

In even the most unlikely of places, leak monitoring is an essential part of the integrity assurance: the author led a team to examine the criticality of a submersible rig and found that the greatest vulnerability arose from the induction of explosive vapours from offshore rigs. In the UK, an explosion at an underground water treatment plant at Abbeystead, forest of Bowland, Yorkshire (23 May 1984) led to 8 deaths and 30 injuries – all arising from an unsuspected methane build-up. On record are apartment house explosions due to gas leakage arising from foundation settlement and cracked mains.

Experience shows, and fault tree analysis can confirm, that structural integrity may often be influenced by vapour build-ups and for such reasons, leakage testing is an important function of such monitoring.

6.2 Location and evaluation

A pressure (or vacuum) container from (or to) which leaks may be occurring must first be located. The leak must then be measured and monitored.

The location of buried pipes involves the use of search instruments and possibly, mapping. The evaluation of leaks may be undertaken by:

(1) Gas or odour monitoring.
(2) Direct leak monitoring.
(3) Internal leakage detection.

6.2.1 Underground services location

Traditionally trial holes and trenches have been used to locate underground pipes and cables; experience based on the tens of thousands of damaged and dangerous excavations made in this manner confirms the benefits from scientifically managed subsurface surveys using modern location instruments.

Techniques used to survey underground services without disturbing the soil are based on the need for a buried pipe or cable in which an electric current can be located or traced. This current generates a detachable alternating magnetic field which is called the signal. Each of the instruments used detects these signals in at least one of four ways:

(1) By direct or indirect connection from a signal generator, or by using the generator to induce a signal onto the buried pipe or cable from the surface.
(2) By picking up the 50–60 Hz hum radiated by a large proportion of power cables.
(3) By picking up 'natural' radio frequency signals from otherwise non-radiating power and telecommunications cables.
(4) By inserting a signal transmitting probe into sewers or non-metallic ducts.

The equipment incorporates several significant technical advances, notably the twin-coil search antenna which enables the engineer to pinpoint a particular service amongst congested services, and makes depth estimation possible.

It is claimed by one leading firm of underground services location surveyors that they can locate:

(1) Live and dead electric cables of all voltages to an accuracy of $\pm 5\%$ in plan and $\pm 10\%$ depth up to 10 metres in good conditions,
(2) Post Office cables to the same accuracy,
(3) Water and gas pipes (including those made from PVC) to a plan accuracy of $\pm 5\%$ and depth accuracy of $\pm 10\%$ in depths of up to 15 metres,
(4) Drains and sewers to a plan accuracy of $\pm 5\%$ and depth accuracy of $\pm 10\%$ in depths up to 30 metres.
(5) Buried metal covers, e.g. manhole and hydrant covers, in depths up to 1 metre.

An additional feature is that this service can be extended to check sites for consolidation; cavities (man-made or natural), e.g. mine shafts, culverts; fissures; and cellars.

6.2.2 'Electrolocation' GPR search instruments

After an alternating current has been applied to a buried pipe or cable an alternating magnetic field is set up. This is called the 'signal' (Fig. 6.1).

'Electrolocation' GPR search instruments [1] were developed for use by their own subsurface survey teams. These are capable of overcoming problems of interference and error which arise in areas of congested buried services or environments of high electrical interference (Fig. 6.2). The selectivity of this range of instruments is such that they are highly effective in power stations as well as in industrial plants, utilities and city centres.

This is achieved by the twin-coil search antenna which greatly improves the signal response. The antenna reads strongest at right angles to the line of the buried service (Fig. 6.3), and can be rotated to enable the operator to follow the

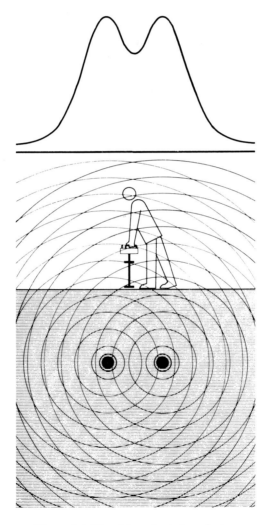

Figure 6.1 *Signal from buried conductors.*

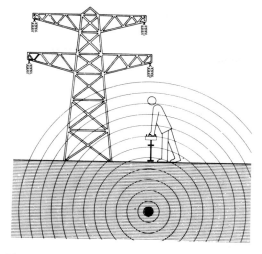

Figure 6.2 *Possible electromagnetic interference.*

line of the pipe or cable and to locate trees or branches. A motor provides push-button depth determination.

Signal filters help to distinguish between different services which all produce a wide variety of signals, ranging from a strong 50/60 Hz hum to a weak radio frequency signal. GPR instruments incorporate low-Q steep cut-off filters designed to accept frequencies which are useful for location and identification while rejecting all other signals. Spectrum analysis further aids such discrimination.

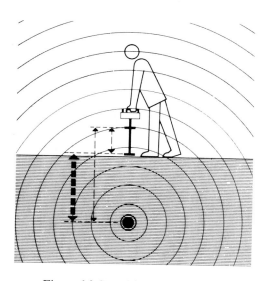

Figure 6.3 *Signal from buried cable.*

Alternating currents which apply the signal may be either universal or toroidal, and may include a phase identification. Continuous or pulsed identifying signals may be applied by the universal signal injector by direct connection (Fig. 6.4) or indirectly by earth/ground with electrodes. An indirect injection of signal can be made from the ground surface.

A toroid signal injector coupled to the VSG output can be clamped to cables (maximum diameter 88 mm). Signals injected in this manner can be traced and positively identified on a particular live or dead cable in cable racks or other areas of bunched cables.

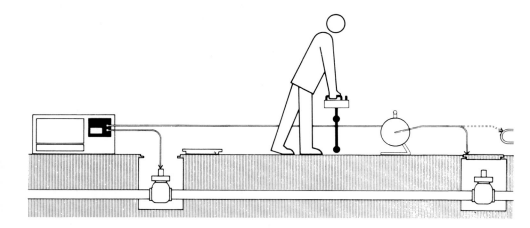

Figure 6.4 *Pipe location probing.*

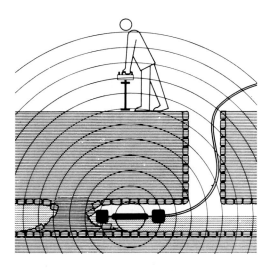

Figure 6.5 *Transmitting 'pig' in blocked drain of sewer.*

The phase identity injector enables the universal signal injector to put a signal into a domestic power socket. This enables a phase signal to be added to the identifying signal to assist in cable location.

Drain probes (Fig. 6.5) are small watertight transmitting probes (pigs) which can be propelled along the pipe by a high pressure water jet, by drain rods or floated from a tether. They are rugged self-contained units of which the standard is 42 mm diameter, 470 mm long and suitable for use in depths of 7 m in sewers and non-metallic drains. A mini-probe also 42 mm diameter but only 260 mm long is available for use in more restricted situations but is limited to depths of 5 m. A flexible probe comprising a radiating transmitting antenna fitted into a small wheeled nosepiece and attached to the end of a 30 m long semi-rigid nylon hose can be used where access is further limited (battery and electronic power pack are housed at the operator's end of the flexi-probe).

Four purpose-designed instruments are available:

GPR 104	for drain location
GPR 204	for pipe location
GPR 304	for cable location
GPR 404	for pipe and cable location

6.2.3 Metal locator

A loop coil in the locator generates a magnetic field which produces a current in metals (both ferrous and non-ferrous) buried underground. A resulting eddy current can be signalled in the instrument and thus provides the basis for buried metal location.

In a typical instrument [2] the oscillatory circuit functions at 7.5 kHz to produce a 30 cm minimum, 60 cm maximum detection capacity. Two 9-volt dry batteries provide the power and earphones or a buzzer may be used to sense the location.

6.2.4 Non-metallic pipe location

Pipes made from materials such as PVC or asbestos can be located by imposing a sound wave. The locator [2] is composed of a sound wave oscillator (based on a Wein bridge system), a sound wave vibrator, and a sound wave locator as shown in Fig. 6.6. It is based on the theory of high sound wave propagation in water.

The sound vibrator is attached to a convenient manhole (on the ground or underground) as the non-metallic pipe detection point. A low frequency sound wave (150 – 800 Hz) is generated in the water of the non-metallic pipe by using the sound wave oscillator. This transmits through the water or pipe and directly radiates sound waves from the centre of the pipe to the earth's surface. The

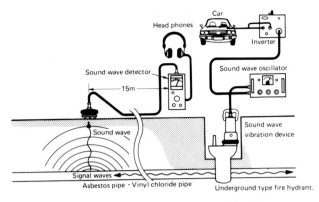

Figure 6.6 *Location system for non-metallic pipes.*

sound wave oscillator sensor receives the oscillation sound waves and the signal is amplified; and when the indicator meter registers the maximum amplitude it is located near to the underground non-metallic pipe.

6.2.5 Underground services location: case studies

(1) Mapping

It was reported [2] that in the course of the provision of new sewers problems arose for the following reasons:

(a) information from the local council was sparse, the plan supplied to the site engineer being as shown in Fig. 6.7;
(b) the junction had a high traffic flow and the digging of trial holes was not realistic;
(c) adjacent industrial and commercial property might have services cut off during working hours;
(d) depths of the services were critical.

A survey carried out without the use of trial holes showed that in reality the whole area was full of pipes and cables.

(2) Leaks

Hydrant and sprinkler main systems on the typical factory site usually have a certain amount of leakage. This is normally compensated for by a jockey jump, which is set to cut in automatically and boost pressure when it drops below the pre-determined level.

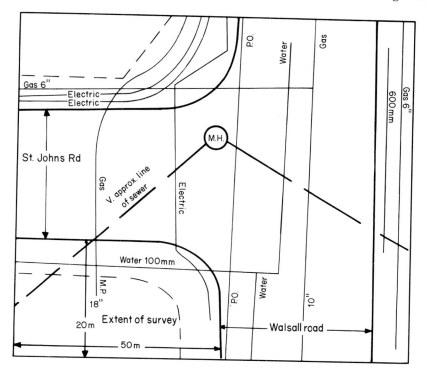

Figure 6.7 *Assumed underground pipe and cable survey [2].*

Pedigree Pet Foods at Melton Mowbray, UK found that the jockey pump on their system was continuously operating in order to maintain the necessary pressure of 827.4 kN/m² (120 psi). This indicated a serious leak, and by closing valves and checking pressure readings the suspect area was identified.

To pinpoint the leak accurately the surveyor first located the line of the water main. Then, using sensitive acoustic leak location instruments, the surveyor was able to take comparative readings along the line to determine the accurate position of the leak. The work was completed in one day causing no disruption to the factory schedule and a suspect point was established. Upon excavation, the pipe was found to have a 13 mm ($\frac{1}{2}$ in) diameter hole.

(3) Middle East

Streets of Middle Eastern countries contain large numbers of electric, telephone and water utilities laid in a somewhat arbitrary fashion. Nowhere in the world are penalties for damage to these so high. Prison sentences, once a routine occurrence for a contractor causing damage, are now becoming less widespread, the penalties being restricted to financial claims. A claim of 1 000 000 SR (£140 000) could be made by a utility company for damage to one service.

Contractors on mammoth drainage schemes, common in the Middle East, are required to meet strict schedules and are forced to use machinery to its maximum advantage. Much unique excavation equipment has been developed to deal with the compact sandstones and coral rock native to the Middle East, and, not surprisingly, such machinery can slice through even the most heavily armoured cable with undue ease.

With temperatures of 60 °C (140 °F) usual during the day, the standard practice of hand digging around utilities becomes tediously slow and diversion, the other resort, can be very expensive and take a great deal of time to obtain the necessary permission.

Unknown utility networks then are proving a major problem in excavation works, and avoiding breakages and designing excavation to avoid any conflict with existing services is of paramount importance for a contractor to achieve maximum efficiency. Standard and special location techniques pioneered by Subtronic have been invaluable to many companies in overcoming this problem. By accurately locating and identifying services prior to excavation, the engineer can fix his trench line and manhole position with precision, instead of arbitrarily drawing a line on a plan. Whole sewer networks can be planned with regard to the existing services, and excavation can proceed with confidence.

6.3 Odour detection

Humans react in different ways to different smells. Some are considered pleasant, whilst others will be objectionable. Equally some odours that are pleasant smelling in low concentrations become very unpleasant at higher levels.

The smell receptors in humans are located in the olfactory epithelium in the roof of the nasal cavity and whilst we are some way from fully understanding the physiology of odour it is generally accepted that the odoriferous molecules must first disslove in the mucus and then be absorbed on the membrane of the cilia.

Measurement of odour concentration is carried out by a method that appears arbitrary and yet as there are considerable quantitative and qualitative differences between the odour sensations generated by different compounds and as we are often if not always dealing with a mixture of different substances producing a composite odour, it is difficult to use a standard that is not based on the human nose.

The intensity of odour is expressed in relation to its threshold concentration, namely the concentration at which it is just perceptible to the human nose or to the average of a chosen set of noses – a reference panel.

Most odours contain a relatively low concentration (mg/l) of various chemicals which, often below the toxic threshold, are nonetheless pungent and nauseating.

Production odours are usually a mixture of chemical compounds in various concentrations, each with its own odour threshold. Some malodorous compounds will cancel out; other reinforce each other. Odour from a given industry is not consistent because feed material, moisture content, temperature may vary. Attempts to analyse odours from meat rendering processes have resulted in the identification of some 50 000 different chemical compounds.

6.4 Gas detection

One of the functions of a gas detector is to determine and measure the amount of gas present after a leak. One such detector known as the IRGA 20 [3] has twin infrared sources and a gas-filled Luft-pattern detector. This is filled at sub-atmospheric pressure to provide the best selectivity.

The standard cell assembly (Fig. 6.8) consists of a single-section analysis cell, a single-section reference cell and make-up tubes and gas filter cell as required. The length of the analysis cell is chosen to suit the measuring range required, long cells being used for high sensitivity applications, short cells for low sensitivity. A two-section analysis cell assembly is available to special order for certain low and medium sensitivity applications.

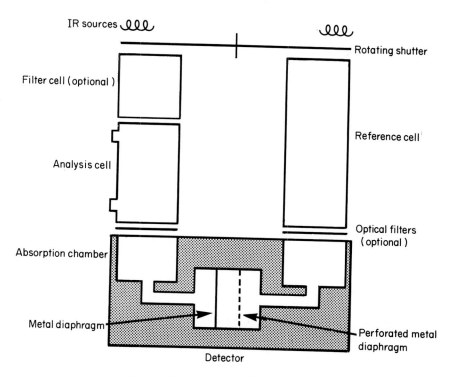

Figure 6.8. *Luft-pattern IR detector.*

The non-absorbing regions of the optical system are continuously purged by means of a built-in pump and chemical absorbent system which circulates a stream of CO_2-free air in a closed loop. A small sampling pump is built in as standard but can be by-passed if not required.

The all-solid-state electronic system includes a two-range amplifier with additional circuitry which enables linearization to be introduced on both ranges. Phase sensitive rectification of the signal ensures optimum signal/noise level and tolerance of vibration. Continuously variable controls are provided for amplifier gain and electrical adjustment of the scale zero. A switched control is also provided for coarse, step-wise adjustment of the amplifier gain.

IRGA 20 is housed in a closed case with rubber feet and, in this form, is suitable for operation on the bench. When removed from its case the instrument can be mounted in a standard 438 mm (19 in) rack if supported on chassis slides.

The standard version is fitted with a flow indicator, needle valve and gas switch to select either sample or test gas. Separate sample and test gas entries are provided. The test gas entry is located on the front panel while the sample gas entry is located at the rear of the instrument.

An alternative version of the IRGA 20 is available to special order for applications where the built-in flowmeter or test gas injection valve are inappropriate. These components and the test gas entry are omitted from this model.

The standard model has a moving coil read-out meter. An alternative version has a digital read-out meter fitted. When the digital meter is used only one measuring range is normally provided.

A switch is provided so that the instrument, when not required for immediate use, can be held in a standby condition. In this condition the sampling pump and rotating chopper motor are switched off. When switched back to the 'operate' condition the instrument can be used immediately.

The IRGA 20 utilizes the fact that each infrared absorbing gas possesses a unique absorbtion spectrum. Two beams of infrared radiation of equal energy are interrupted by a rotating shutter which allows the beams to pass intermittently but simultaneously through an analysis cell assembly and a parallel reference cell and thence into the detector.

The detector consists of two sealed absorption chambers separated by a thin metal diaphragm which, with an adjacent perforated metal plate, forms an electrical capacitor. The two chambers are filled with the gas to be measured so that the energy characteristic of this gas is selectively absorbed.

The reference cell is filled with a non-absorbing gas. If the analysis cell is also filled with non-absorbing gas equal energy enters both sides of the detector. When the sample gas is passed through the analysis cell, the measured gas present absorbs some of the energy to which the detector is sensitized resulting in an imbalance of energy causing the detector diagragm to be deflected and thus changing the capacitance. This change is measured electrically and a corresponding reading is obtained on the meter. An electrical output is also provided for connection to a recorder.

The sensitivity of this detector is given in Table 6.1.

Table 6.1

Gas	Lowest range (ppm)
Carbon dioxide	0–30
Carbon monoxide	0–200
Nitrous oxide	0–30
Nitric oxide	0–200
Sulphur dioxide	0–50
Ammonia (dry)	0–200
Ammonia (wet)	0–500
Methane ⎫ Other hydrocarbons ⎬	0–200
Freon 12 (arcton)	0–100
Sulphur hexafluoride	0–300
Water vapour	0–2000
n-Hexane	0–200
Other paraffins	0–200
Aromatics	0–500
Aliphatic alcohols	0–500
Acetone	0–500

6.4.1 Gas leak detectors

To 'sniff' gases, colorimetric devices are used where the colour of a reagent changes in the presence of known gases. 'Auer' detector tubes are widely used, each being selectively reactant to gases such as ethyl alcohol, ammonia, n-butane, butyl alcohol, chlorine, acetylene, benzene, carbon dioxide, carbon monoxide, etc.

The aspirator or 'sniffer' consists basically of a bulb and pump body into which the detector tube is inserted. When the aspirator bulb is compressed air is drawn through the detector tube and in the presence of leakage gases produces a characteristic discoloration of the reagent. Other methods based on colorimetric reactions include the use of a lead acetate dampened filter paper which turns brown on exposure to hydrogen sulphide (H_2S); a machine for continuous monitoring passes wetted paper through an orifice exposed to the atmosphere.

Other gas 'sniffing' monitors use physical properties such as: (a) electron capture cells with a radioactive source such as tritium absorbed in titanium; and (b) changes of thermal conductivity as by the change of current when passing through a heated platinum element. Electron capture cells are used for automatically testing for leaks from transistors and scaled electronic components; thermal conductivity forms the basis of refrigerant and cable leak testers.

The GPO type 24 cable leak testing gun is a hand-held leak detector which is sensitive to low concentrations of tracer gas which is injected together with the

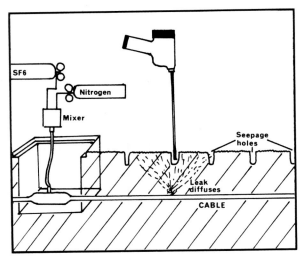

Figure 6.9 *GPO cable leak 'sniffting'.*

normal pressurizing nitrogen into the GPO cable (Fig. 6.9). The approximate location of a buried cable leak is established by the existing method of monitoring pressure and plotting a graph. Ideally a monitoring point should be fitted at every cable junction box. When the area is located a tracer gas is injected along with the pressurizing gas. This must be done the day before the leak location is carried out due to the flow resistance of the cable and the limit to the pressure which can be applied on the cable. A concentration of 5% SF_6 in nitrogen is all that is required.

The day after pressurization, the line of the cable is located and seepage holes sunk with a steel bar approximately 25 mm (1 in) diameter. It is recommended that the holes be the same distance apart as the depth of the cable in the soil. SF_6 which has diffused through the soil accumulates in the hole and is easily detected using the long probe fitting on the Leakgun. When SF_6 is detected, the holes showing the highest concentrations mark the point for excavation. Alternatively if only one hole shows SF_6 extra holes can be made. After excavation the actual leakage point on the cable is readily determined with the Leakgun.

For exposed cables, the pressurization procedure is the same but the cable can subsequently be probed directly. For ducted cables air is blown through the duct from an air line which is drawn along the duct. When the air line passes the leak a large change in SF_6 concentration is measured in the air emerging from the duct.

6.4.2 Equipment leaks: trace gas

Although the simplest method of leak detection is that used to examine pneumatic tyres, i.e. to immerse the articles under test in water, applying air or gas pressure to the article and detecting leaks by observing escaping air bubbles, it is of limited application because of the possibility of damage to the article under test, impracticability and because the limit of leak detection using water is not very high.

An alternative method of finding leaks based on the detection of halogen vapours using the 'Arcton' range of compounds which can be detected by instruments such as the halide lamp or electronic leak detector [4]. Electronic detectors can for example locate leaks of 'Arcton' 12 (dichlorodifluoromethane) as small as 0.02 g/yr (6 × 10^{-4} oz/yr). Instruments of this type detect the presence of halogens and are more sensitive to chlorine than fluorine.

'Arcton' 12 is a vapour at ambient temperature and since it is non-flammable, non-corrosive and of low toxicity it is ideally suited for use in leak detection operations. The main applications for leak detection are as follows:

(1) Detection of leaks in systems where 'Arcton' vapour is already present. Examples include refrigerators, air conditioners, aerosol filling equipment and vapour pressure instruments, etc. Testing is relatively simple.
(2) Vessels, pipes and other enclosures which have to be leak tested during manufacture or periodically whilst in service. In this case the 'Arcton' vapour has to be introduced into the vessel prior to test.

Other gases and mixtures which can be used for leak detection are listed in Table 6.2 [5].

Table 6.2 Leak detection mixtures. The gases used for leak detection depend on the sensitivity required and the type of detector employed

	Gases and mixtures	Detector type
Helium and helium mixtures	0.5 to 10% helium balance air or nitrogen	mass spectrometer thermal conductivity detector
Halocarbon-12	0.5 to 2% halocarbon-12 balance air or nitrogen	electron capture detector thermal conductivity detectors flame detector
Sulphur hexafluoride and sulphur hexafluoride mixtures	0.1 to 10% sulphur hexafluoride balance air or nitrogen	electron capture detectors thermal conductivity detectors
Krypton-85 mixtures in balance gas to suit application		radiation detectors

Halide lamp detector

This works on the principle of passing the test vapour over a copper element heated to about 400 °C (752 °F). In the presence of halogen vapour the normal flame turns an intense blue colour. This is caused by the volatilization of copper chloride formed by the chemical reaction of the test vapour with the copper element.

Several types of lamp are available but sensitivity and rapid recovery from contamination is greater when a relatively hot flame is used. Halide lamps commonly operate on the pressure lamp principle using butane paraffin or methylated spirits as feul. These lamps are widely used by the refrigeration industry. Minimum detection level is in the range 20 to 50 ppm of 'Arcton' 12.

Electronic detectors

Most detectors indicate the level of 'Arcton' in the atmosphere by means of a pointer moving on a scale or by a flashing light or audible alarm which can be set to record a specific 'Arcton' level. The sensitivity of the instrument depends upon its specification, typical figures being between 0.05 to 14 g/yr.

Typical sensors are:

(1) Leak-A-Larm – battery operated $9 \times 20 \times 5\,cm^3$ ($3\frac{1}{2} \times 8 \times 2\,in^3$) excluding probe. Sensitivity better than 14 g (0.5 oz) per year. During use in the absence of leaks the instrument buzzes and then emits a high pitched note when a leak is found [6].

(2) Bacharach – variety of detectors available covering different duties. Typical model is battery operated Leakator $11.4 \times 5 \times 19\,cm^3$ ($4\frac{1}{2} \times 2 \times 7\frac{1}{2}\,in^3$) excluding probe. Sensitivity in terms of 'Arcton' 12 is 14 g (0.5 oz) per year. During use a light glows in the probe in the absence of leaks. Light is extinguished in the presence of a leak [7].

(3) Robinair – transistorized Duo-Range refrigerant leak gun. Battery operated $16.5 \times 18\,cm^2$ ($6\frac{1}{2} \times 7\frac{1}{4}\,in^2$). Dual range to determine area of leak, then the precise location. Flexible nozzle for difficult access. Audible warning of leak. Sensitivity 14 g/yr (0.5 oz/yr) [8].

(4) FB Detectors – several models available. Basis detector is $20 \times 9 \times 13\,cm^3$ ($8 \times 3\frac{1}{2} \times 5\,in^3$) and models for battery and mains operation are available. Maximum sensitivity on 'Arcton' 12 is 0.6 g (0.5 oz) per year. The presence of leaks is shown by a flashing lamp. The instrument has an adjustable sensitivity and alternative models are available which indicate leaks by a scale reading or audible alarm [9].

(5) SF6B Detectors – basic detector gives a variable pitch audible warning or a scale reading. Maximum sensitivity is $6 \times 10^{-4}\,oz/yr$ or 'Arcton' 12. The response time varies from 0.5 to 2 seconds. The instrument will operate with a low background of refrigerant vapour. An accessory can be purchased which will differentiate between leaks of 'Arcton' 12 and 'Arcton' 11. This operates on a sample basis rather than continuously and the instrument takes 2 minutes to analyse the sample. Battery and mains electric versions are available and both require a source of nitrogen as a carrier gas [10].

6.4.3 Ethylene leak monitoring: laser detectors

Conventional ethylene detectors are 'slow' with a response time of 3 to 6 seconds, which can be excessive if a dangerous leak occurs. Since escaping ethylene gas might occur as jets, many conventional point indicators would be needed to cover a whole plant. With this in mind laser scanning was developed by Dr John Cramp of ICI Limited and incorporated in an absorption gas infrared system called 'Eagle' [11]. Laser detectors have been used before for environmental monitoring. For example, operating in the visible range they are fired at the tops of stacks to detect air emissions. But these are very big, cumbersome machines that need a special high power (kV) supply and consume a lot of energy. Eagle is robust, it uses eye-safe infrared light and works from mains electricity.

6.4.4 Benzene-in-air monitoring

A limitation to the exposure of employees to an airborne concentration of benzene in excess of 1.0 ppm in air, as an eight-hour time weighted average (TWA), or to a concentration in excess of 5 ppm as averaged over any 15 minute period, has been established by the Occupational Safety and Health Administration (OSHA) [12].

A Hewlett-Packard Gas Chromatograph Model 5830/40A equipped with a Model 7671 automatic liquid sampler uses a two-column system which when used for the benzene analysis provides fast analysis (benzene retention = 7–8 min), isothermal operation, and reduction of the possibility for interferences due to materials co-eluting with the benzene [13].

A portable pump samples the air and collects the benzene on charcoal tubes. For benzene the sampling rate to determine the ceiling level is 200 ml/min for 10 min, for a total of 2.0 l of air sampled. For finding the TWA level the rate is 20 ml/min for 8 h (480 min) for 9.6 l sampled or 50 ml/min for 3.3 h for 9.90 l sampled.

Table 6.3 shows data for a series of four National Bureau of Standards calibrated (SRM 2661, benzene on charcoal) samples analysed using a method in which 1.0 ml of CS_2 was added to the charcoal from the NBS tube. The data indicate that the peak areas are reproducible, over a wide range of benzene concentrations.

6.4.5 Infrared gas concentration analyser

Gas concentrations may be monitored continuously by the use of this analyser which is sensitive to CO, CO_2, CH_4 and other constituents [14].

As illustrated functionally in Fig. 6.10, the analyser is a non-dispersive infrared photometer operating on the single-beam principle with modulated

Table 6.3 Peak area reproducibility using NBS standards

Amount of benzene (NBS certified) (ng)	Peak area average of six analyses	Relative standard deviation (%)	Response factor (area/ng)
15 ± 1	23 867 ± 106	0.44	1591
68 ± 2	104 133 ± 208	0.20	1531
264 ± 10	399 067 ± 321	0.08	1512
1018 ± 37	1 533 500 ± 2121	0.14	1506

Average = 1535 ± 39
(2.54%)

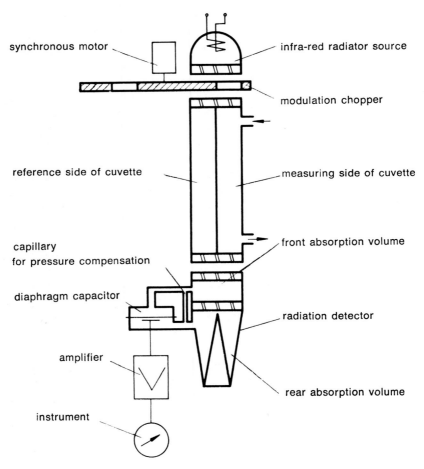

synchronous motor

infra-red radiator source

modulation chopper

reference side of cuvette

measuring side of cuvette

capillary for pressure compensation

front absorption volume

diaphragm capacitor

radiation detector

amplifier

rear absorption volume

instrument

Figure 6.10 *Functional diagram of 'UNOR' gas analyser.*

radiation. The selective radiation detector of the optical-pneumatical type is sensitized to the constituent to be measured by its gas filling; it consists of two absorption volumes, pneumatically separated but optically in series. In view of the greater layer thickness of the rear volume, its wavelength-dependent absorption lines (fine structure of bands) have a larger half-value breadth than those of the front volume. Therefore, the front volume mainly absorbs the energy of the centre of each absorption line, while the rear volume absorbs the remaining energy of the flanks. The detector is dimensioned so that both portions of energy are equal, resulting in pressure pulses of the same amplitude by heating of the gas volumes.

6.4.6 Explosive gas (Ammonia) detection and alarm unit

Ammonia is cheap as compared with many other refrigerant gases and is extensively used in industry, particularly in refrigeration systems associated with food and drink processing and storage. But ammonia is explosive, toxic and corrosive. A relevant Code of Practice is set out in the British Standards Specification 4434.

As an alternative to fully flameproof equipment in compressor plant areas, remotely mounted gas detectors linked to electrical shutdown procedures, may be adopted in ummanned plant areas. This philosophy has been endorsed by many professional bodies, insurance companies, and industry itself, aiming to avoid the high costs of disaster, and to satisfy the Health and Safety at Work Act.

Ammonia (NH_3) is slightly heavier than air at NTP. Its lower explosive limit (LEL) is 16% v/v, and its upper explosive limit (UEL), is 25% v/v. Only between these levels will ammonia ignite. Below the LEL there is insufficient gas to promote combustion and above the UEL there is insufficient oxygen to support combustion.

The R006 Refrigeration Safety System [15] is designed to give an indication of the presence of ammonia before an explosive gas/air mixture is accumulated, and for this purpose the instruments are calibrated 0–4% v/v or 0–10% v/v. This system detects concentrations of ammonia gas at pre-explosive levels by ensuring optimum location of gas detectors and sensors. These are super-sensitive electrocatalytic diffusion type sensors, each of which incorporates two platinum elements representing the arms of a Wheatstone bridge network. One element is impregnated with specially selected catalysts and under gassing conditions the electrical resistance of this element varies and unbalances the bridge network. The out of balance voltage is amplified in the appropriate detection module and is utilized to operate the switching circuit at the appropriate selected alarm level. The second element is rendered insensitive to gas and acts as a temperature compensating device.

Ammonia gas alarm systems are designed and arranged to detect ammonia over the range 0–4% v/v NH_3/air and to initiate audible alarms, with visual indication of alarm acceptance at two preset levels (adjustable, but typically set at 1% v/v and 2% v/v) (Fig. 6.11).

At first gas alarm condition, the system automatically switches on flameproof fans to disperse the ammonia build up. At second gas alarm condition, the system isolates all electrical supplies to the plant, except fans, emergency lighting, and the gas detectors themselves. To cope with fire, panic or other emergency situations, operation of break-glass pushes (suitably located at entrances and exits) sounds the alarm and brings about all second gas alarm control conditions. Audible and visual alarm is raised if a fault should occur on any sensor or its cabling, and system inhibits are fitted to facilitate normal operation of the plant.

6.4.7 Ammonia monitoring

Present methods of measuring ammonia concentrations in air are generally monitored using devices such as the midget liquid impinger or the long path infrared or chemiluminescent analyser. A new system [16] for measuring ammonia concentrations in air has been developed in the USA, using optical waveguides. It involves the deployment of coated optical waveguides at monitoring sites and a laboratory measurement of the chemical reaction between the coating and any ammonia in the atmosphere.

The optical waveguides are small quartz rods, 20 mm long and 1 mm diameter. These are sensitized to react selectively with ammonia by incorporating a chemical in the coating on the surface of the rod. The coating is about 0.1 mm thick. The waveguides are exposed in special holders for about 24 hours

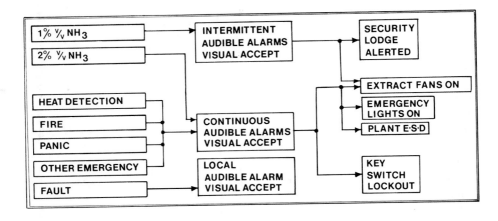

Figure 6.11 *Ammonia detection incorporated in an alarm system.*

at the various locations where ammonia concentrations are being monitored. Reaction between ammonia and the chemical coating brings about a colour change or light scattering and the resulting change in light transmission through the waveguides is measured spectrophotometrically in a central laboratory. The waveguides are carried to and from the exposure sites in sealed vials.

The light transmission of the coated quartz waveguides is measures in a specially developed 'gradient light analytical detector'. The concentration of ammonia can be calculated from the measured change in light transmission before and after exposure to ammonia. Transmittance is defined as the ratio of final to initial measured light transmission for a given exposure time. It is shown that there is a relationship between transmittance and ammonia concentration multiplied by exposure time. Calibration curves for batches of waveguides are obtained by exposing a number of waveguide collector/sensors from each batch to known concentrations of ammonia for different times.

It is claimed that the sensitivity, small size, cheapness and simplicity of the sensitized waveguides and instrumentation make the technique ideal for large-scale monitoring of ammonia concentrations in air. Concentrations as low as 1 ppb have have been assayed.

6.4.8 CO leak: measuring and warning

Highly toxic yet odourless and colourless, carbon monoxide is found in many industrial gases such as blast furnace gas, producer gas, etc., sometimes in rather high concentrations. It is also contained in pipelines and vessels and could become dangerous only by leakage. The CO-laden exhaust gases of motor cars escape in large quantities directly into the atmosphere which creates an increased risk of CO poisoning in garages and motor traffic tunnels. The first symptoms of CO poisoning are headache, giddiness and stupor. The maximum permissible concentration of CO is specified as 50 ppm.

A CO detector designed for use in garages, tunnels, work and store rooms acquires the gas sample (atmospheric air) by a built-in diaphragm pump through room air filter, cleaned in fine dust filter and delivered through flow indicator into infrared CO analyser.

The analyser is equipped with a limit switch (with two threshold valves) which can be set to any value. When the concentration exceeds this value, a visible signal is automatically released on the UNOR and on a signal panel. At the same time, ventilators can be started by suitable switching relays. The second contact is set between 150 and 250 ppm. When this value is exceeded, the second visible signal on the signal panel and, in addition, a hooter is sounded. The warning lamp A1 18 and the additional warning lamp incorporated in the UNOR continue to signal CO alarm until the concentration falls below the set CO limit. The ventilators are switched off automatically only when the CO gas concentration has fallen to harmless values.

6.5 Cable pressure monitoring

Pressurized cables have been used widely in the US telephone industry to protect against moisture from holes or breaks in the cable sheath. A continuous flow system using air dryers is commonly used, maintaining a pressure of $0.045 \, \text{N/mm}^2$ ($5 \, \text{lb/in}^2$) for underground cables, $0.018 \, \text{N/mm}^2$ ($2 \, \text{lb/in}^2$) for aerial cables.

Air is fed directly from an air pipe through a simplified shut-off manifold with pressure sensors placed between each manifold location. Breaks in cables can be identified to any 2362 m (6000 ft) manifold section and an analysis of the air flow at the manifolds at both ends of a pressure drop can more closely locate the fault.

Strategically sited low pressure limit switches were first used as go/no go leakage alarms from which Chatlos Systems Inc. developed a cable pressure transducer for cable pressure monitoring. This consists basically of a pressure sensor, converting cable pressures into resistance values calibrated in 0.15 lb steps from 0.95 lb. It can be 'dialled up' from a central office, the resistance value checked and the equivalent pressure value recorded. Severity of a leak at any point of trouble can thus be determined without actually sending someone out. Cost savings achievable with this technique have resulted in the pressure transducer being widely used throughout the US telephone industry.

Considerable time was required to read the pressure values at points along a cable route, pressure transducers could be read only perhaps once a day. Automatic cable pressurization monitoring/alarm systems were therefore developed. Such systems operate automatically from a central office and can each monitor in the region of 500 pressure transducers, reporting directly to a teletype via the telephone network to furnish a simplified printout (Fig. 6.12) continuously recording alarm point failures and associated analogue data.

6.6 Leak detection cable

Leakages from above-ground and below-ground storage tanks can be monitored to reduce the hazards and losses from liquids and gases, including all petroleum products [17]. The 'Petro-Probe' system can be installed either vertically, or at an angle, close to liquid storage tanks, or used to form a warning barrier between hazardous gases or liquids and water courses, homes, subways, etc.

Each Petro-Probe contains:

(1) One perforated 10 cm (4 in) diameter × 3.6 meter (12 ft) (other lengths available) probe tube, complete with infiltration prevention sleeve.
(2) 47.6 m (25 ft) Leak 'X' sensor cable for liquid hydrocarbons and other hazardous liquids.

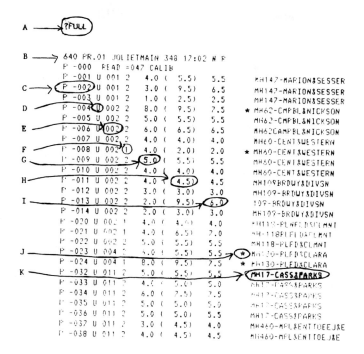

A   Request for a full report listing all transducers
B   Location of the 640 monitor
C   Serial number of the transducer
D   Type of transducer U = underground
                        A = aerial
E   Cable circuit group
F   Priority — determines to which stations this transducer will
    report when there is a fault
G   Alarm value in PSI — can be reprogrammed at any time from
    the keyboard
H   Yesterday's reading (PSI)
I   Today's reading (PSI)
J   ★ indicates there is an alarm condition
K   Address of the transducer

Figure 6.12 *Typical printout generated by the Chatlos 640 monitoring system.*

The leak 'X' petroleum leak detection cable is a twisted pair of woven conductors, insulated with a patented degradeable dielectric affected by liquid phase hydrocarbons. Degradation of the dielectric creates an electrical property change in the cable; thus when connected to a Leak 'X' monitor, it will provide a leak detection system for any vessel containing liquid phase hydrocarbons, such as, above- and below-ground storage tanks and pipelines. The degrading condition of the dielectric plus the electrical properties of the cable, facilitate fast and accurate leak or fault location.

The cable is designed to be vulnerable to all liquid petroleum derivates and will deteriorate at different time intervals depending upon the type, temperature, quantity of petroleum product, and the general environmental conditions prevailing.

In addition Leak 'X' extrudes a non-degradeable, dielectric cable (control cable) to carry the signal between the sensor cable and the monitor. The control cable has the same electrical properties as the sensor cable, for consistency in fault or leak location procedures (Table 6.4).

Table 6.4 Compound cable: typical specification

Specifications	Sensor cable	Control cable
Conductors		
Number	2	2
Size	20 AWG	20 AWG
Material	Tinned copperweld	Tinned copperweld
Strands	7 per conductor	7 per conductor
Tensile strength		
(2 conductors)	61.2 kg (135 lbs) $\pm$ 10%	61.2 kg (135 lb) $\pm$ 10%
Resistance		
(1 conductor)	30 ohm $\pm$ 10% per 305 m (1000 ft)	30 ohm $\pm$ 10% per 305 m (1000 ft)
Dielectric		
Thickness	Standard 25 mil per conductor 15 mil (used as outer cover)	25 mil/per conductor with 35 mil cover
Electrical	2000 volts (across conductors)	200 volts (across conductors)
Material	Basic block copolymer of styrene and butadiene	Standard sensor cable with PVC cover
Hardness	Marketing standard 65 or 90	
Temperature range	-84 to $+66$ °C	-84 to $+66$ °C
(standard product)	(-120 to $+150$ °F)	(-120 to $+150$ °F)
Armour protection	Loose interlocking alluminium alloy	Not applicable
Abrasion	Taber : 0.325 cc/kc	Good
resistance	NBS : 83% of standard	
Cathodic effect	Nil	Nil

6.7   Leakseeker: thermal conductivity element [18]

With a capability to detect, measure and quantify, the high performance thermal conductivity (TC) unit can be used for monitoring leaks of hydrogen, helium, ammonia, argon, carbon dioxide, methane, ethylene oxide, hydrocarbon and refrigerant gas.

The hand-held Leakseeker unit uses a new advanced design thermal conductivity detector and sampling pump. It provides greater audio and meter output amplification. The user simply selects the appropriate working range, sets the Leakseeker to zero and uses the probe to search for leaks on the system. When a leak is located, it is indicated both on the meter and by an audio signal, emanating from the instrument.

Intrinsic safety is incorporated which has led to a BASEEFA Certificate (Numbered EX74063/B) for use in hazardous atmospheres. A sister model, the Leakseeker TC is also available for use in non-hazardous atmospheres.

Because the Leakseeker is as portable as it is versatile, it can be used across a variety of applications, from offshore drilling production platforms to petrochemical plants, boiler houses, tank farms, process plants, pipelines, refrigeration plants and shipping.

Sensitivity is expressed in terms of leak rate sensitivity and concentration sensitivity. The concentration sensitivity is a function of the detector design. The leak rate sensitivity depends on the flow rate into the detector probe which is only $0.05 \, \text{cm}^3/\text{s}$. Response time and recovery time are maintained at 1 second. Characteristic values for sensitivity are given in Table 6.5. Values to provide a conversion between units in different measurement systems are given in Table 6.6.

6.8   Fluid line leak monitoring

The traditional way of shutting down a fluid line in the event of pressure loss is to use a simple device such as a switch directly operated by a pressure element to sound an urgent alarm.

Current good practice is to provide an entirely separate indication of the pressure so that an operator may look at the present value and initiate action accordingly. Alternatively an alarm device on the signal representing the pressure can give confirmation and the two signals provide the necessary action.

Another method is offered by using strain gauges. Since they can sense forces with virtually no movement, their electronic circuits operate at very low levels while the usual considerations of components and electrical reliability are hardly applicable. The attraction of such transducers for alarm/shutdown systems is that they are small enough to screw directly into a flow line of virtually any size and thus connection requires no more than the single electrical cable which can conveniently terminate in a plug and socket. Thus installation is greatly simplified.

Table 6.5

Gas	Response sense	Full scale deflection at maximum sensitivity	
		Leak rate (atm cm³/s)	Concentration sensitivity (%)
Hydrogen	+ VE	7.5×10^{-5}	0.15
Helium	+ VE	9×10^{-5}	0.18
Methane	+ VE	2.4×10^{-4}	0.48
Carbon dioxide	− VE	3.2×10^{-4}	0.64
R12 refrigerant	− VE	1.1×10^{-4}	0.22

Detection limit is 5% of maximum sensitivity, e.g. hydrogen 3.75×10^{-6} cm³/s or 0.0075%.
For sensitivity of TCS multiply above values by a factor of 2.

Table 6.6 Conversion table for leak rate units

Multiply by factor to convert to	lusec	Torr l/s	atm cm³/s	oz/yr R12
lusec	1	10^{-3}	1.3×10^{-3}	8.8
Torr l/s	10^3	1	1.3	8.8
atm cm³/s	760	0.76	1	6.7×10^{-3}
oz/yr R12	0.11	1.1×10^{-4}	1.5×10^{-4}	1

The silicon semiconductor type strain gauge which fits directly onto a metal diaphragm gives an adequate electrical output from a d.c. input with almost no electronic signal processing circuitry and thus the output of such a strain gauge can be used directly for alarm purposes. Erroneous action is only likely to occur due to a break in the circuit or loss of power supply, both of which would result in a zero output, easily distinguished from a zero pressure signal. The simplicity of the system, particularly for multipoint monitoring, is illustrated in Fig 6.13. A single pressure sensor can incorporate two independent strain gauges so that

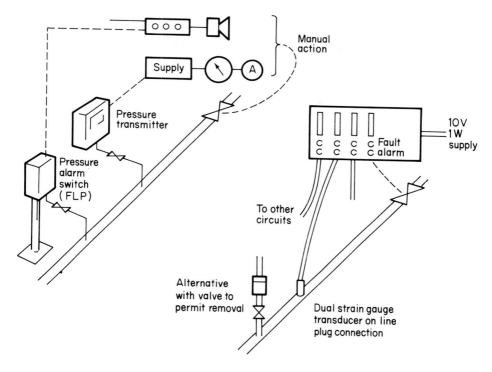

Figure 6.13 *Strain gauge/pressure leak transducer system.*

one inherently monitors the other and can give advance warning of the need for maintenance.

6.9 Ultrasonic leak testing

Leaks in any pressure or vacuum system can be located by utilizing the ultrasonic energy generated by molecular collisions when gas is forced through a small orifice. The ultrasonic frequences involved are well outside the audible range thus enabling the instrument to be used in areas of extremely high audible noise level.

Ultrasonic energy is picked up by a high frequency ceramic microphone. The signal is amplified and used to provide a visual indication on a moving-coil meter as well as an audible signal on headphones [19]. The headphones also provide an effective screen for the ambient noise, thus enabling the operator to distinguish more easily the noise produced by the leak. The microphone is highly directional and enables the operator to detect a leak at long range.

6.9.1 Applications

The applications of this instrument are almost umlimited and the list given below is only representative.

● General leak detection in chemical plant
● Testing reforming plant and high pressure gas mains
● Steam condensers
● Refrigeration equipment
● Heat exchangers
● Corona discharge
● Pressurized telephone cables
● Compressed air systems
● Watertight compartments in ships
● Pressurized compartments in aircraft
● High pressure pneumatic and hydraulic servo systems
● Ovens etc. under pressure or vacuum
● Boilers on test
● Pipelines under construction or repair
● Seating of valves in enclosed or open systems
● Pneumatic systems of road vehicles
● Pneumatic control systems
● Pressure and vacuum brake systems
● Air operated contactors on electric traction control
● Valve operation in compressors and internal combustion engines
● Bearing noise
● Air leaks from pneumatic tyres
● Radiation from faulty electrical appliances

6.10 Detection of water leaks

6.10.1 Water leak detector WL-200

Water which leaks from a buried pipe generates noises with contributions from: (a) flow rate; (b) impact forces; (c) frictional effects; and (d) vibration. When these noises are transmitted to the surface the tonal quality of each is changed by the transmission path – such factors as the magnitude of the leak, soil quality, pipe material, burial depth are involved.

The noises can be analysed in terms of their frequencies, from which the pipe-based noises usually dominate.

The WL-200 water leak detector [20] has been developed on this basis; it filters out all other sounds and 'listens' only to noises in the probable pipe-frequency bands. According to the material of the pipe the operator selects one of the three steps of the filtered frequency bands: 'high', 'low' or 'wide'. The tone-quality of leak noise can be grasped more clearly by cutting the noises with a sharp filter. Leak noises of low frequencies generated from pipes made of vinyl chloride and polyethylene can be detected.

6.10.2 Meterized water leak detector WL-91

All kinds of leaks from fluids under pressure in pipes can be quantified by use of the meter incorporated in this detector. Water leaks are detected by the sound in the headphones and confirmed by a large indicator meter. The WL-91 uses a relatively heavy sensor to accurately catch the faint sounds made by leaking water and filter out external noises in the amplifier, making it possible to detect leaks easily.

Automatic mechanisms, such as a limiting circuit for automatic electronic regulation of excessive input and a fixed voltage diode, stabilize variations in voltage. Most sounds made by leaking water are in the range 200–800 Hz. The WL-91's amplifier thus gives optimum passage to sounds in this range and cuts out high and low frequency sound ranges with a filter circuit. By narrowing the frequency band to avoid extraneous noises the sounds from leaking water are made clearly audible.

6.10.3 Water pressure recorder

Water leakage can be regulated by controlling pressure and by making a pressure distribution chart through which it is possible to find irregularities in water leakage in a system.

An electric version of a water pressure recorder uses a strain gauge in the transducer attached to the water pipe as the basis for pressure measurement. Recording is made on a continuous paper roll through an electric discharge.

6.10.4 Water mains: nitrous oxide leak detection

This is a 'seeding' diagnostic method whereby a nitrous oxide tracer gas is injected into the suspect length of main and the main is brought to full test pressure. The operator then proceeds along the length of main, commencing at the injection point, probing at intervals of approximately 3.7 m (12 ft) (less if necessary) or at joint positions, with the detection apparatus. Where a leak occurs the equipment will register the tracer gas as indicated in Fig. 6.14.

The following points are recommended to facilitate the search and to ensure a satisfactory result:

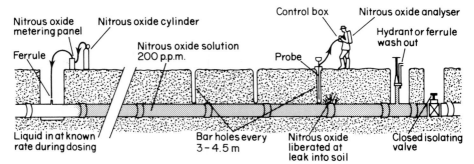

Figure 6.14 *Nitrous oxide leak detection method.*

(1) Equipment should be available on site to pressure test the main.
(2) The position of the main must be known and, if possible, the position of all joints etc. Barholes are to be made adjacent to the main along its length to the invert level of the main.
(3) If a water main is at fault an adequate supply of water will be required to fill the main and also facilities for disposing of the water after use.

6.11 Pressure decay leak detector

As a replacement to the 'bubble under water' test this detector monitors rate of change of pressure as well as differential pressure between the system under test and a reference (Fig. 6.15) [21].

No special services are needed other than a suitable electrical mains supply and compressed or bottle air supply. Once supplies are connected, the reference and test items are in turn connected to the detector, known as the Qualitek, the test pressure is set by a rotatable control and the fill time adjusted to suit the component by digital selectors.

To instigate the automatic test cycle the 'Initiate' push button is operated. The test and reference items are then pressurized to the pre-set test pressure for the pre-set fill time. The Qualitek then measures electronically any pressure difference between the test and reference items and computes the rate of change in pressure. This operation is achieved in only half a second.

If the pressure difference remains within the pre-set limits during the test time then the green 'accept' lamp will illuminate. If the pre-set limits are exceeded then the red 'fail' lamp will illuminate. In the event of there being a very large leak which is outside the instrument's measuring range the 'gross leak' lamp will illuminate.

A differential pressure transducer is the basic detector element. Overall the equipment has a sensitivity of 0.01 mbar/s (pressure decay 2×10^4 atm cm^3/s leak rate for 10 cm^3 volume or, 1×10^{-3} atm cm^3/s leak rate for 100 cm^3 volume). Response time is 0.5 s.

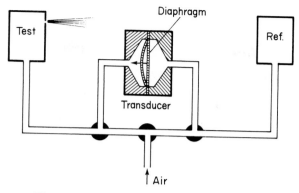

Figure 6.15 *Differential pressure leak detector.*

6.12 Gas appliance continuous flow leak detector [22]

A continuous test is applied to gas appliances and components to British and European standards enabling leaks to be found and eliminated until a pass is obtained. The detector is shown diagrammatically in Fig. 6.16.

In the standby condition, there is a continuous flow from the test port. When a test item is connected to the leak detector it is filled rapidly through a bypass until no flow is registered through the bypass valve. The valve closes automatically, leaving the test item connected to an internal reference volume by a sensitive linear orifice flow meter. The system then stabilizes, showing the leak rate on the readout meter and indicating 'pass' or 'fail' lamps according to the setting of the 'fail level' control. If the leak level is greater than full scale, the bypass valve will open and close on a slow cycle, the 'fail' lamp will remain on.

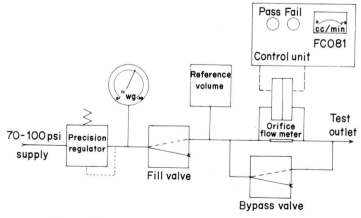

Figure 6.16 *Gas appliance continuous flow leak detector.*

6.13 Internal leakage measurement and analysis

Portable instruments to measure valve leakage, developed at the David W. Taylor Naval Ship Research and Development Center [23] are based on acoustic emission and called acoustic valve leak detectors (AVLD). These are suitable for gathering 'quick look' data, and for gathering field data on valve leakage, as well as data on background noise in piping systems

The detector monitors acoustic emission in a 25 kHz bandwidth centred at 25 kHz, or sweeps a frequency range from 10 kHz to 100 kHz with a bandwidth of 10 kHz. Monitoring in the fixed frequency mode can be accomplished by reading a meter or by listening to the acoustic emission signal heterodyned down to the audio range.

In the sweep mode, the output is generally displayed on an X–Y plotter. The sweep is slow enough (10 s) so that the signal is averaged to some extent. The AVLD also has a differential input capability to accommodate a second transducer for subtraction of structure-borne background noise.

Data derived by a study of internal leakage generally involve a plotting of leakage rate versus signal amplitude at a specific frequency to determine how well the two variables correlate in terms of accuracy, resolution, and repeatability.

6.13.1 Sources of acoustic emission

Three factors influence the amplitude and spectral distribution of acoustic emission signals from an internal leakage:

(1) The transducer's sensitivity and spectral response.
(2) Cavity and mechanical resonances in the valve and its associated piping.
(3) Noise source itself.

In addition to these there may be a background of structure-borne or airborne noise which may interfere with the leakage signal.

The source mechanisms may be divided into four types:

(1) Fluid dynamic sources, i.e. acoustic emission generated by unsteady flow.
(2) Cavitation, including rapid bubble growth where low pressure transients in the flow cause the formation of vapour or gas filled cavities.
(3) Flashing of superheated water into steam.
(4) The mechanical movement of valve parts and pipe walls and structure-borne noise due to local pressure variations near the boundary in the flowing fluid.

Even though these sources have been singled out and treated separately, the generating mechanisms interact and the acoustic emission observed is a sum from all sources. The contributing mechanisms may not be individually identifiable. It is considered likely that some of the source mechanisms are better indicators of leakage rate than others.

6.13.2 Cavitation sources of acoustic emission

Cavitation occurs whenever the local pressure drops below the vapour pressure for the liquid or any gas dissolved in that liquid. The degree to which cavitation occurs in leakage through an orifice is:

(1) Directly proportional to the pressures across the orifice.
(2) Inversely proportional to the difference between the local static pressure P, and the sum of partial pressures of vapours and dissolved gasses P_c. The cavitation number σ is generally accepted as an indication of the tendency of a fluid to cavitate and is given by

$$\sigma = \frac{P - P_c}{\frac{1}{2}\rho U^2}$$

where U is the fluid velocity and ρ the fluid density. The acoustic power radiated by cavitation resulting from flow through orifices is generally taken to be proportional to the square of the pressure drop across the orifice.

Cavitation occurs very irregularly in seawater valves operating at low pressure and leaking to atmosphere. This variability lies partly in the variation of air content of the water ingested into the system, partly on the temperature of the water at the time it is depressurized through the leak, and partly on the leak path itself which may generate negative pressures.

Cavitation in hydraulic systems is usually more predictable. Most hydraulic systems use an air interface at atmospheric pressure or higher and a temperature above ambient. Under these conditions the oil reaches saturation so that leakage occurs to a lesser pressure or higher temperature; air comes out of solution in a 'cavitation' event. This is not strictly cavitation. The cavity persists (i.e. does not collapse) and contains air and oil vapour rather than only oil vapour. Small bubbles persisting after 'cavitation' manifest themselves as foam which makes the oil compressible and decouples the fluid from the metal pipe (acoustically).

Cavitation events in water last from 1 to 50 μs, emitting frequencies in the 20 kHz to 1 MHz region. Oscillation of bubbles entrained by air in leak is another source of sound, usually as 'whistles' in the audio and ultrasonic ranges. Resonant frequencies for bubbles (1 to 0.1 mm diameter) range from 5 to 40 kHz [24].

6.13.3 Acoustic characteristics: air leaks

Characteristic curves for air leaks can be constructed as follows. First, graphs of acoustic amplitude versus frequency are recorded for leakage rates ranging from 82 ml/min to 3000 ml/min. From each spectral plot the leakage rate is then plotted against the highest acoustic amplitudes occurring near 55 kHz and 175 kHz. The three signatures shown for a 5.08 cm (2 in) air ball valve with a razor cut seat in Fig. 6.17 are typical graphs of maximum acoustic amplitude versus frequency. They show clearly that the signature amplitude is reduced by about two-thirds when the leakage through the same basic path is reduced by about two-thirds.

Although some spectra are recorded at frequencies as high as 500 kHz most are recorded over the range 0–250 kHz. Peaks occur in the frequency ranges 50–60 kHz and 170–180 kHz. Leak rate plotted versus peak acoustic amplitude for the frequency band 50–60 kHz for three damaged value seats is shown in Fig. 6.18. This shows that the peak acoustic amplitude for all three leak paths is related to leak rate and that it increases with increasing leak rate. The tightness of the data point grouping is an indicator of the accuracy, resolution and repeatability with which high pressure air valve leakage may be quantified acoustically. Curves which fit the data are of the general form

$$\log y = a + b \log x$$

where x is the peak acoustic amplitude and y the leak rate. Values for the intercepts and slopes were calculated for each curve and are listed in Table 6.7 together with the standard deviations of the data.

Data taken at 50–60 kHz had the smallest standard deviation which suggests that an instrument tuned to 50–60 kHz might give better accuracy, repeatability and resolution than an instrument tuned to 170–180 kHz.

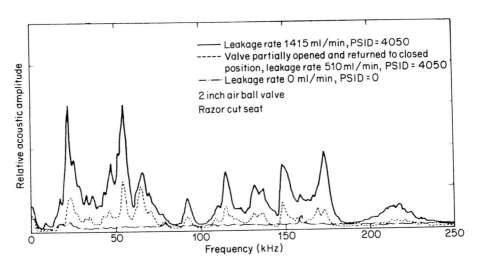

Figure 6.17 *Three leakage signatures.*

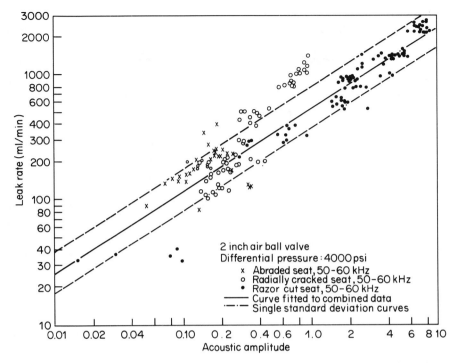

Figure 6.18 *Leak rate plots for three leak paths (50–60 kHz)*.

Table 6.7 Characteristics of curves fitted to leak rate versus peak acoustic emission
amplitude data for high pressure air valve leakage

Frequency band (kHz)	Intercept (a)	Slope (b)	Standard deviation of log (leak rate)
50–60	2.759	0.6627	0.1595
170–180	2.834	0.5190	0.2361

Other considerations include the availability of transducers with strong
resonances at the frequency bands of interest and the necessity for attenuation
of structure-borne noise (particularly at higher frequencies).

6.13.4 Acoustic characteristics: steam leaks

A test on a 90 mm ($3\frac{1}{2}$ in) globe valve which leaked steam past the seat provided
a graph of leak rate versus peak acoustic emission amplitude of the averaged
spectra as shown in Fig. 6.19. These high acoustic emission amplitudes occurred
in the frequency band 45–50 kHz. To obtain these data the transducer was
located at the top of the upstream valve flange.

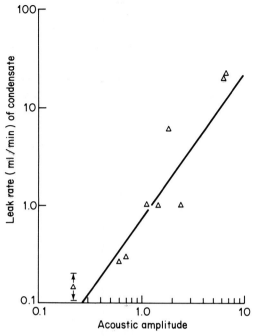

Figure 6.19 *Leak rate in a steam valve (45–50 kHz)*.

Curves which fitted had an equation of the general form

$$\log y = a + b \log x$$

where x is the acoustic amplitude, and y the leak rate. The resulting intercepts, slopes and standard deviations are tabulated in Table 6.8 along with similar results obtained in the frequency range 35 to 45 kHz.

Compared with high pressure air valve data, the standard deviations of the steam valve data are fairly large, due possibly to the reduced number of data points (arising from difficulty in controlling the leakage rate), being data not recorded at the same temperatures and pressures, and inaccuracies of steam flow measurement.

Table 6.8 Characteristics of the curves fitted to leak rate versus peak acoustic emission amplitude for steam valve leakage

Frequency band (kHz)	Intercept (a)	Slope (b)	Standard deviation
35–45	−0.987	1.108	0.600
45–50	−0.795	1.573	0.274

Nevertheless it was considered that there is an acoustic emission signature associated with steam valve leakage and that it increases with increasing leak rate. From this it was reasonably concluded that the ultrasonic instrument was capable of quantifying steam valve leakage rates acoustically with useful accuracy, resolution and repeatability. It was further considered that this technique could be extended to discriminate between defects such as those arising from valve damage compared with those due to wire-drawn seats.

6.13.5 Acoustic characteristics: hydraulic valve leaks

Acoustical measurements of leakage rate were made through a number of two-position one-way hydraulic spool valves: acoustic amplitude spectra were recorded for numerous leak rates, at three different backpressures. Plots of leakage rates versus the peak acoustic amplitudes revealed that these maximum valves arose within narrow frequency bands of about 20, 50 and 160 kHz. High background noise which tended to mask the signal was found to be related to backpressure. It was possible to differentiate one leaking manifold valve from other non-leaking valves (Fig. 6.20).

Fig. 6.21 is typical of numerous acoustic amplitude spectra recorded. The leak rates at the three backpressures of 0, 0.135 and 0.45 N/mm^2 (0, 15 and 50 psi) are very similar, being 20 ml/min, 21 ml/min, and 19.8 ml/min, respectively. At zero backpressure, the signatures typically exhibit a large peak near 25 kHz, with little or no signal above the noise at frequencies much higher than about 65 kHz. Conversely, at higher backpressures, most of the acoustic energy appears at frequencies of more than 100 kHz.

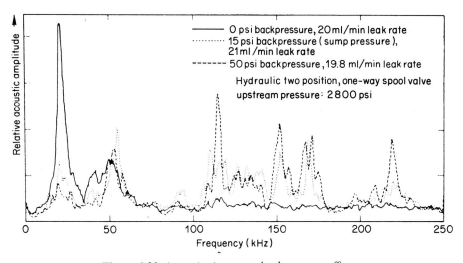

Figure 6.20 *Acoustic signature: backpressure effect.*

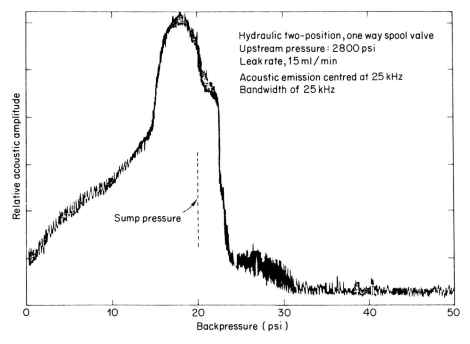

Figure 6.21 *Acoustic emission: hydraulic curve.*

Idiosyncracies which occurred with other tests showed that the effect of back-pressure on leakage noise is related to sump pressure, at which the hydraulic oil may or may not have reached equilibrium saturation with air. Large increases in leakage noise observed at low frequencies and low pressures are probably due to the air suddenly coming out of solution.

Plots of leak rate versus peak acoustic mission amplitude near 20 kHz for three different backpressures with the single valve in the test stand are shown in Fig. 6.22 together with data from an x-valve manifold. These curves have the general form

$$\log y = a + b \log x$$

where x is the acoustic amplitude, and y the leak rate. The resulting slopes, intercepts and standard deviations for the single valve data are listed in Table 6.9. Similarly data for the six-valve manifold is listed in Table 6.10.

For the single valve the data recorded at zero backpressure and 20 kHz provided the best sensitivity and the smallest standard deviation. This is consistent with the characteristic for the corresponding strong peak of acoustic emission amplitude. High backpressure again resulted in the poorest sensitivity and the largest standard deviation. The data suggest that acoustic measurements at 20 kHz are probably preferred over a range from 10 ml/min to at least

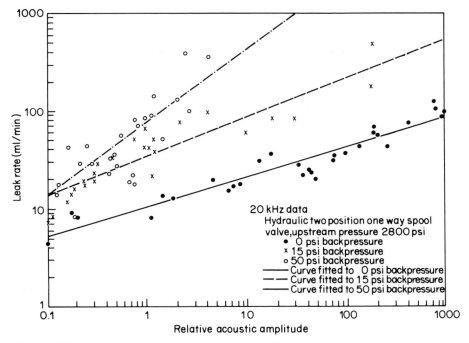

Figure 6.22 *Acoustic amplitude versus leak rate (20 kHz) for three backpressures on a hydraulic valve.*

Table 6.9 Characteristics of curves fitted to leak rate versus peak acoustic emission amplitude for hydraulic blocking valve leakage

Frequency band (kHz)	Back pressure		Intercept (a)	Slope (b)	Standard deviation of log (leak rate)
20 kHz	0	0	1.021	0.311	0.107
20 kHz	0.135	15	1.513	0.438	0.173
20 kHz	0.45	50	1.840	0.775	0.238
160 kHz	0	0	1.207	0.487	0.193
160 kHz	0.135	15	1.322	0.356	0.227
160 kHz	0.45	50	1.295	0.822	0.282

1000 ml/min. Detection relationships which remain to be established include those between acoustic signal and backpressure, the significance of long-term storage of hydraulic oil under air pressure, the relationship between leak path and acoustic signals, and the effect of oil temperature.

Table 6.10 Characteristics of curves fitted to leak rate versus peak acoustic emission amplitude for hydraulic six-valve manifold

Frequency band (kHz)	Location	Intercept (a)	Slope (b)	Standard deviation of log (leak rate)
20	1	1.628	0.2982	0.0990
20	2	1.841	0.2974	0.0729
20	3	1.876	0.3320	0.0495
55	1	1.381	0.5074	0.2802
55	2	1.891	0.3016	0.0845
55	3	1.971	0.2501	0.0809

6.13.6 Acoustic characteristics: water valve leaks

Tests on a variety of valve ball and seat combinations and conditions involving differential pressures across the valve of 0.045–0.072 N/mm² (5 to 8 psi), revealed the presence of several types of acoustic emission apparently unrelated to leakage rate. Some emission was due to air coming out of solution, some due to the existence of air pockets in the leakage flow path immediately downstream from the valve, other emissions resembled random whistles and periodic clicks.

Fig. 6.23 shows the plot for three differential pressures. There was an adequate signal-to-noise ratio only below frequencies of about 75 kHz. Acoustic emission amplitude increased dominantly with increasing differential pressure suggesting that higher differential pressures produce acoustic signals related to leak rate that are stronger than signal fluctuations not related to leak rate.

In general there was poor correlation between the amplitude of the acoustic signal and the signature leakage rate at constant pressure. The fact that acoustic emission levels are being measured near threshold, and that there are low seating forces on a relatively massive valve, may explain this poor correlation between signal level and leakage rate.

The audio frequency spectrum in the range 0–20 kHz shows that there are usually acoustic emission signals at low frequencies and that the structure-borne background noise spectrum is such that significant signal-to-noise ratio begins to occur at about 8 kHz.

6.14 Gas and pipeline testing

High level pipeline testing by the British Gas Council is today virtually standard for all new high pressure gas and oil pipelines [25]. The technique originated in

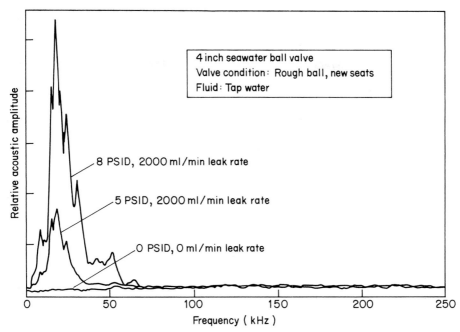

Figure 6.23 *Typical spectra of leakage through a 10.16 cm (4 in) seawater ball valve.*

the USA where long-distance large-diameter pipelines are hydrostatically tested up to the minimum specified yield of the pipe material.

High level testing aims to minimize the chance of operational failure by revealing defects at the time of testing. It is reasoned that if there are defects the high pressure will cause defects to grow to significant proportions, open up and leak. During tests on 10 000 miles of pipe almost 72% of all defects detected were found above 80% of minimum specified yield.

Traditional hydraulic testing is based upon the 'duty' of a pipeline, commonly 1.5 times or twice the designated working pressure, but always within the elastic range of the pipe material. High level testing is based upon the physical attributes and characteristics of the pipe material irrespective of the particular duty which a pipeline is designed to perform. Conventional tests prove that a pipeline is capable of performing a given duty at the time of test but they may not reveal defects that escape detection during the laying of the pipeline.

By testing to higher pressures so as to stress the pipe material to minimum specified yield or above, significant weaknesses (where they exist) in a pipeline will be revealed at the time of test without in any way damaging acceptable sections of the line. It is also thought that testing to such levels results in a form of stress relieving by causing small defects which have not grown to significant proportions to yield locally.

Correctly called high level testing, increasingly the term 'yield testing' has come to be used to describe this testing procedure. This is a little unfortunate as it tends to give the impression that the entire length of line under test is yielding and that it is the reaching of yield point which produces the benefits claimed. In fact the yield point is used as a convenient limit which can be defined with reasonable ease in the test, but essentially the purpose of the test is to subject the pipe to as high a pressure as possible consistent with the present and future security of the pipeline.

When a pipeline under test shows signs of yielding not all of the pipelengths in the section will be yielding simultaneously. Inevitably there will be variations in the stress imposed due to differences in elevation in the test section and in the yield strength, wall thickness and diameter of individual pipes. Research indicates that in pipelines tested to as much as 110% of specified minimum yield no pipe yields more than 1%; only about 0.25% of the test length yields more than 0.5% and over 99% of the length either does not yield or yields less than 0.25%.

High level testing up to 90% minimum specified yield usually follows procedure similar to that adopted for normal hydrostatic testing. Thereafter a pressure – volume curve is plotted – quantity of water injected into the pipeline on the *x*-axis against pressure on the *y*-axis. Pumping is stopped when either the gradient of the pressure – volume plot halves or the pressure equivalent of the specified percentage of minimum yield is reached, whichever happens first.

For a period up to two hours afterwards post-yielding usually occurs, resulting in a drop in pressure. Provided that this is within specified limits, pressure is restored to its original maximum and maintained for the 'hold' period. It is usual to hold the pipeline at test pressure for a period of not less than 24 hours. Should a burst occur, pre-yielded pipe is used to replace the defective pipe and the section is re-tested.

Experience in the United States indicates that 50% of revealed defects become evident on pressurization, 16% in the first hour of the 'hold' period, 18% in the next four hours and the remaining 16% in the final 19 hours of the 24 hour test.

During the last decade high level testing has become standard practice in the oil and gas industries; 90% of minimum specified yield being usual in the oil industry and 105% or sometimes higher in gas practice. However, insurers still treat testing beyond 100% of minimum specified yield apprehensively and have shown some reluctance to provide cover. To some extent this arises because of physical change having taken place in the pipeline and also on the general reasoning that the test is designed to reveal defects.

References and list of companies

1   Electrolocation Ltd, 129 South Liberty Lane, Bristol BS3 2SZ, UK.
2   'Subtronic underground service location', Subtronic Ltd, 1 Lucas House, Craven Road, Rugby, Warks CV21 3JQ, UK (Tel. 0788-70241/2); also 5 Foxhill Road, West Haddon, Northampton NN6 7BQ, UK (Tel. 078-887-270).

3   GP Instrumentation Division, NEI (Grubb Parsons) Instrumentation Ltd, Whitley Road, Longbenton, Newcastle-upon-Tyne, NE12 9SB, UK (Tel. 0632-669091).
4   Imperial Chemical Industries Ltd, Mond Division, The Heath, Runcorn, UK.
5   BOC Special Gases, Deer Park Road, London SW19 3UF, UK (Tel. 01-542-6677).
6   Refrigeration Spares Ltd, 31 Harrow Road, Leytonstone, London E11 3PT, UK (Tel. 01-555-1321).
7   Shandon Southern Instruments Ltd, Camberley, Surrey GU16 5ET, UK (Tel. 0276-63401).
8   Kent-Moore Ltd, 19–21 Stockfield Road, Acocks Green, Birmingham B27 6AJ, UK (Tel. 021-706-4262).
9   Hi-Fi Ltd, Hi-Fi Works, Station Road, Cradely Heath, Warley, West Midlands, B64 6TN, UK (Tel. 021-559-2446).
10 Analytical Instruments Ltd, Green Lane, Fowlmere, Royston, Herts, UK (Tel. 076-382317).
11 Taylor, C. (1981), 'Laser will search plant to detect leaking gases'. *Process Engineering*, August, 29.
12 (1977), 'Occupational exposure to benzene', *Federal Register*, Part VI, **42**, No. 103, 27452.
13 Rooney, T. A. (1983), 'Automated determination of benzene absorbed on charcoal tubes', Application Note AN 228-6, Hewlett Packard Co., Avondale, Pa, USA.
14 'UNOR' Gas Monitors, H. Maihak AG, 2 Hamburg 60, Semperstrasse 38, FRG (Tel. 040-27161).
15 G. C. Davis & Co. (Gas) Ltd, Binary House, Park Road, Barnet EN5 5SA, UK (Tel. 01-440-7161).
16 (1977), 'New air monitoring method', Techlink Report No. 2381, Subject Code 8.16.32, Techlink Unit, Technology Reports Centre, Orpington, Kent BR5 3RF.
17 Detection Devices Ltd, 1843 Merivale Road, Ottawa K26 1ES, Canada; also AMEECO Ltd, Bentalls, Basildon, Essex, UK (Tel. 0268-284561).
18 'Leakseeker' TCS, Hazard Control Ltd, 61 High Street, Barnet, Herts EN5 5UR, UK (Tel. 01-449-3974).
19 'Type 8902A Leak Detector', Dawe Instruments Ltd, Concord Road, Western Avenue, London W3 0SD, UK (Tel. 01-992-6751).
20 Subtronic Ltd, 1 Lucas House, Craven Road, Rugby, Warks, UK (Tel. 0788-70241, Telex 311794 Chacom G); also Fuji Sangyo Co. Ltd, 1-11 Izumi-cho, Kanda, Chujoda-Ku, Tokyo, Japan (Tel. (03) 862-3196, Telex 02657448 DETECT J).
21 A. I. Industrial, London Road, Pampisford, Cambridge CB2 4EF, UK (Tel. 0223-834420, Telex 817536).
22 Furness Controls Ltd, Beeching Road, Bexhill, East Sussex TN39 3LJ, UK (Tel. 0424-210316).
23 Dickey, J., Dimmick, J. and Moore, P. E. (1978), 'Acoustic measurement of valve leakage rates', *Materials Evaluation*, January, 67–77.
24 Flynn, H. G. (1964), 'Physics of acoustic cavitation in liquids', in *Physical Acoustics*, Vol. B, Ed. W. P. Mason, New York, Academic Press.
25 (1979), *British Engine Bulletin*, December, British Engineer Insurance Ltd, Longridge House, Manchester M60 4DT, UK.

7
Structural vibration

7.1 Introduction

Structures of doubtful integrity may vibrate with excessive amplitudes. For such structures the change in 'flexibility' would appear to be the dominant consequence of deterioration. Known technically as 'modal analysis' this characteristic is the relationship between the applied force and the resulting vibration.

Techniques of vibration monitoring are well developed and for many structures the use of vibration monitoring techniques can be recommended without restriction. Machine vibration monitoring features very high frequencies and low vibratory amplitudes. Structural vibrations can also be influenced by harmonic frequency effects, but overall these are considerably smaller in magnitude than those for machine vibration monitoring.

7.1.1 Causes of vibration

Structural vibrations are damped forced vibrations. This means that the applied forcing disturbance will produce a magnified vibration according to the ω/ω_0 frequency ratio (Fig. 7.1), where ω is the frequency of the forcing disturbance and ω_0 the natural frequency of the structure.

'Damping' is likely to be significant in structural vibrations. This is the energy dissipating effect of fluid friction, internal friction, fixation looseness, etc. A highly damped structure reduces vibrations quickly, a lightly damped structure fails to stop dangerous vibrations.

Defects which may cause excessive vibration in structures are:

(a) Mechanical slackness.
(b) Loose foundations.
(c) Ineffective load-bearing members.
(d) Excessive environmental loadings – wind, rain, tidal action, earthquake, settlement.

These will produce the following typical vibratory modes:

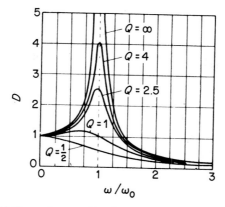

Figure 7.1 *Dynamic magnifier for a range of damping conditions.*

(1) Bending vibration (aircraft wings, bridges, pipes).
(2) Flexural/plate-mode vibration (aircraft fuselages, bridges, ships, walls).

7.1.2 Natural modes

It is possible to calculate the natural frequencies of the components of a structure provided their flexural mode is known. Examples of boundary conditions and mode-shape for various configurations of single beams of uniform section and uniformly distributed load are shown in Fig. 7.2.

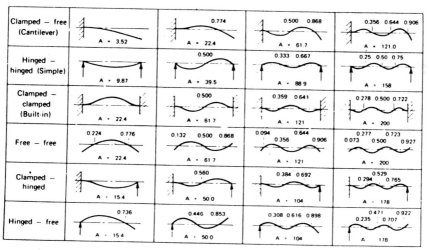

Figure 7.2 *Examples of boundary conditions and mode shapes for various configurations of single beams of uniform section and uniformly distributed load.*

The natural frequency (f_0) for each of these modes is given by

$$f_0 = 2\pi A \left(\frac{EI}{\rho L^4}\right)^{\frac{1}{2}}$$

where

A = constant
E = Young's modulus (modulus of elasticity)
I = second moment of area of beam cross-section
ρ = mass per unit length
L = beam length

Thus for the free–free mode the first (natural) frequency has $A = 22.4$ and nodes located at $0.224L$ and $0.776L$. For the first overtone $A = 61.7$ and nodes occur at $0.132L$, $0.500L$, and so on.

Similar two-dimensional effects for plates are shown in Fig. 7.3.

7.1.3 Pipe and tubing vibrations

Apart from its frequency as a loaded beam, a pipe has longitudinal and hoop frequencies which resonate as a function of pressure fluctuations.

Housner reports an investigation of the effects of wind velocity on an oil pipeline crossing a desert. He found the frequency of forced vibrations from pumping to be of the same order as pipe resonant frequency.

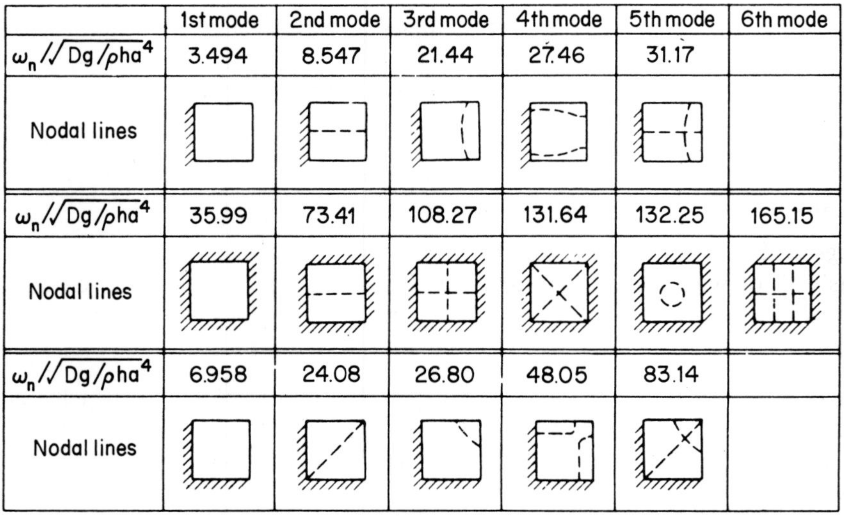

	1st mode	2nd mode	3rd mode	4th mode	5th mode	6th mode
$\omega_n / \sqrt{Dg/\rho ha^4}$	3.494	8.547	21.44	27.46	31.17	
Nodal lines						
$\omega_n / \sqrt{Dg/\rho ha^4}$	35.99	73.41	108.27	131.64	132.25	165.15
Nodal lines						
$\omega_n / \sqrt{Dg/\rho ha^4}$	6.958	24.08	26.80	48.05	83.14	
Nodal lines						

$\omega_n = 2\pi f_n$ ρ = weight density a = plate length

$D = Eh^3 / 12(1-\mu^2)$ h = plate thickness

Figure 7.3 *Examples of modal line configurations for square plates under various edge conditions.*

A good working equation for the frequency of vibration of thin-walled tubes for which the wavelength of the pressure variations is much greater than the tube diameter is given by

$$f_o = \frac{1}{2\pi} \left(\frac{4E}{D^2 \rho} \right)^{\frac{1}{2}}$$

where E is the modulus of rigidity of pipe material, D is the pipe bore, and ρ is the density of the pipe material. For a steel pipe with $E = 20 \times 10^{10} \text{N/m}^2$, $\rho = 7.85 \text{ Mg/m}^3$, the following frequencies and pipe diameters are obtained:

Pipe bore (m)	2	1	0.5	0.25	0.1
Resonant frequency (kHz)	0.81	1	3.24	6.48	16.2

7.1.4 Structures

The discrete frequencies of complex structures are evaluated by the combined use of a finite element technique to simplify the mass and stiffness distribution together with a computer to perform the interrelated calculations. Such evaluations are made as normal practice. A typical example for part of a ship's hull is shown in Fig. 7.4. Programs are available from organizations such as Lloyd's Register of Shipping and Pafec Ltd.

7.2 Vibration monitoring

Three stages are involved in the overall monitoring of a structure

Stage 1: Monitoring, by means of sensors or transducers, the signals produced by the machine or system. This is the means by which interpretative data are acquired, hence the more general term 'data acquisition'.

Stage 2: Signal/data processing by means of meters, visual displays or 'print-outs', all of which may be coupled with alarm systems of lights, sounds or cutout switches. In principle, this stage manipulates the raw signals to provide significant vibration data.

Stage 3: Condition assessment. This is the 'decision stage' by which comparison with reference data on ideal 'signatures' provides information on which the system condition can be assessed and operative decisions made. Concepts of reliability feature largely at this stage.

7.2.1 Vibration sensor selection

A suitable vibration sensor ('transducer' or 'pickup') involves: (a) displacement

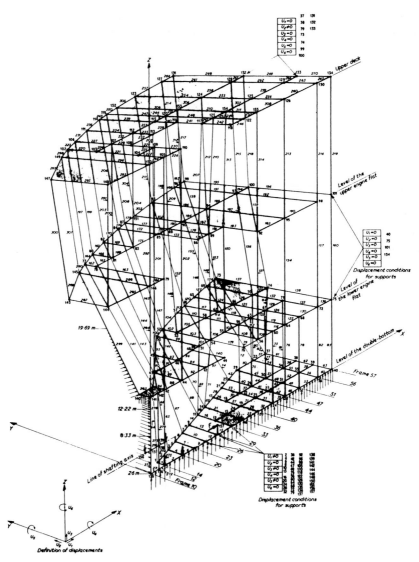

Figure 7.4 *Deformation calculation of the after part of a ship's hull – equivalent elastic model.*

sensed by a proximity transducer with an output signal proportional to displacement; (b) velocity sensed by a seismic-type transducer; or (c) acceleration sensed by an accelerometer. The strongest influence on selection of the appropriate parameter is the frequency at which measurement is important; in any event, a large signal is required so that the signal-to-noise ratio is high. (Noise is electrical 'noise' or interference with the primary signal due to stray currents which arise from various electrical interference effects.)

As a general rule, displacement measurements record large signals at low frequencies, while acceleration measurements are effective at high frequencies (see Table 7.1). The vibration parameters sensed differ in selectivity, as shown in the octave frequency spectra (Fig. 7.5) for the same machine.

Table 7.1

Frequency range (Hz)	Preferred vibration parameter
1–50	Displacement
50–1000	Velocity
1000+	Acceleration

For many basic observations a stroboscope examination provides the first level of investigation. The stroboscope permits rotating or reciprocating objects to be viewed intermittently and produces the optical effect of slowing down or

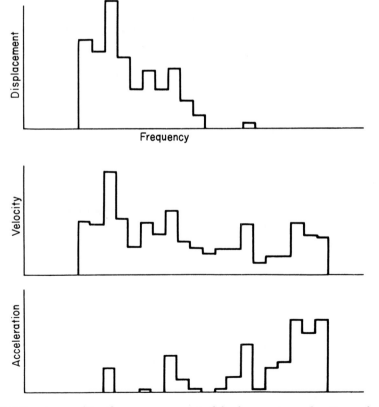

Figure 7.5 *Spectra resulting from measurement of displacement, acceleration and velocity of rotating machinery in the frequency range 0–1 kHz.*

stopping motion. For instance, an electric fan revolving at 1800 rpm will apparently stand still, if viewed under a light that flashes uniformly 1800 times per minute. At 1799 flashes per minute the fan will appear to rotate at 1 rpm, and at 1801 flashes per minute it will appear to rotate backwards at 1 rpm.

The amplitude of vibration can be assessed if a fine reference line is scribed on the vibrating part. This technique has been used to confirm the calibration of vibration calibrators, and automotive engineers have used it to measure crank-shaft whip and vibration.

7.2.2 Frequency analysis

A vibration may be monitored by comparing the frequency 'spectrum' of the structure at different time intervals.

The analysis of a vibration is said to be in the frequency domain if it displays signal amplitude versus frequency. On the other hand, a plot of signal amplitude against time (a waveform plot) is said to be in the time domain.

Energy in a vibration (or sound emission) is distributed over a range of frequencies consisting of the fundamental frequency and combinations of the harmonics.

With this technique the defective component is identified by resolving the vibration signal into its constituent frequencies and relating these to the known discrete frequency of the component: In this way the component's 'signature' or 'thumbprint' is established.

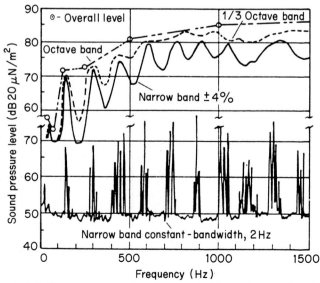

Figure 7.6 *Roller contact sound analysed with filters of various bandwidths.*

Vibration signals are filtered electronically and resolved into frequency bands which may be: (a) octave; (b) fractional octave; or (c) narrow bandwidth. Finer analysis (bandwidths of the order of 1 Hz) improves the sensitivity and the ability to identify defective components. A typical comparison of the spectra derived from different filtering resolutions is shown in Fig. 7.6. The effect of the different types of filter is as follows:

(a) Octave and fractional octave divide the frequency spectrum into a series of bands which have a constant width when plotted logarithmically.
(b) Narrow band may be either a constant percentage of the centre frequency or a constant frequency width.

As shown in Fig. 7.7, other than for those with a constant frequency width, the discrimination of all filters is much coarser at high frequencies and therefore less effective for analytical purposes.

By applying microprocessor techniques to digitized signals the data can be analysed in 'real time' through a Fourier transform algorithm programmed into the computer.

7.2.3 Waveform analysis

Known as 'time domain', this is probably the most appropriate system for the monitoring of structures. In its elementary form it may involve an 'Askania' vibrograph. In more sophisticated forms it can involve process-capture data and oscilloscopes.

(a) 'Askania' vibrographs

Structural vibrations may be recorded by the use of this hand-held instrument,

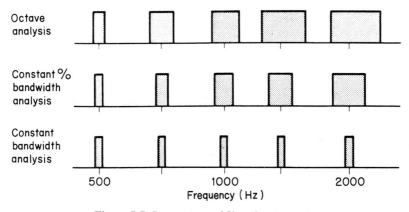

Figure 7.7 *Comparison of filter discrimination.*

shown in Fig. 7.8. The housing (1) with feeler tube (3) and stylus (11) forms one main part of the vibrograph, while the spring drive (21), together with the battery housing (30) and the time markers (18, 19) forms the other main part. The feeler tube (3) contains a light-weight feeler rod (4) which moves in longitudinal direction and is usually pressed against the object under test by means of the tension spring (7). The spring drive (21) operates the driving roller (17) for the waxed tape (15), the speed of which is fixed at 40 mm/s by a governor.

Recording tapes are evaluated for amplitude and frequency; at a low frequency the phase angle may also be determined. Frequency is based on paper speed, provided there is no slipping. For accurate evaluation the recording length of one second is determined by means of the time marks, e.g. 38 mm. Then the number of deflections – either the zero passages in one direction or the reversal points on one side – within this space is counted. The figure thus obtained represents the frequency in hertz.

Deflection is determined by measuring the double amplitude on the tape and dividing it by twice the magnification factor (the standard factor is 5 : 1, then deflection is one-tenth of the measured double amplitude).

(b) 'Askania' waveform analysis

Manual enveloping of the waveform printouts can provide basic structural vibration data.

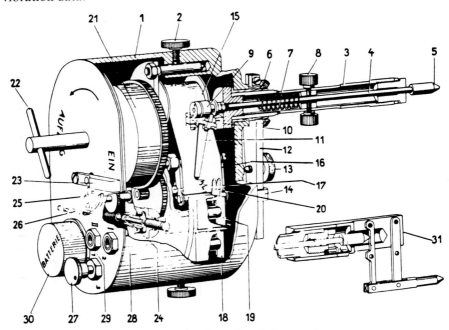

Figure 7.8 *'Askania' hand vibrograph.*

Typical wavetraces are shown in Fig. 7.9 (two and three component vibrations). The trace in Fig. 7.10 is made up of three components. As applied to a helicopter drive using the 'Askania' vibrograph, from the time base four cycles of $4R$ (i.e. 1 revolution of the main rotor) are established and this distance is used as unit time. The highest frequency is determined as a rotor order (i.e. if 11.3 cycles of this frequency occur in unit time, the rotor is $11.3R$). This rotor order is then enveloped and the double amplitude measured (i.e. the distance A).

Enveloping the highest frequency reveals the next lowest frequency. This frequency is determined as a rotor order and the double amplitude is measured. In

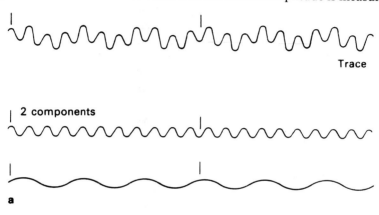

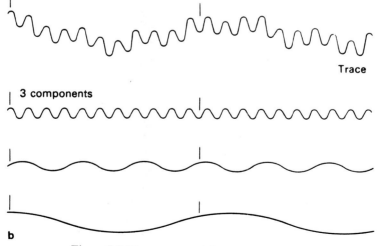

Figure 7.9 *Wavetraces and their components.*

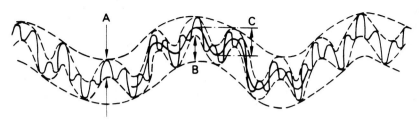

Figure 7.10 *Envelope three-component wavetrace.*

this case the double amplitude is the distance at C. The process of enveloping and measuring the amplitudes of the rotor orders continues until the lowest order is reached down to the frequency limitation of the recorder.

An example of a sequence of component estimation for structural vibration in a helicopter is shown in Fig. 7.11 (R = main rotor speed; T = tail rotor speed; E = engine speed). At 17 500 engine rpm the rotor speed is 221.4 rpm (3.69 Hz); with four rotor blades the significant forcing frequency is $4 \times 3.69 = 14.76$ Hz (sometimes called the fourth main rotor order).

7.3 Holographic and speckle pattern vibration monitoring

Holographic interferometry (first described by Powell and Stetson in 1964) provides a means of recording the vibration pattern of two-dimensional surfaces.

One developing method is based on the combination of the holographic interferometer with a televison system such that the advantages of coherent optics can be used for data acquisition followed by a more convenient system of electronic data processing. With this system it is necessary to reduce the spatial frequencies below those used for optical/photographic holography and at the same time use the resulting speckle pattern for sampling and analysis.

The speckle system has been developed to display surface vibration modes in real time. At present the surface mode patterns are obtained on a television monitor using a combination of optical and electronic techniques. The system operates by illuminating the surface of interest with laser light and imaging it onto a vidicon camera screen, together with a uniform reference wavefront derived from the same laser. The resultant video waveform is processed electronically before the image is displayed on a monitor and one obtains a fringe pattern on the image indicating the amplitude and distribution of vibrational motion present on the object.

Operation is dependent on the speckle effect in the image, which provides informaton bearing spatial frequency within the resolving capabilities of the vidicon. This speckle occurs whenever a scattering surface is illuminated with highly coherent light and results in a very grainy or speckled image. Because of the finite resolving power of the lens, any point in the image will receive light

Example
Trace to be analysed
A time of 1 rev of main rotor (or 4 cycles of 4 *R*) is established

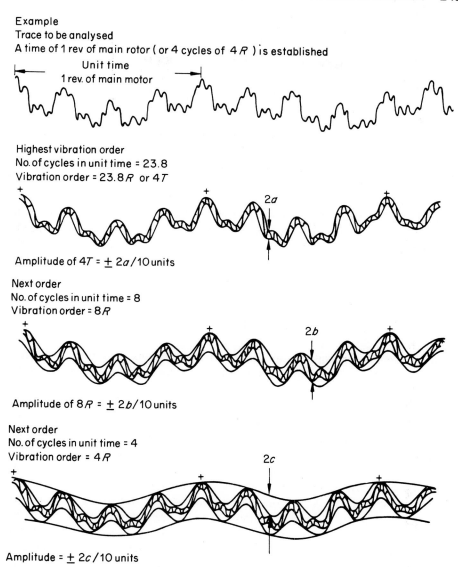

Unit time
1 rev. of main motor

Highest vibration order
No. of cycles in unit time = 23.8
Vibration order = 23.8 *R* or 4 *T*

Amplitude of 4*T* = $\pm$ 2*a*/10 units

Next order
No. of cycles in unit time = 8
Vibration order = 8 *R*

Amplitude of 8*R* = $\pm$ 2*b*/10 units

Next order
No. of cycles in unit time = 4
Vibration order = 4 *R*

Amplitude = $\pm$ 2*c*/10 units

Figure 7.11 *Example of three-component graphical resolution.*

from a finite area of the imaged surface. This finite area will contain many large random fluctuations in light, on an optical scale, by the definition of a scattering surface. Thus one obtains a random vector diagram for the resultant intensity at the image point which varies from one image point to another.

Although the amplitude and phase vary from point to point in the image they remain fixed in time for any particular image point, the phase being directly related to the position of the corresponding object point. Localized interference effects may be observed if another coherent wavefront is superimposed on the image.

The scanning action of the television camera converts spatial frequencies in the image into temporal frequencies in the electrical output. Thus, parts of the image containing high spatial frequencies may be separated by inserting a high pass filter between the camera and the monitor where the picture is reconstituted. The resultant image appears brightest for the nodal regions. In practice only a few of the maxima and minima will be seen due to random noise generated in the system. Resolution is poor due to the CCTV system having a maximum resolution of 600×600 picture elements. However, the major advantage of the system is the production of a high contrast nodal pattern in real time.

The use of pulsed laser holography to monitor stress corrosion cracks on an aircraft wing is described in Section 5.20.1.

7.4 Vibration data manipulation

The incorporation of a microprocessor in vibration monitoring equipment makes it possible to introduce mathematical processes which enhance periodic effects and progressively remove 'noise' and disturbances. Such methods include averaging, Cepstrum analysis, coherence, convolution, correlation, cross-correlation, cross-power spectrum, Kurtosis, etc.

7.4.1 Cross-spectrum

Periodic signals may be described spectrally in terms of the root-mean square (RMS) values of their various frequency components (the frequency spectrum), while random vibration signals are best described in terms of mean square spectral density functions. This is because random signals produce continuous frequency spectra and the RMS value measured within a certain frequency band will therefore depend upon the bandwidth.

The cross-spectrum measures the similarity between two signals in the frequency domain, it is the Fourier transform of the cross-correlation and expresses the similarity as a function of frequency, conjugate multiplication being applied to calculate the cross-spectral density function.

Only frequencies which are common to both waveforms are contained in the cross-spectral analysis. For example, the vibration of the flaps of an aeroplane wing can be analysed by means of a vibration pickup (Fig. 7.12) which can detect the output waveform caused by the vibration generated by the flaps. A cross-spectrum of this waveform and the input will show the major frequency generated. This is a common application for almost every airframe manufacturer, often referred to as 'flutter testing'.

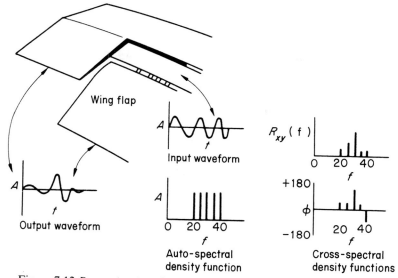

Figure 7.12 *Determination of vibration effects in aircraft wing structure.*

7.4.2 Autospectrum analysis

A set-up for monitoring the vibration from sources X and Y is shown in Fig. 7.13. An accelerometer is placed at X to determine whether the vibrations measured at Y are generated at X or at some other source.

A typical power or autospectrum (G_{yy}) at point Y (Fig. 7.14) identifies three major vibrations: a broad and fairly strong band from approximately 10 Hz to 50 Hz; a very narrow but strong component at 120 Hz; a weaker, but significant, tonal component at 500 Hz.

This power spectrum can be 'normalized' by calculating the RMS value of the power at each particular frequency. This can be done quickly by a computer.

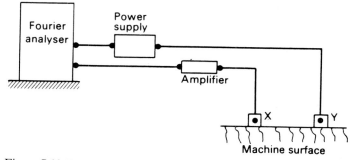

Figure 7.13 *Test set-up to determine source of vibration in a machine.*

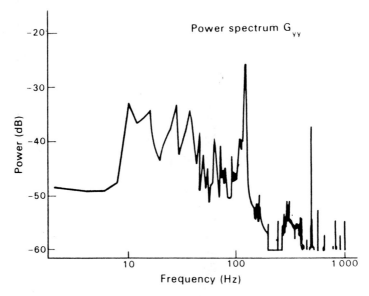

Figure 7.14 *Power spectrum of vibration measured by the accelerometer at point Y.*

The cumulative value at each frequency produces a 'normalized' result which is relative to the degree of random distribution and comparable with a Gaussian distribution. In the normalized integrated noise-power spectrum (Fig. 7.15), 30% of the power is distributed in the broad band below 60 Hz, 45% is in the 120 Hz spectral line, while the 500 Hz line contains less than 5% of the total power.

Autospectrum G_{xx} at point X (Fig. 7.13) revealed that energy below 60 Hz is not significant except for a small line at about 50 Hz and a stronger one at 60 Hz. The low energy content suggests that the 10 and 60 Hz signals may be

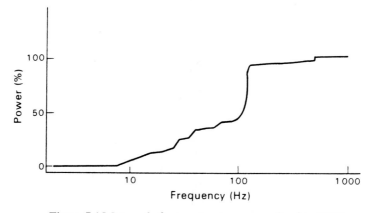

Figure 7.15 *Integral of power spectrum normalized to 100%.*

due to external sources, but this is only conjecture without some additional knowledge of the transfer mechanism; the 60 Hz component is much stronger relative to 120 Hz in the point X signal than in the point Y signal, which does not help source interpretation. The 120 Hz and 500 Hz lines, on the other hand, are clearly present in the point X spectrum and can in part be attributed to the source being measured. However, even here there is some difficulty in establishing the reliability of the result, since the ratio of the 120 Hz to 500 Hz tones is 10 dB in the point Y spectrum and 6 dB in the point X spectrum.

A cross-relationship between X and Y must be established in order to determine the degree to which the monitored power spectrum G_{yy} is caused by the assumed source. The cross-power spectrum is such a relation. The cross-spectrum differs from the two power spectra in that it is the cross-conjugate product between the linear spectrum at X and the linear spectrum at Y, rather than the self-conjugate product as is G_{yy}.

7.4.3 Cross-correlation

The cross-correlation of two waveforms is a measure of their similarity, measured as a function of the time shift between the two waveforms. Cross-correlation is obtained by multiplying two signals (rather than one signal alone, as with autocorrelation). The resulting graph (Fig. 7.16), although similar to that for autocorrelation, has a different comparative significance.

The concept of cross-correlation can be understood from the principle of echo-ranging as used in radar, sonar or geophysics. Thus, in Fig. 7.17 if the abscissa is either time or distance (they are directly related in echo-ranging) then:

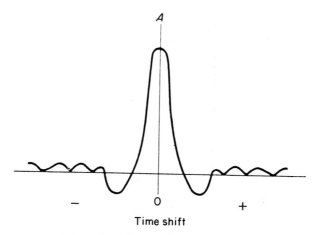

Time shift

Figure 7.16 *Cross-spectrum waveform.*

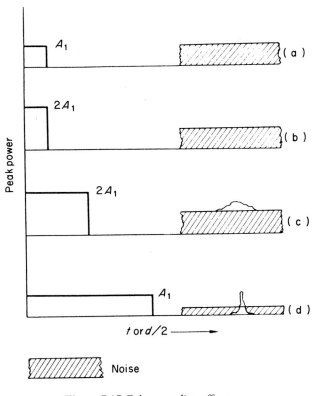

Figure 7.17 *Echo-sounding effect.*

(a) The left-hand portion of the graph (Fig. 7.17(a)) shows the amplitude and duration of a transmitted pulse, the characteristics of which are shown to receive an echo at time t (this corresponds with twice the distance, since the pulse must travel to the target and then return).

(b) The right-hand portion shows the received echo (which in this case is not distinguishable from the noise of the receiver).

Logical efforts to receive and identify the echo by using a transmitted pulse with a larger peak amplitude (Fig. 7.17(b)) or increasing the signal duration (Fig. 7.17(c)) either produce no echo or one of such duration that it loses much of its resolution. If a signal of long duration were cross-correlated against the echo, a response like that in Fig. 7.17(d) would appear. The long transmitted signal has been 'compressed' by the receiver, since the cross-correlation function for the transmitted signal is very narrow if the bandwidth of the signal is wide. Furthermore, the noise-level has been greatly attenuated since it does not cross-correlate with the transmitted signal except by chance. The cross-correlation function of two waveforms contains just the frequencies common to both.

Cross-correlation (as in echo-ranging) can easily detect a signal even though it is immersed in noise of 100 times its amplitude. Fig. 7.18(a) shows a 'chirp' signal and Fig. 7.18(b) its autocorrelation taken from 20 dB noise.

This cross-correlation technique can be used to study signal transmission paths. When several transmission paths are observed, the transmission path that is used can be determined by cross-correlation techniques. Fig. 7.19(a) shows a source of vibration generating some complex signal through three different paths. The lengths of the paths and their transmission characteristics, if different, will determine which path contributes the most to the signal detected by the pickup, as shown in Fig. 7.19(b).

A typical automobile study to identify the source of vibration at the driver's seat would be to cross-correlate signals from the axles. A further technique of fault location is a triangulation method (Fig. 7.20) to record separate measurements with respect to a static sensor, the time delay being proportional to transmission path effective distance.

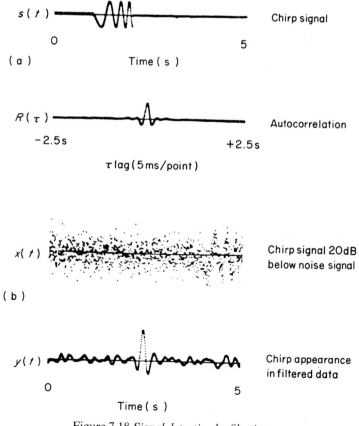

Figure 7.18 *Signal detection by filtering.*

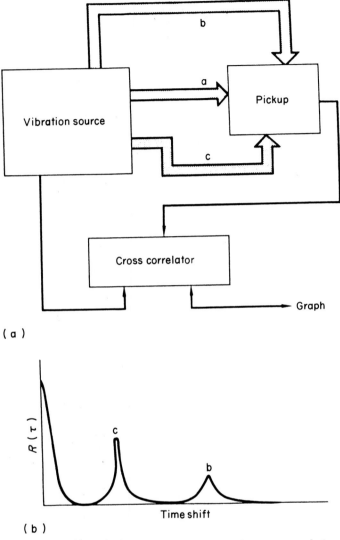

(a)

(b)

Figure 7.19 *Complex signal path determination by cross-correlation.*

7.5 Impedance and structural response

Applied forces and the resulting motions at different points in a structure are related by their 'mechanical impedance.' Measurements can be used to:

(1) Determine natural frequencies and mode shapes.
(2) Measure specific material properties such as damping capacity or dynamic stiffness.

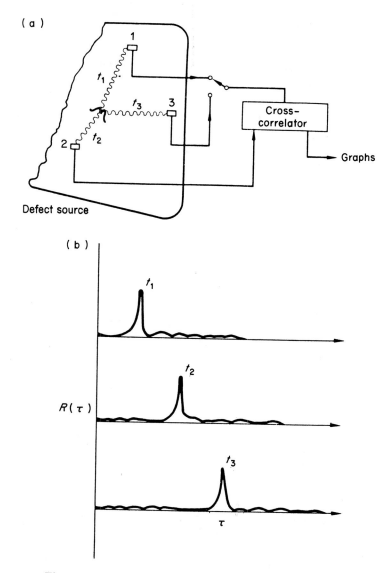

Figure 7.20 *Triangulation to determine source of vibration.*

(3) Provide the basis of an analytical model. From measurements of the impedances of individual components or substructures it is possible to predict the behaviour of combined systems, in a manner completely analogous to the study of complex electrical circuits.

There is an analogy between electromechanical and electro-acoustic technologies. Mechanical impedance and mobility (for simple harmonic motion) are defined as the complex ratios of force vector to velocity vector, and velocity vector to force vector respectively. This is shown in Table 7.2 where, in addition, similar ratios involving acceleration and displacement are given.

Table 7.2 Terminology for complex dynamic ratios of force and motion. F is force, m mass, a acceleration, v velocity, d stiffness

Physical quantity	Notation	Unit	Reciprocal of physical quantity	Notation	Unit
Dynamic mass (apparent weight)	F/a	Ns²/m	Acceleration through force	a/F	m/(Ns²)
Mechanical impedance	F/v	Ns/m	Mobility (mechanical admittance)	v/F	m/(Ns)
Stiffness	F/d	N/m	Compliance	d/F	m/N

The terms given in Table 7.2 are taken from the American Standard USAS S2.6-1963: Specifying the Mechanical Impedance of Structures.

As both force and motion are vectors in space as well as in time care should be taken to define directions of motion relative to the direction of force when this is not obvious from the measurement conditions or from the calculations.

Point impedance relates to force and motion. Values are measured at the same point and in the same direction; the ratios are termed driving point values, or point values for short. Transfer impedance relates to force and motion measurements at different points or at the same point with an angle between them.

The ratios given in the first and second column of Table 7.2 really represent (as functions of frequency) the difficulty or ease, respectively, with which a structure can be set into motion. By measuring, for example, the mechanical impedance of points on a structure, knowledge is gained about its response to vibrational forces at different frequencies. Similarly, a measurement of the motion of the structure, after it has been placed on a vibrating support, may be compared to its mechanical impedance to obtain information about the forces which act on the structure.

To solve vibrational problems, mechanical impedance may have to be measured and a narrow band frequency analysis carried out to obtain detailed knowledge about the response ability of the structures involved, and of the actual responses or forces. After the combination of this information the need for, or the possibility of, corrective measures may be evaluated.

7.5.1 Practical impedance measurement

An impedance head is commonly used to measure the point impedance of a structure. This is an electromechanical device which contains two types of electromechanical transducer. One transducer measures the force which is applied to the structure, while the second measures the motion of the point to which the force is applied. The actual construction of an impedance head is shown in Fig. 7.21. It consists basically of an accelerometer and a force gauge. These are made up of pairs of lead zirconate titanate piezoelectric discs with conductors sandwiched between them and leading out to output sockets in the housing. A seismic mass of tungsten alloy is mounted above the accelerometer discs. The driving platform below the force gauge is machined from beryllium to obtain high stiffness and low mass (typical weight 1.0 g).

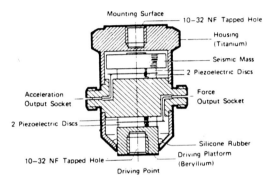

Figure 7.21 *Schematic view of impedance head (Bruel & Kjaer).*

7.5.2 Impedance and mobility of structural elements

A structure may be represented by a theoretical model which consists of masses, springs and dampers. If the structure is complicated and if the response must be duplicated exactly over a large frequency range the number of elements needed may be very large. However, for simple systems, and even for complicated structures in a limited frequency range, the response may be represented sufficiently well by a few elements.

The force F needed to set a pure mass m into vibration is proportional to the acceleration a:

$$F = ma \tag{7.1}$$

The force required to deflect a spring with stiffness k is proportional to the relative displacement d of the two ends of the spring:

$$F = kd \tag{7.2}$$

Finally, the force is proportional to the relative velocity v of the two ends of a damper with damping coefficient c for pure viscous damping:

$$F = cv \tag{7.3}$$

For sinusoidal motions, acceleration, velocity and displacement measured at a given point are related by the relationships

$$a = j\omega v = -\omega^2 d \qquad (7.4)$$
$$v = j\omega d = (1/j\omega)a \qquad (7.5)$$
$$d = (1/j\omega)v = (1/-\omega^2)a \qquad (7.6)$$

here $\omega = 2\pi f$ and f is the frequency of vibration.

The graphical signatures for the three elements are given in Fig. 7.22 together with their mechanical impedances and mobilities. While the mass is free in space, the spring and damper each require one end fixed in order for the absolute motion of the excited end to correspond to the relative motion between their ends. They are considered massless. In cases where both ends of a spring or damper move, then it is the difference between the absolute motions of their ends which must be substituted into equations (7.2) and (7.3).

The impedances and the mobilities of the elements are best illustrated in log–log diagrams with frequency $f = \omega/2\pi$ as the abcissae.

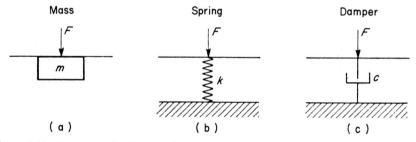

Figure 7.22 *Signatures for the basic elements:*
(a) mass = m (kg), mechanical impedance $Z = j\omega m$, mobility $M = 1/j\omega m$;
(b) stiffness = k(N/m), $Z = k/j\omega$, $M = j\omega/k$;
(c) damping coefficient = c (Ns/m), $Z = c$, $M = 1/c$.

(a) Single sprung mass system

In simple systems the mass can be considered to be placed on one spring which has a stiffness value equal to the sum of stiffnesses of the supports. Due to a misconception this is called a series system while it is, in fact, a parallel system where the force is shared between the mass and the spring as indicated in Fig. 7.23. Here the force is applied to a moving plane to which both the mass and the spring are attached.

As the motion is common to the two elements their impedances can be added to obtain the point impedance:

$$Z = Z_m + Z_k = j\omega m + k/j\omega$$
$$= j(\omega m - k/\omega) \qquad (7.7)$$

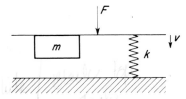

Figure 7.23 *A mass supported on a spring shown in the correct way.*

At low frequencies ω is very small and Z equals $k/j\omega$ approximately as $j\omega m$ can be neglected. At high frequencies ω is large and Z equals $j\omega m$ approximately. At a frequency f_R where $j\omega m = -k/j\omega$ a resonance occurs because $Z = 0$ and $\omega = \omega_0 = \sqrt{(k/m)}$.

The impedance can be plotted from equation (7.7) but it is less time consuming to combine the curves graphically. The mass impedance has a positive phase angle of $90°$ (j) and the spring impedance has a negative phase angle of $90°$ ($-j$ or $1/j$) relative to the force.

The curves can be obtained by subtracting the lowest value from the highest value at each frequency, but a more straightforward method is to construct a so-called 'impedance skeleton'. This is shown in Fig. 7.24(b) where the spring and mass lines are combined up to their intersection at f_R where they counteract each other to produce a vertical line for $Z = 0$. The impedance curve can then be drawn to the desired accuracy by determining two or more points from the spring and mass curves and drawing a curve through the points from the skeleton values at $0.1f_R$, f_R and $10f_R$ (Fig. 7.24(c)).

With the base excited, the velocities at the base and at the mass are different, i.e. both point and transfer values do exist. As the force on the mass is equal to the force at the base the system can best be evaluated from the mobilities of the mass and the spring. These are taken directly by inversion of the impedance curves of Fig. 7.24 (see Fig. 7.25(a)). From these curves the point mobility skeleton and the point mobility curve can be constructed in a similar manner to the impedance curve of Fig. 7.24(c) (see Fig. 7.25(b)).

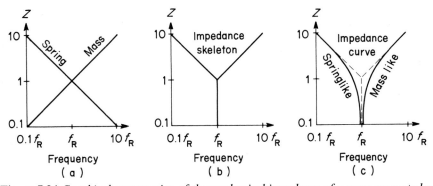

Figure 7.24 *Graphical construction of the mechanical impedance of a mass supported on a spring.*

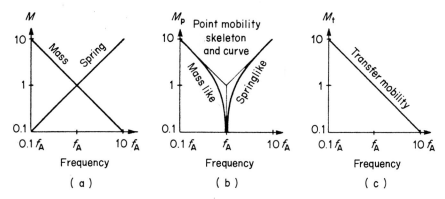

Figure 7.25 *Point mobility and transfer mobility for a base excited system.*

This is a mobility plot. The minimum at f_A represents an antiresonance, i.e. an infinite force would be required to produce any motion at all.

Transfer mobility experiences no discontinuity. As the force transmitted through the spring remains constant and equal to F the velocity v_t of the mass remains the same as it would be for a mass suspended in space. Hence the transfer mobility is a straight line with the same slope and position as a point mobility curve for the mass alone. The motion of the mass being reduced rapidly suggests that at high frequencies to all practical purposes the spring can be considered as placed on a rigid support as in Fig. 7.22(b).

(b) Mass–spring–mass system

The frequently encountered system, shown in Fig. 7.26, has the force divided between the mass m_2 and the spring-supported m_1 (the sprung mass). Point impedance is found from the combination of the mass impedance line (shown for three different masses in Fig. 7.27), and the point impedance skeleton of the sprung mass which has antiresonance at f_A.

The point impedance skeleton is obtained by inversion of the point mobility skeleton of Fig. 7.25(b), as shown in Fig. 7.28. The resulting point impedance skeletons and curves for the three values of m_2 are given in Fig. 7.29.

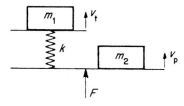

Figure 7.26 *A mass–spring–mass system.*

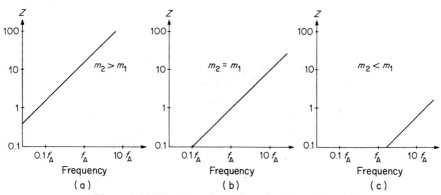

Figure 7.27 *Mass impedance lines for three values of m_2.*

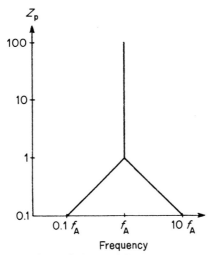

Figure 7.28 *Point impedance skeleton of the sprung mass shown in Fig. 7.25.*

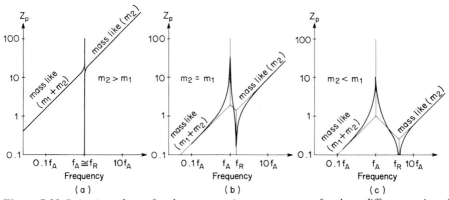

Figure 7.29 *Point impedance for the mass–spring–mass system for three different ratios of m_2/m_1.*

Impedance is the combination of Figs 7.24 and 7.28 by keeping the highest value and allowing the values to go to infinity at the antiresonance and to zero at the resonance. This produces the peak–notch response curve. At the anti-resonance the point impedance switches instantaneously from an infinitely high mass value to an infinitely high stiffness value which is negative (180° phase shift) with respect to the impedance of m_2. As the high stiffness value is reduced with increasing frequency the point impedances of the sprung mass and m_2 compensate each other at the resonance to produce zero impedance.

Figs 7.29(a), (b), (c) show that the curves commence as mass like impedances with respect to $m_1 + m_2$. After the peak–notch they continue as impedances with respect to m_2 as m_1 is now decoupled. Between the peak and the notch there is an interval in which the impedance is spring-like.

To maintain a constant transfer velocity V_t below and over the antiresonance, force F must be $j\omega(m_1 + m_2) V_t$ and, hence, the transfer impedance value continues as a straight line with slope $+1$ across the antiresonance (see Fig. 7.30).

At the resonance the point impedance curve shows that no matter which velocity V_p (and thereby the force input to the sprung mass) is chosen the total input force is zero; hence the transfer impedance also goes to zero. Above the resonance the point impedance slope changes to $+1$, compared to -1 before the resonance. As the total force required to keep V_p constant, and thereby keep the force input to the sprung mass constant, must experience a similar change of slope the transfer impedance slope will increase from $+1$ to $+3$. From the impedance skeletons in Figs 7.29 and 7.30, the mobility skeletons can be obtained by simple conversion as

$$M = 1/Z \tag{7.8}$$

By adding subsystems to the system in Fig. 7.26 or by letting itself be part of a larger model, the total response can be evaluated by combining either impedance or mobility skeletons following the foregoing rules.

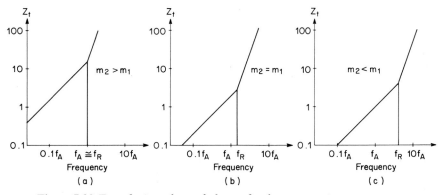

Figure 7.30 *Transfer impedance skeletons for the mass–spring–mass system.*

(c) Dynamic damping

Structures are normally lightly damped. This low damping produces high mechanical amplification factors Q. In simple viscoelastic systems only

$$Q = c/\sqrt{(km)} \qquad (7.9)$$

The amplification factor represents the factor with which to multiply or divide the intersection values between mass and stiffness lines in the mobility diagram to obtain the mobility values of the resonances and the antiresonances respectively, and vice versa for the impedance values. This is illustrated in Figs 7.31 and 7.32 where the mobility curve from Fig. 7.25(b) and the impedance curve from Fig. 7.29(b) have been redrawn for Q values of 4, 10 and 25.

Q values of 4 to 10 are often experienced, e.g. for masses placed on rubber isolators. Other isolating materials used in compression may provide Q factors

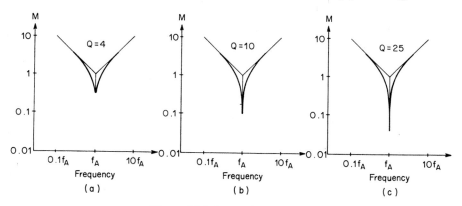

Figure 7.31 *Damped mobility curves.*

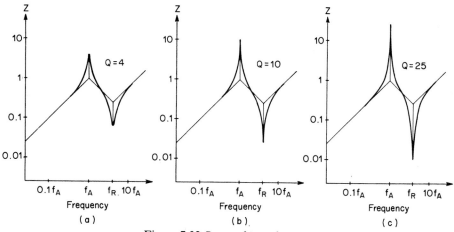

Figure 7.32 *Damped impedance curves.*

around 10 while many other mechanical engineering or civil engineering constructions are found with Q values in the range from 10 to 25.

Sandwich constructions with viscoelastic layers in shear, specially designed dampers or materials with high integral damping, provide Q values considerably lower than 4. When the Q value is 0.5 or smaller, the system is said to be critically damped (see Fig. 7.33), i.e. after a forcing function has been discontinued the vibration amplitude will die out without any oscillations. All systems with higher Q will oscillate at their resonance frequency for a shorter or longer period after excitation depending on the Q value.

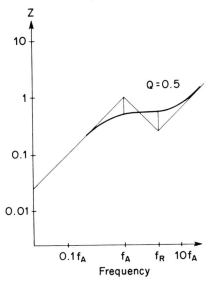

Figure 7.33 *Critically damped impedance curve.*

7.6 Measurement of transfer function and modal analysis

Low cost minicomputers and computing techniques such as the fast Fourier transform have given rise to powerful new 'instruments' known as digital Fourier analysers. These machines quickly and accurately provide the frequency spectrum of a time-domain signal and it is now relatively simple to measure the transfer function and to perform a modal analysis [2].

Techniques also identify the modes of vibration of an elastic structure from measured transfer function data [3, 4]. Once a set of transfer (frequency response) functions relating points of interest on the structure have been measured and stored, they may be operated on to obtain the modal parameters, i.e. the natural frequency, damping factor, and characteristic mode shape for the predominant modes of vibration of the structure.

The modal responses of many modes can be measured simultaneously and complex mode shapes can be directly identified. It is not then necessary to isolate the response of one mode at a time, i.e. the so called 'normal mode' testing concept.

7.6.1 Frequency response measurement

Frequency responses (transfer functions) are measured by:

(1) Simultaneous measurement of an input and a response signal in the time domain.
(2) Applying a Fourier analysis.
(3) Dividing the transformed response by the transformed input.

This digital process enjoys many benefits over traditional analogue techniques in terms of speed, accuracy, and post-processing capability [5].

One of the most important features of Fourier analysers is their ability to form accurate transfer functions with a variety of excitation methods. Three popular methods for exciting a structure for the purpose of measuring transfer functions are: random, transient, and sinusoidal excitation. A shaker is used to excite the mechanical structure under test, as illustrated in Fig. 7.34.

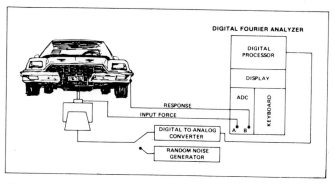

Figure 7.34 *The general test set-up for making frequencey response measurements with a digital Fourier analyser and an electrodynamic shaker.*

(1) Random excitation

Three types of broad-band random excitation can be used:

(a) pure random;
(b) pseudo-random; and
(c) periodic random.

Fig. 7.35 illustrates each type of random signal. Pure random excitation has a Gaussian distribution and is in no way periodic, i.e. does not repeat. A pseudo-random excitation avoids the leakage effects of a non-periodic signal.

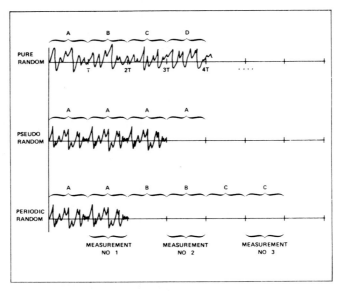

Figure 7.35 *Comparison of pure random, pseudo-random and periodic random noise. Pure random is never periodic. Pseudo-random is exactly periodic every T seconds. Periodic random is a combination of both, i.e. a pseudo random signal that is changed for every ensemble average.*

Periodic random waveforms combine the best features of pure random and pseudo-random, but without the disadvantages; they satisfy the conditions for a periodic signal, yet change with time to excite the structure in a truly random manner.

(2) Sinusoidal testing

A controlled sinusoidal force is applied to the structure, and the ratio of output response to the input force versus frequency is plotted. Sinusoidally measured transfer functions can be digitized and processed with the Fourier analyser or can be measured directly.

(3) Transient testing

There are two basic types of transient tests:

(a) impact;
(b) step relaxation.

A hand-held hammer with a mounted load cell is a typical impact excitation. The load cell measures the input force and an accelerometer mounted on the structure measures the response. The process of measuring a set of transfer functions by mounting a stationary response transducer (accelerometer) and moving the input force around is equivalent to attaching a mechanical exciter to the structure and moving the response transducer from point to point. In the former case, we are measuring a row of the transfer matrix whereas in the latter we are measuring a column [4].

Step relaxation is a transient test in which an inextensible, light-weight cable is attached to the structure. This preloads the structure to some acceptable force level. The structure 'relaxes' when the cable is severed, and its transient response as well as the transient force input, are recorded.

7.7 Case studies

7.7.1 Nuclear reactor: tiebar wear

A special-alloy steel tiebar charges fuel into and discharges it from advanced gas-cooled reactors. Gas flow may cause this tiebar to strike the fuel-supporting grids and braces. It is important that this movement should not cause wear and reduce significantly the strength of the tiebar.

Tests by the Nuclear Power Co. (Whetstone) Ltd measured movement in two directions at right angles to a number of planes along the length of the fuel stringer. The resulting signals were recorded on magnetic tape for subsequent analysis. Digital signal processing techniques were used to describe the energy associated with the movement and provide a basis for simulated laboratory-based endurance tests to examine the tribology of the wear process.

Peak velocity distribution and the radial position probability density function were derived for the combined signals from two transducers at right angles. Since only movements associated with impacts were of interest, a more sophisticated program was developed to permit calculation of velocity of impact, azimuth of impact and angle of impact normal to the radius. These three parameters were stored and the data accumulated from multiple impacts were used to derive probability density functions. DC offset of the signals varied slightly from day to day and between transducers and had a critical influence.

7.7.2 Chemical industry: heat exchangers

A unique vibration was detected in a new platformer installation involving an air dryer. When flow in the heat exchangers reached a particular velocity the last two heat exchangers in the series developed a tremendous noise.

Vertical vibration profiles for different flow conditions (Fig. 7.36) showed that the fundamental frequency of vibration as detected in (a) was exactly the same as in (b). This proved conclusively that the source of the noise was aerodynamic and unrelated to the compressor.

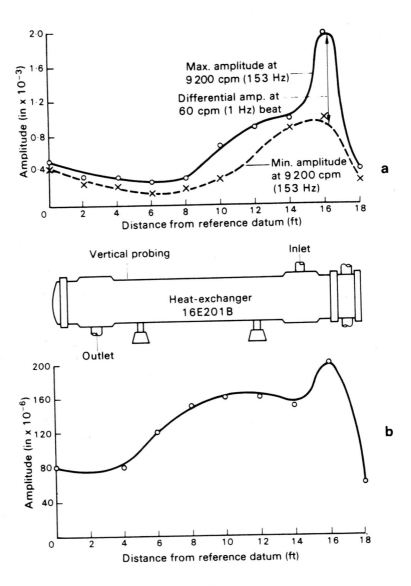

Figure 7.36 *Heat exchanger vibration profiles:*
           (a) fundamental frequency, 153 Hz
                compressor, 126 Hz
                compressor flow, 8.05 (air)
                exchanger temperature: inlet/outlet 15 °C
                vibration outlet flange: max. 0.00762 mm, min. 0.00635 mm;

           (b) fundamental frequency, 153 Hz
                compressor, 110 Hz
                compressor flow, 7.2 (air)
                exchanger temperature: inlet/outlet 15.5 °C.

The mass of air in the system under the specified pressure and flow conditions was calculated and compared with that for hydrogen (which it was proposed to use). This showed that with hydrogen, because of its entirely different thermodynamic characteristics, the aerodynamic vibration would disappear. Hydrogen was used, and the heat exchanger noise disappeared.

7.7.3 Chemical industry: tube collisions

Although regular vibration monitoring was not at the time employed, a failure reported by Taylor is of interest.

A desulphurizer overhead condenser at Conoco's Humber refinery had been operating successfully for about a year. Shortly after a shutdown and a subsequent start-up the overhead product went off-specification and indicated a leak in the condenser. A fracture was found at the shell-side edge of the tube-sheet in a tube located near an inlet nozzle and supported at the maximum span length within the shell. Recent collision had occurred between tubes on the maximum span lengths under both inlet nozzles and at the U-bends.

Analysis of the broken tube, inspection of the bundle, and calculations based on current vibration theories showed that the failure was caused by excessive inlet velocities. These had been caused partly by an increase in plant throughput, and partly by the recycling of stored naphtha, leading to greater overhead volumes. The combined effect doubled the excitation load, leading to a resonant condition and fatigue failure.

7.7.4 Machine tool base

Modal plots of a surface provide a means of locating positions in need of stiffening so that excessive vibration does not impair product quality. These modal plots reduce the cost of large machine tool bases by ensuring that stiffeners and sectional changes are only introduced where they are needed.

A cooperative study between the US Steel Corporation, the Snyder Corporation and the University of Cincinnati resulted in the successful analysis of a welded steel base to replace a cast-iron machine base. Both bases were placed on flexible supports with a natural frequency of less than 5 Hz; frequency tests were run and mode shape was plotted.

An electromechanical exciter was used with its pin held against the test structure. Two devices measured structural motion, one a velocity pickup and the other a piezoelectric acceleration transducer. A mechanical impedance transfer function analyser automatically analysed and plotted signals from the pickups, which were filtered, converted to logarithms and subtracted, and plotted by an X–Y recorder together with the phase between the force and motion signals.

Two machine tool bases, cast-iron and welded steel, were made as geometrically similar as fabricating techniques would allow. Dynamic testing showed little difference in response characteristics, the welded steel having greater stiffness, the cast-iron base having better damping.

Fig. 7.37 shows a typical mode shape, the areas outside the 'node' line being 180° out of phase with those inside. Using these shapes for the 'steel box', it was found exactly where and how the steel base could be improved. Judiciously placed stiffeners produced the desired dynamic performance. Welded machine tool bases need not reproduce the configuration of the castings they replace as they have their own stress distribution characteristics, different from those of castings. In the test performed, the major dynamic weakness from plate vibration in the centre section was corrected by introducing an additional Z-shaped internal rib right through this section.

7.7.5 Structural problems: pipes

A portable type SD 330 analyser was used by the Structural Dynamics Corporation to solve a piping problem. Natural gas is piped to the Michigan, Great Lakes area from Texas and stored to provide the winter supply for Detroit and the surrounding area in numerous re-injection wells where the gas is pumped into storage cavities some 1128 m (3700 ft) underground.

Piping at the well in question, and at four others, was vibrating and causing considerable concern since the gas pressure is at 121 bar (1750 psi) and a break would have serious consequences.

It was determined that the peaks were occurring at 11 Hz, the second harmonic of the engine speed. The solution was either to add considerable reservoir capacity to remove the pulsations or to add more mass to the tiedown straps for the pipe at each well-head.

About five minutes of set-up time was all that was needed to identify the vibration. The engineers involved left Kansas City at 7.00 am, flew about 700 miles, analysed five sites and were back in Kansas City by 7.00 pm the same day.

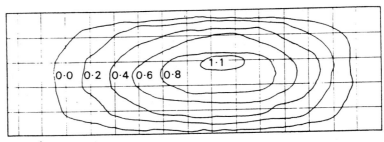

Figure 7.37 *Mode shape plot of top of steel wing base at lowest resonant frequency (unmodified).*

7.7.6 Low-level vibrations in buildings

A measuring system developed by Bruel & Kjaer used an accelerometer with integral preamplifier and ultra-high sensitivity to measure vibrations below 1 g over a frequency range 1–1000 Hz. A mechanical impedance system was used for measurement. A schematic diagram of this accelerometer is shown in Fig. 7.38 and the principal characteristics of the system are given in Table 7.3.

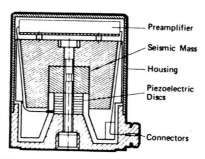

Figure 7.38 *Bruel & Kjaer high-sensitivity accelerometer.*

Table 7.3 Characteristics of Bruel & Kjaer high-sensitivity measuring system

Accelerometer sensitivity	10.0 ± 0.2 V/g
Accelerometer mounted resonance	2500 Hz
Preamplifier gain	nearly zero
Preamplifier output impedance	$< 500\,\Omega$
Preamplifier max. output current	1.4 mA
Preamplifier max. output voltage	10 V
Frequency response (acc. + preamp.) (10% limits)	0.2–1000 Hz
Noise level (2–1000 Hz)	$< 40\,\mu$V
Maximum temperature transient sensitivity (3 Hz LLF)	$4 \times 10^{-5} g/°C$
Special connector gives charge output via	1000 pF
Power supply	2 mA, 28 V DC
Weight	approx. 600 g

The sensitivity of the accelerometer is about 100 to 1000 times that of most ordinary types of accelerometer. The noise level sensitivity of the preamplifier is 5 to 10 times greater than that of conventional preamplifiers because of a special filtering arrangement. Consequently, the minimum vibration level that can be measured with the accelerometer is approximately 10 μg, which is about the same order of magnitude as the background vibrations in quiet areas, free from the influence of traffic vibration and other disturbances.

A typical application to part-structure of reinforeced concrete is shown in Fig. 7.39 with the equivalent mathematical system model of a clamped–hinged–

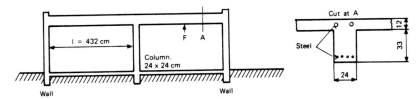

Figure 7.39 *Dimensions of structure.*

clamped supported beam (Fig. 7.40). Calculations and measurements produced the following values:

Mode	1	2	3	4	5
f(measured) (Hz)	22	36	70	95	160
f(calculated) (Hz)	25	36	80	99	167

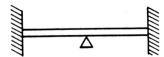

Figure 7.40 *Clamped–hinged–clamped beam.*

7.8 Electricity pylon failures: line vibration and its effect [6]

In Trainor *et al.* [6] some work on vibration effects was reviewed [7–11]. The general field, including galloping, aeolian vibration, and subspan oscillation, has been summarized [7]. Beards [8] concentrated on the control of aeolian vibration; others [9, 10] focused on conductor galloping. Johns [11] reviewed wind loading of general structures and included references on the vibration of conductors and the wind excitation of transmission towers. Several other review articles relate to aeolian vibration. Fleishmann and Sallet [12, 13] covered the unsteady flow phenomena related to vortex shedding. A comprehensive paper by Tsui [14] summarizes modern advances in nonlinear mechanics and fluid dynamics as they are applied to aeolian vibrations. A practical method for calculating the peak response of a conductor was included.

Galloping motions are low-frequency, vertical, or horizontal oscillations. They correspond to mode shapes having from 1 to 4 loops per conductor span and natural frequencies from 0.1 to 0.8 Hz. The term conductor galloping generally refers to the self-excitation of these modes to very large amplitudes; self-excitation is usually initiated by an icing storm. Aeolian vibrations, conversely, are high-frequency vertical oscillations with mode shapes of 50 to 250 loops per span. They are caused by vortex shedding and have a frequency range of about 10 to 50 Hz. The upper frequency limit depends upon air turbulence and conductor tension and can be greater for long spans over water or very flat terrain. The third major type of vibration, subspan oscillation, is a wake-induced phenomenon affecting the opposite pairs of a multiconductor bundle. Single or double vibration loops occur in the spacer subspans at frequencies of about 1 to 3 Hz.

Both aeolian vibrations and galloping have an indirect influence on tower dimensions. Phase-to-phase clearances are determined by galloping amplitudes [15]. In addition, conductor tensions are usually set at a limit of 20% of ultimate tensile strength (UTS) to control the aeolian response. This limit determines the magnitude of sag, which, in turn, dictates tower height [16]. Thus, successful efforts to construct more economical and more compact transmission lines depend somewhat on improving the control of line vibrations [17, 18].

Most galloping does not cause tower damage. Measurements during galloping on a line with self-supporting lattice towers [19] have shown that the insulator tension fluctuates from about 60% of the tension under no-wind conditions. Such loads can be considered as being transferred statically to the towers because the fundamental frequency of a lattice tower ranges from about 1.5 Hz to 4 Hz, i.e. substantially above the range of conductor galloping frequencies.

Qualitative investigations by White [20], however, have shown that certain structural arrangements can nevertheless convert galloping motions into destructive forces that cause tower collapse. A running-angle suspension insulator assembly connected to any type of tower and a dead-end insulator assembly connected to a very flexible tower are examples of hazardous arrangements. The potential for any new type of insulator assembly to be affected adversely by galloping requires investigation by model or full-scale tests [21].

Aeolian vibrations can also occasionally lead to structural damage. For example, they have caused the crossarms of towers on several lines in Saskatchewan to suffer severe fatigue damage. According to Mitchell [22], individual crossarm members were observed to vibrate violently although the conductors were fitted with Stockbridge dampers and the bending amplitude of the conductors appeared to be within acceptable limits. In this case the problem was resolved only by the addition of bracing, which raised the resonant frequencies of individual crossarm members above the frequency range for aeolian vibration.

7.9 Offshore structures

Structural vibration monitoring of offshore platforms is intended to establish that sufficient strength remains to withstand the largest wave forces envisaged. Inspection is normally carried out by divers. Attention is focused on the inclined members which provide the shear connection between the legs. Wind and wave loads are resisted mainly by axial tension and compression forces transferred by these shear connections to the legs and bracing members (and not by bending of the legs). Only a small percentage of such members can be examined annually, usually by concentrating on those welds which have been calculated to be most susceptible to fatigue.

Platforms are positioned in deep water where the environment is more extreme. Diver inspection becomes more severe, costly and dangerous so that inspection programmes in deep water cover fewer members than would be inspected in shallower waters. Undetected loss of load-carrying capability of a member presents a major problem. Other problems may result from a collision or accident.

Restricted operation pending complete repair and shutdown and evacuation when waves exceed a stipulated height reduce immediate risk. Neither procedure checks that the platform is fit for re-occupation. An operating limit is imposed as a result of known damage – a decision influenced by the thought that the damage may have spread, that the load may redistribute, that there may be other undetected damage.

Continuous assurance that the condition of the structure has not seriously deteriorated is provided by augmenting the inspection and damage surveillance of primary load-carrying members by vibration monitoring [23].

7.9.1 Base line dynamics

Assurance of maximum accuracy involves complete understanding of the dynamics of the particular platform and acquisition of a good base line for comparison. Sensitivity of the platform to damage at various locations must be calculated; this provides a basis for identifying damage associated with changes in vibration spectra. Such calculations involve:

(1) Platform response calculation by element (beam element) analysis.
(2) Direct measurements of the base line signature.

7.9.2 Vibration measurement intervals

Significant changes in mass or in support stiffness (as would arise from scouring) would be logged if measurements were made every 6 months; on platforms which have not settled to their final operating configuration (deck mass, number of conductors and risers etc.) more measurements are needed. The initial base line signature and subsequent measurements should be made at the end of a diving inspection period when information on the integrity of the platform is at a maximum.

7.9.3 Portable/real-time instrumentation

Measurement/recording visits with on-shore processing of data are the cheapest method in the short-to-medium term if the monitoring is regarded simply as a more frequent periodic check on platform strength. On-board instrumentation is necessary if structural integrity is to be assured before the platform can be put into service; this means that the data can be processed in real time and assessments of integrity made on the spot.

Interpretation is a highly skilled operation in which experience of platform vibration analysis is essential. A preliminary (but fairly definite conclusion) can be reached after 6 hours of measurement and analysis; final confirmation takes about 72 hours.

7.9.4 Vibration monitoring

Wind and wave motion continually vibrate offshore platforms. Additional excitation arises from onboard activity. Structures are lightly damped so that almost all energy is concentrated either at the wave frequency or at the natural or resonant frequencies of the structure.

Failure of one member in an efficient structure will cause a change in stiffness. This will affect the magnitude of the responses and possibly alter the natural frequencies. Response changes will be larger for the more structurally important member. All the quantities involved (e.g. wave input forces, response amplitudes) are stochastic; accordingly the relationships between cause and effect can only be described in statistical terms.

(1) Wave response

Direct wave action produces a response at a frequency determined by the wave period. Monitoring involves the measurement of response *amplitude* relative to input wave excitation. Measurement difficulties limit the potential accuracy of such a comparison.

(2) Platform resonance

Platform resonances are at frequencies governed by mass and platform stiffness (including foundation stiffness). Severance at one end of a shear member in a braced panel increases the forces and deformations in other members. Local loss of stiffness reduces some natural frequencies of the structure; such reductions can be measured. First and second groups of natural frequencies are easily obtained from measurements made with only natural excitation. Failure of a particular member produces larger changes in some frequencies than in others. Location of the failed member can be deduced approximately from the pattern of these changes. Frequencies of the third mode group can also be obtained for some structures and provide further confirmation of a failure and increased confidence in its location [24–26]. Fig. 7.41 shows the first three sway mode shapes for a typical platform.

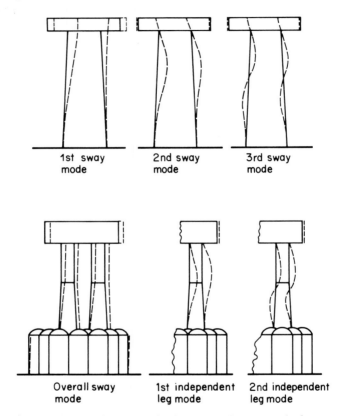

Figure 7.41 *Overall sway modes for steel and concrete platforms.*

Natural frequencies contain information about structural integrity, amplitude changes help in vibration mode shape derivation which assists defect location.

(3) Monitoring capability

Accurate techniques have now been developed for the measurement of structural vibration on steel platforms and for the deduction of vibration mode shapes and foundation characteristics [23]. Offshore measurements made on 18 different structures have shown that on many steel platforms the changes in the natural frequencies caused by the failure of a single, structurally important member are large enough to be detected. Such detection has been unambiguous.

Vibration monitoring by frequency/mode measurement is less suitable for structures such as squat platforms with base dimensions comparable to height. In such platforms, it may not be possible to identify any more than the fundamental modes; this reduces the detection sensitivity. Platforms with strong nonlinear response due to pile/soil or pile/leg interactions may not be suitable although nonlinearities which develop at only relatively large amplitudes can be monitored at smaller amplitude vibrations.

7.9.5 Node weld cracking: underwater inspection

Completely severed primary members can be detected from vibration changes without underwater inspection. Partial failures, such as cracked node welds, make it necessary to use higher modes. With such higher modes the vibration is concentrated in small groups of members (even in single members). Fig. 7.42 shows the typical modes involving in-plane vibration of a K-braced frame. Such modes may be analysed by cross-spectral analysis using acceleration measurements simultaneously on all members.

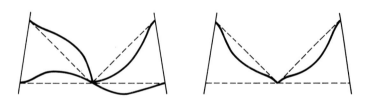

Figure 7.42 *Vibration modes of a K-frame.*

7.9.6 Crack penetration

Penetration of the wall of one member by a crack produces a leakage or 'flooding' and the entering water increases the mass and reduces the natural frequency. Water leakage depends on the depth at which the member is located, its angle of inclination, the position and size of the crack. At 100 m depth a structural member will flood to about 90% of its volume; a typical bracing member with a crack 10 mm long and 0.1 mm wide will fill in a matter of weeks, even if the crack is open for only 10% of the time. At 10 m depth a member may take in only 50% of its volume of water in several months with the crack open all of the time.

Mode shapes before and after flooding of one of the inclined members of the K-frame are shown in Fig. 7.43 and a typical 5 to 10% reduction in frequency. Flooding is positively responsible for such changes and is different from an increase in mass from other causes such as marine growth.

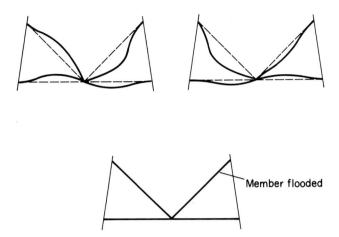

Figure 7.43 *Vibration modes of K-frame with flooded member.*

7.10 Concrete platforms

Procedures in 1978 did not assess the platform strength directly; strength was inferred from the lack of local damage.

Tests on a multileg platform showed that many vibration modes associated with overall sway and the vibration of individual legs can be measured and identified by adapting the instrumentation developed for steel jackets. Such measurements may monitor some aspects of foundation conditions, integrity of the legs and their attachment to the base. Vibration monitoring of concrete structures may always be less sensitive and lacking in diagnostic power than on steel jackets.

7.10.1 Vibration response measurement

Overall monitoring using vibration response measurement involves sets of orthogonal accelerometers positioned at two levels – one near the top of the legs, and the other as close as possible to water level [27, 28]. Accelerations are recorded simultaneously following excitation from wind and wave. A typical acceleration/time record for a steel platform is shown in Fig. 7.44. Analyses then give an autospectrum (Fig. 7.45) and a cross-spectrum (Fig. 7.46) from accelerometers at two levels of a platform. Variation of phase angle between the two levels varies with frequency as shown in Fig. 7.46.

Wave excitation and fundamental vibration contained in the first two groups of peaks are in phase. Relative motion between the two levels can be deduced from such information and the natural modes of vibration identified.

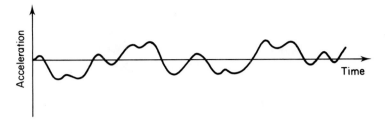

Figure 7.44 *Acceleration-time record.*

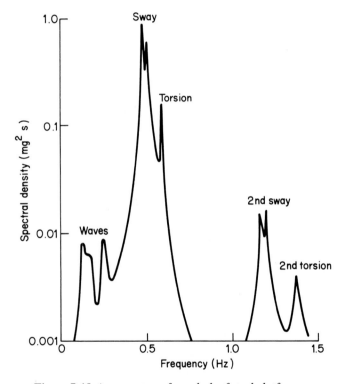

Figure 7.45 *Autospectrum from deck of steel platform.*

An acceleration autospectrum for a concrete platform is shown in Fig. 7.47.. Underwater measurements for node weld inspection use accelerometers attached temporarily with magnetic clamps; a typical autospectrum is shown in Fig. 7.48. Even under the relatively calm conditions necessary for diving the autospectra respond clearly to the excitation.

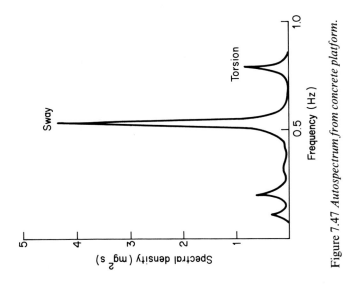

Figure 7.47 Autospectrum from concrete platform.

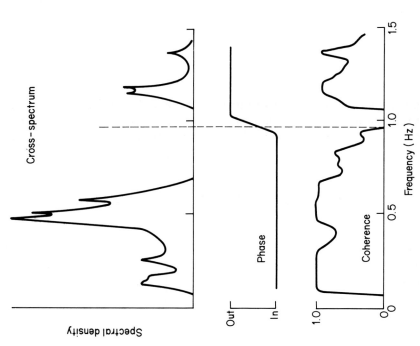

Figure 7.46 Cross-spectrum between two levels of steel platform.

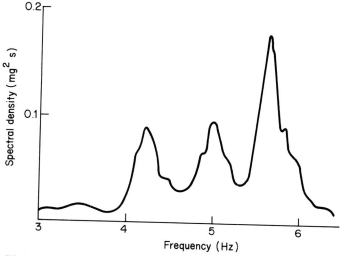

Figure 7.48 *Autospectrum from underwater member of steel platform.*

7.10.2 Fatigue life

Platforms have operated for several years in the North Sea such that fatigue failure is a major possibility. Fatigue can be induced by:

(1) Primary forces – these are the structural response due to the wave action. This type of fatigue occurs at low natural frequencies.
(2) Localized forces – these are shocks due to wave slam or vortex shedding. In this case fatigue occurs at high frequencies.

Historical analysis of the primary forces in the members at the significant frequencies can be expressed in appropriate statistical terms. This indicates the stress history of a joint in any primary member and provides an assessment of the residual life of a platform [28]. Platform displacements are not useful for monitoring fatigue due to wave slam or vortex shedding. Sensors on individual members or groups of members are necessary for this.

7.11 Dynamic analysis of structures (SEAPOS)

A system of computer calculation applied to finite elements within a structure has been developed by Structural Dynamics (Offshore) Ltd with a capability for solving a variety of structural problems: static and dynamic; elastic, plastic, creep and swelling; small- and large-scale deflections. Several types of analysis can be used, for example:

(1) Static – to solve for displacements, stresses and strains caused by applied loads.
(2) Mode/frequency – to solve for resonant frequencies and mode shapes characterizing a structure.
(3) Nonlinear transient dynamic – the time history response of an arbitrary structure to a known force and/or forcing function.
(4) Linear harmonic response spectrum – the steady-state solution under a set of harmonic loads of known amplitude and frequency, where several frequencies are to be analysed.

The SEAPOS software system represents the entire platform by a set of connected units, so allowing a piecewise super-element approach to the whole solution. In each case the individual areas are mathematically modelled as a system of node points joined by finite elements such as beams, plates and shells (Fig. 7.49). Masses, inertias, stiffness and damping factors are then given to these elements and distributed along the nodes so that the physical characteristics of any point can be described in detail.

Structural motion is represented through the dynamic degrees of freedom at various node points, from which a set of matrix equations is derived from the

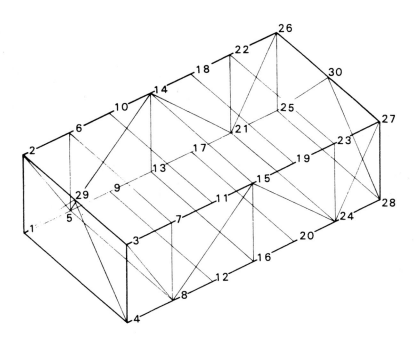

Figure 7.49 *SEAPOS geometry input, checking and record. Simple modular structure with numbered modes.*

finite element model. These equations are then solved on a large-core, batch process machine (a CDC6500). The batch job results are received on the SEAPOS software system which utilizes a more flexible and interactive time-sharing machine to give immediately interpretable information. The linear harmonic response spectrum is calculated on-line to indicate significant modes; both natural and forced motion of the structure may be displayed graphically. The effect of forcing the structure at different frequencies is shown as acceptance frequency plots (Fig. 7.50). Applied to a computer-aided design, it is possible to display the dynamic deflection profile as in Fig. 7.51.

Effects of local structure design changes can be calculated on-line and displayed immediately. This means that a sequence of potential remedial measures can be quickly evaluated for engineering cost effectiveness, without the need for a major re-run on the large batch machine.

The SEAPOS system overall capability includes the efficient solution of plastic, small- and large-scale deflection and specific nonlinear transient dynamic problems, as well as offering an effective system for static analysis, if required.

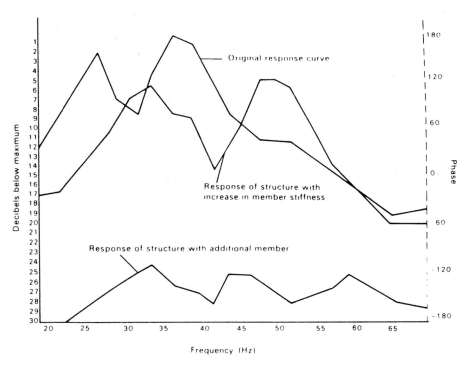

Figure 7.50 *Effect of forcing the structure at different frequencies.*

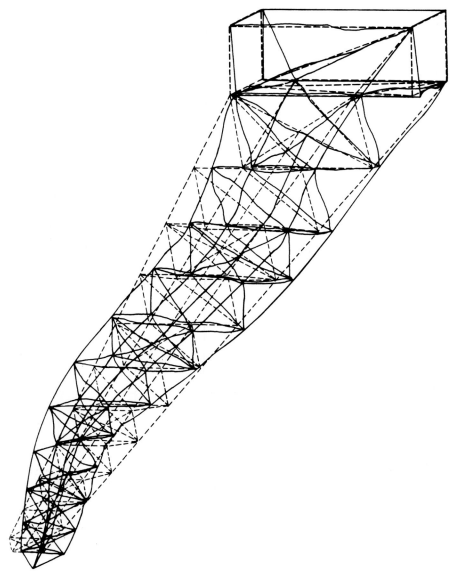

Figure 7.51 *Resonant mode of a flare boom: the displacement scale is arbitrary but gives direct information on the forced response of the mode.*

References

1 Koury, A. J. (1975), Maintenance Technology Department, Naval Air Systems Command, Washington DC 20361, USA. Paper MFPG-33, April 21–23, 1981, AIR-4114C.

2 Ramsey, K. A. (1975), 'Effective measurements for structural dynamics testing', *Sound & Vibration*, November, 24–35.

3 Richardson, M. and Potter, R. (1974), 'Identification of the modal properties of an elastic structure from measured transfer function data', *Proc. 20th ISA Conf.* Albuquerque, NM, May.

4 Potter, R. and Richardson, M. (1974), 'Mass, stiffness and damping matrices from measured modal parameters', *Proc. ISA Conf. and Exhibit,* New York City, October.

5 Roth, P. R. (1971), 'Effective measurements using digital signal analysis', *IEEE Spectrum,* April, 62–70.

6 Trainor, P. G. S., Popplewell, N., Shah, A. H. and Pinkney, R. B. (1983), 'Static and dynamic behaviour of mechanical components associated with electrical transmission lines', *Shock & Vibration Digest,* **15**, No. 6, 27–38.

7 Ramamurti, V., Sathikh, S. and Chari, R. T. (1978), 'Transmssion line vibrations', *Shock & Vibration Digest,* **10**, No. 11, 27–31.

8 Beards, C. F. (1978), 'Damping overhead transmission line vibration', *Shock & Vibration Digest,* **10**, No. 8, 3–8.

9 Dubey, R. N. and Sahay, C. (1980), 'Vibration of overhead transmission lines III', *Shock & Vibration Digest,* **12**, No. 12, 11–14.

10 Dubey, R. N. (1978), 'Vibration of overhead transmission lines', *Shock & Vibration Digest,* **10**, No. 4, 3–6.

11 Johns, D. J. (1982), 'Wind excited behaviour of structures III', *Shock & Vibration Digest,* **14**, No. 7, 23–38.

12 Fleishmann, S. T. and Sallet, D. W. (1981), 'Vortex shedding from cylinders and the resulting unsteady forces and flow phenomena – part I', *Shock & Vibration Digest,* **13**, No. 11, 9–22.

13 Fleishmann, S. T. and Sallet, D. W. (1981), 'Vortex shedding from cylinders and the resulting unsteady forces and flow phenomena – part II', *Shock & Vibration Digest,* **13**, No. 12, 15–24.

14 Tsui, Y. T. (1982), 'Recent advances in engineering science as applied to aeolian vibration: an alternative approach', *Electr. Power Syst. Res. (Switzerland),* **5**, No. 1, 73–85.

15 Rawlins, C. B. (1981), 'Analysis of conductor galloping field observations – single conductors', *IEEE Trans. PAS,* **100**, No. 8, 3744–53.

16 Havard, D. G. and Nigol, O. (1981), 'Research on overhead conductor vibration', *Ont. Hydro Res. Rev.,* No. 3, 39–45.

17 Cassan, J. G. (1974), 'Ontario Hydro's overhead-transmission research program', *Ont. Hydro Res. Q.,* **26**, No. 2, 1–6.

18 Havard, D. G., Paulson, A. Ş. and Pohlman, J. C. (1982), 'The economic benefits of controls for conductor galloping', *Proc. CIGRE 29th Session,* Vol. 1, (22–02).

19 Anjo, K., Yamasaki, S., Matsubayashi, Y., Nakayama, Y., Otsuki, A. and Fujimura, T. (1974), 'An experimental study of bundle conductor galloping on the Kasatori–Yama test line for bulk power transmission', *Proc. CIGRE 25th Session,* Vol. 1, (22–04).

20 White, H. B. (1979), 'Some destructive mechanisms activated by galloping conductors', *IEEE Power Engng Soc. Winter Mtg,* February, New York, (A79–106–6).

21 Souchereau, N., Sabourin, G., Cayer, P. and Tsui, Y. T. (1978), 'Validation of a Chainette Tower for a 735 kV line', *Proc. CIGRE 27th Session,* Vol. 1, (22–04).

22   Mitchell, J. (1976), 'Steel tower vibration problems', *Can. Electr. Assoc. Fall Mtg,* Calgary, Alberta, October.
23   Begg, R. D. and McKenzie, A. C., (1978), 'Monitoring of offshore structures using vibration analysis', paper 17. Offshore Structures Conference, City University, London, September.
24   Loland, O. and McKenzie, A. C. (1974), 'On the natural frequencies of damaged offshore oil platforms', *Mech. Res. Commun.,* 1, 353–4.
25   Vandiver, J. K. (1975), 'Detection of structural failure on fixed platforms by measurement of the dynamic response', OTC 2267. Offshore Technology Conference, Aberdeen, Paper 2267.
26   Loland, O., McKenzie, A. C. and Begg, R. D. (1975), 'Integrity monitoring of fixed steel offshore oil platforms', *BSSM/RINA Conf.,* Edinburgh, September.
27   Loland, O. and Dodds, C. J. (1976), 'Experiences in developing and operating integrity monitoring systems in the North Sea', OTC 2551. Offshore Technology Conference, Aberdeen, Paper 2551.
28   Begg, R. D., McKenzie, A. C., Dodds, C. J. and Loland, O. (1976), 'Structural integrity monitoring using digital processing of vibration signals', OTC 2549. Offshore Technology Conference, Aberdeen, Paper 2549.

Further reading

Ahmad, M. B. *et al.* (1984), 'Self-supporting tower under wind loads', *ASCE J. Struct. Engng,* 110, No. 2, 370–84.
Archbold, E., Buch, J. M., Ennos, A. E. and Taylor, P. A. (1969), *Nature,* 222, 263.
Bhatti, M. H. (1982), 'Vertical and lateral dynamic response of railway bridges due to nonlinear vehicles and track irregularities' *PhD Thesis,* Illinois Institute of Technology.
Collacott, R. A. (1978), *Vibration Monitoring and Diagnosis,* London, Longmans/Godwin.
Iwatsubo, T. *et al.* (1983), 'Reliability analysis of structures under seismic excitation', *Bull. JSME,* 26 (222), 2233–8.
Karaolides, C. K. and Kounadis, A. N. (1983), 'Forced motion of a simple frame subjected to a moving force', *J. Sound Vib.,* 88, No. 1, 37–45.
Macovski, A., Ramsey, S. D. and Schaefer, L. F. (1971), 'Time lapse interferometry and contouring using television systems', *Appl. Optics,* 10, No. 12.
Melke, J. and Kraemer, S. (1983), 'Diagnostic methods in the control of railway noise and vibration', *J. Sound Vib.,* 87, No. 2, 377–86.
Paidoussis, M. P. *et al.* (1983), 'Dynamics of a cluster of flexibly interconnected cylinders', *J. Appl. Mech. Trans. ASME,* 50, No. 2, 421–8.
Plunkett, R. and Roy, A. K. (1983), 'Wave attenuation in damped periodic structures', Department of Aerospace Engineering and Mechanics Report, Minnesota University, 12 July.
Rayleigh, Lord, *Scientific Papers of Lord Rayleigh,* Vol. 1, New York, Dover, pp 411–96
Ross, T. J. (1983), 'Direct shear failure in reinforced concrete beams under impulsive loads' Report AFWL-TR-83-24, US Air Force Weapons Laboratory, Kirtland AFB, NM, USA.
Rutkowski, M. J. (1983), 'Vibration characteristics of a coupled helicopter rotor-fuselage by a finite element analysis', Report No. NASA-A-9053/NASA-TP-2118, NASA Ames Research Center, Moffett Field, Ca, USA.
Sano, M. (1983), 'Pressure pulsations in turbo-piping systems', *Bull. JSME,* 26 (222), 2129–35.

Stetson, K. A. (1970), *Optics and Laser Technology*, **2**, No. 3, 179–81.

Younis, C. J. and Tadjbakhsh, I. G. (1984), 'Response of sliding rigid structures to base excitation', *ASCE J. Engng Mech.*, **110**, No. 3, 417–32.

8
Strain gauge techniques

8.1 Introduction

Force, torque, pressure and stress are measurable by using strain gauges. By introducing the properties of the material strain measurements may be related to the applied force, torque, pressure or stress.

The whole range of such transducers is based on the property that a material's electrical resistance or conductance changes with change of length under stress. Precise electronic magnification is needed. Spurious effects can arise. It is the skill and quality of the manufacturer which leads to precision measurement.

8.2 Strain gauges

8.2.1 Basic principle

The 'strain gauge' was developed in the late 1930s by two researchers in the USA. Working independently of each other, Simmons at California Institute of Technology and Ruge at Massachusetts Institute of Technology developed a strain gauge consisting of a length of wire glued to the test object so that changes in length (strains) on the surface are transferred to the wire. These length changes cause alterations in the resistance of the wire which can be measured by comparatively simple electrical circuitry. The modern strain gauge,

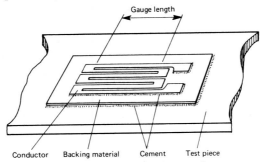

Figure 8.1 *Electrical resistance strain gauge.*

examples of which are shown in Figs 8.1 and 8.2, works in exactly the same way, with strain being detected by measuring the resistance variations caused by changes in the gauge length of the wire. The strain gauge should be very small and compact, having negligible mass to exert a minimum of influence on the measuring object; it should also be easily connected to the test specimen.

Electrical detection circuits to measure the very small changes in the gauge resistance are comparatively uncomplicated, being variations of the familiar Wheatstone bridge. When suitable compensation circuits are employed or self-compensating gauges used, temperature sensitivity is virtually eliminated.

Typical well known applications for strain gauges include experimental strain and stress measurement on aircraft, boats, cars and other forms of transportation. Strain gauges are also used for the measurement of stress in larger structures, for example apartment buildings and office blocks, pressurized containers, bridges, dams, etc.

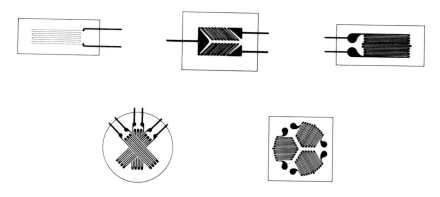

Figure 8.2 *Typical strain gauge configurations.*

Earlier types of strain gauge were made of thin copper–nickel or chrome–nickel alloy wire approximately 0.025 mm (0.001 in) in diameter. Long gauges, to give big resistance changes while occupying small area, are obtained by folding the conductor into a grid pattern similar to that shown in Fig. 8.3. The grid layout has a long gauge length without increasing the transverse sensitivity.

Foil gauges are now in widespread use. In this type of gauge the conductor is made by etching a grid pattern in a thin metal foil only a few microns in thickness. The foil is made of a similar alloy to the wire. The grid may be cut from foil using accurate dies. A typical foil gauge grid is illustrated in Fig. 8.4. The large tabs at each turn of the conductor path reduce sensitivity to strains across the grid, also their large surface ensures that linear conditions exist over the complete active length of the grid. End effects are minimized, and the creep problem greatly reduced.

Semiconductor gauges have been introduced since 1970. They have very high sensitivity and gauge factors some 50 or 60 times greater than for wire or foil

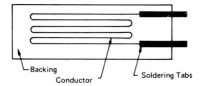

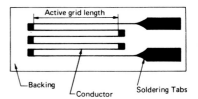

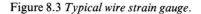

Figure 8.3 *Typical wire strain gauge.*   Figure 8.4 *Typical foil strain gauge.*

gauges. Semiconductor gauges consist of a strip conductor made from a single crystal of silicon or germanium that contains an accurately adjusted amount of impurity to give the characteristic desired.

Semiconductor gauges are more sensitive to temperature variations, and generally not so rugged as wire or foil gauges. They are more suitable for dynamic measurements but can be used for short-term measurements of static strain levels.

8.2.2 Gauge factor

Small changes in the gauge strength of a conductor caused by an applied load, induce changes in the resistance of the conductor (an effect first described by Lord Kelvin); these changes in gauge resistance are detected by the measuring instrumentation. The change in gauge resistance is related to the change in gauge length (strain) by the gauge factor k:

$$k = \frac{\delta R/R}{\delta L/L} = \frac{\delta R}{\varepsilon R} \tag{8.1}$$

where R is the gauge resistance, δR the change in gauge resistance, L the gauge length, δL the change in gauge length, and $\varepsilon\ (=\delta L/L)$ is the strain ($\mu\varepsilon$ = micro strain = $10^{-6}\varepsilon$). Thus with $k = 2$, $\varepsilon = 600 \times 10^{-6}$, $R = 350$ ohm the change of resistance would be 0.42 ohm.

One of the major factors that affects the performance and usefulness of any strain gauge is the material from which the conductor is made. Ideally the conductor should have a high gauge factor, so that small strains give as large changes as possible to the resistance. The specific resistance of the material should also be high, to give the biggest possible changes in resistance when strained, i.e. better resolution. These qualities make the strain gauge sensitive to small strains.

Further, multiples of a given load (strain) must give the same multiple of the resistance change, i.e. the gauge factor must be linear (as implied by equation (8.1)) and must not vary with the degree of loading. Similarly, the gauge factor should not vary with time so that repeated applications of a given load will always give the same resistance change.

8.3 Strain gauge selection

A representative range of available strain gauges is given in Table 8.1 [1].

Table 8.1 Details of a range of typical strain gauges

Gauge Series	DESCRIPTION AND PRIMARY APPLICATION	TEMPERATURE RANGE	STRAIN RANGE	FATIGUE LIFE	
				Strain Level in Microstrain	Number of Cycles
EA	Constantan foil in combination with a tough, flexible, polyimide backing. Wide range of options available. Primarily intended for general-purpose static and dynamic stress analysis. Not recommended for highest-accuracy transducers.	Normal: $-100°$ to $+350°$ F $(-75°$ to $+175°$ C) Special or Short-Term: $-320°$ to $+400°$ F $(-195°$ to $+205°$ C)	$\pm3\%$ for gauge lengths under 1/8 in (3.2 mm). $\pm5\%$ for 1/8 in and over.	±1800 ±1500 ±1200	10^5 10^6 10^8
CEA	Universal general-purpose strain gauge. Constantan grids completely encapsulated in polyimide, with large, integral, copper-coated terminals. Primarily used for general-purpose static and dynamic stress analysis. See p. 2L of Catalogue 200 for details. Fatigue life improved with use of low-modulus solder, e.g. M-Line 242-25S or 358-25S.	$-100°$ to $+400°$ F $(-75°$ to $+205°$ C) Stacked rosettes limited to $+125°$ F $(+50°$ C)	$\pm3\%$ for gauge lengths under 1/8 in (3.2 mm). $\pm5\%$ for 1/8 in and over.	±1500 ±1500 *Fatigue life improved using low-modulus solder.	10^5 10^6 *
MA	Open-faced constantan gauge on a thin special epoxy cast-film backing. Recommended for highest-accuracy transducers. Somewhat brittle nature of backing too delicate for general-purpose use. Only Option S is available.	Precision Static Transducer Service: $-100°$ to $+250°$ F $(-75°$ to $+120°$ C) Dynamic: $-320°$ to $+350°$ F $(-195°$ to $+175°$ C)	$\pm2\%$	±1900 ±1700 ±1500	10^5 10^6 10^7
ED	Isoelastic foil in combination with tough, flexible polyimide film. High gauge factor and extended fatigue life excellent for dynamic measurements. Not normally used in static measurements because of very high apparent-strain characteristic. Many options.	Dynamic: $-320°$ to $+400°$ F $(-195°$ to $+205°$ C)	$\pm2\%$ Nonlinear at strain levels over $\pm.5\%$	±2500 ±2200	10^6 10^7
WA	Fully encapsulated constantan gauges with high endurance leadwires. Useful over wider temperature ranges and in more extreme environments than EA Series. Option W available on some patterns, but restricts fatigue life to some extent.	Normal: $-100°$ to $+400°$ F $(-75°$ to $+205°$ C) Special or Short-Term: $-320°$ to $+500°$ F $(-195°$ to $+260°$ C)	$\pm2\%$	±2000 ±1800 ±1500	10^5 10^6 10^7
WK	Fully encapsulated K-alloy gauge with high endurance leadwires. Widest temperature range and most extreme environmental capability of any general-purpose gauge when self-temperature-compensation is required. Option W available on some patterns, but restricts both fatigue life and maximum operating temperature.	Normal: $-452°$ to $+550°$ F $(-269°$ to $+290°$ C) Special or Short-Term: $-452°$ to $+750°$ F $(-269°$ to $+400°$ C)	$\pm1.5\%$	±2400 ±2200 ±2000	10^6 10^7 10^8
EP	Special annealed constantan foil with tough, high-elongation polyimide backing. Used primarily for measurements of large post-yield strains. The 08 self-temperature-compensation value is the best compromise in the 06 and 13 range normally encountered. Available with Options E, S, SE, L, and LE.	$-100°$ to $+400°$ F $(-75°$ to $+205°$ C)	$\pm10\%$ for gauge lengths under 1/8 in (3.2 mm). $\pm20\%$ for 1/8 in and over.	±1000 EP gauges show zero shift under high cyclic strains.	10^4
SA	Fully encapsulated constantan gauges with solder dots. Same matrix as WA Series, but slightly thinner. Same uses as WA Series but derated somewhat in maximum temperature and operating environment because of solder dots. Often used in transducer service.	Normal: $-100°$ to $+400°$ F $(-75°$ to $+205°$ C) Special or Short-Term: $-320°$ to $+450°$ F $(-195°$ to $+230°$ C)	$\pm2\%$	±1800 ±1500	10^6 10^7
SK	Fully encapsulated K-alloy gauges with solder dots. Same uses as WK Series, but derated in maximum temperature and operating environment because of solder dots.	Normal: $-452°$ to $+450°$ F $(-269°$ to $+230°$ C) Special or Short-Term: $-452°$ to $+500°$ F $(-269°$ to $+260°$ C)	$\pm1.5\%$	±2200 ±2000	10^6 10^7
WD	Fully encapsulated Isoelastic gauges with high endurance leadwires. Used in wide-range dynamic strain measurement applications in severe environments.	Dynamic: $-320°$ to $+500°$ F $(-195°$ to $+260°$ C)	$\pm1.5\%$ — Nonlinear at strain levels over $\pm.5\%$	±3000 ±2500 ±2200	10^5 10^7 10^8
SD	Equivalent to WD Series, but with solder dots instead of leadwires.	Dynamic: $-320°$ to $+400°$ F $(-195°$ to $+205°$ C)	$\pm1.5\%$ See above note.	±2500 ±2200	10^6 10^7
TA	Open-faced constantan gauges on thin reinforced laminated backing. Primarily used in transducers where temperatures exceed capability of MA Series. Option S available.	$-100°$ to $+400°$ F $(-75°$ to $+205°$ C)	$\pm2\%$	±2000 ±1700	10^5 10^6
TK	Open-faced K-alloy gauges on a thin reinforced laminated backing for transducer use. Option S available.	$-452°$ to $+450°$ F $(-269°$ to $+230°$ C)	$\pm1.5\%$	±2000	10^8
TD	Open-faced Isoelastic gauges on a thin reinforced laminated backing for transducer use. Option S available.	Dynamic: $-320°$ to $+400°$ F $(-195°$ to $+205°$ C)	$\pm1.5\%$	±2400	10^7
VK VC VN	Open-faced gauges on a strippable backing. Normally installed with ceramic adhesives for special high-temperature and/or cryogenic measurements.	$-452°$ to over $+800°$ F $(-269°$ to $+425°$ C) depending on specific application	$\pm1\%$	----	----

8.4 Anisotropic etched pressure transducers

Traditionally, miniature silicon pressure transducers have been made similar to most large flat diaphragm strain gauge transducers. When pressure is applied to a flat diaphragm, bending stress is distributed over its surface, changing from compression in one area to tension in another. Strain gauges can be positioned on a diaphragm to provide increasing and decreasing resistance. For more than two generations four-arm Wheatstone bridge transducers have been made this way, by simply cementing gauges to one side of a metallic flat diaphragm. Now for almost a generation transducers have been made with piezoresistive gauges diffused into flat circular diaphragms of silicon, thus providing atomically bonded elements. By using silicon semiconductor technology, miniaturization was achieved. The process reduces the diameter of the circular diaphragms and permits extremely thin diaphragms.

A pressure transducer design can be made dramatically more efficient if the stresses can be concentrated at the locations where the strain gauges are placed. Stress should not be spread over a large area such as on a flat circular diaphragm. In general, if a structure is under bending stress, stress concentration can be achieved by varying the thickness of the bending element.

Diaphragms are shaped to concentrate stress using a combination of plate and beam theory. A circular diaphragm should have two thicknesses. It should be considerably thicker at the outer edge, and have two 'islands' in the middle section, as shown in Fig. 8.5. A distributed load, or pressure on one side results in stress concentrations at points A, B, and C where all the bending occurs. If a stress sensitive material is placed at A, B, and C, a sensor of improved efficiency results [2].

Diaphragms are fabricated from single-crystal silicon. This material is anisotropic, meaning that its physical properties vary according to the direction in which they are measured. Another characteristic of anisotropic materials is that their chemical reaction rates may vary according to crystalline directions.

When a silicon crystal is placed in a caustic bath such as hot hydrazine the material is etched faster in certain directions than others. By controlling etch bath parameters, the ratio of etching rates with direction can exceed 30 to 1. This results in precise control of shape.

The technique provides deep notches and flat bottoms to the thinned sections.

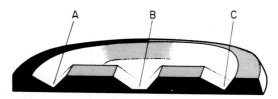

Figure 8.5 *Etched diaphragms: section through notches and islands.*

To obtain optimum transducer performance, the crystal must be correctly oriented, and then using a multiple step lithographic, etching and diffusing process, the part is fabricated. This technology is basic to the semiconductor industry. Varieties of etch patterns and contours can be achieved which permit the concept to be used over a wider range of pressures and sizes.

8.5 Strain gauge adhesives

Since a strain gauge relies on replicating the movement of the points to which it is attached it is a cardinal necessity for the contact adhesion to be faithful. The use of the correct adhesive is as critical as any other selection. Some typical adhesives are listed in Table 8.2.

8.6 Strain gauge (transducer) applications

Strain gauges are not limited to the measurement of strain alone. In fact, some of the most interesting applications of strain gauges have been as sensing elements in specialized transducers, most commonly for measurements of load, deflection, torque and pressure.

Most of the basic types of transducer are produced commercially. However, there are times when the desired characteristics are not available in a commercial device. Four specialized transducers that were designed and built at Batelle Memorial Institute are described here:

(1) Highly sensitive deflection device: this device was capable of sensing intervals of deflection as small as 25.4×10^{-6} mm (10^{-6} in). It was constructed by slitting a small portion of a spring steel cantilever beam from both sides toward the neutral axis, leaving a small amount of metal near the axis. On both sides, at the top of the tines, epoxy decks were cemented to provide a basis for bonding strain gauges. A four-arm bridge circuit was used. In operation, the tines spread apart under very light loads, providing very high amplification of the strains.

(2) Thin deflection device: this device provided a linear deflection–strain output. It was constructed of spring steel, 25.4 mm (1 in) diameter by 2.032 mm (0.080 in) thick and equipped with a full bridge composed of four 1.588 mm (1/16 in) gauge length bakelite gauges. Contact with the cantilever portion of the device was made through a small bearing peened into the end of the lever.

(3) Miniature pressure detector: this detector was used for insertion in the throat of a nozzle to indicate slight pressure variation. It employed a 0.0762 mm (0.003 in) thick beryllium–copper diaphragm as the elastic member. An active gauge was placed on the diaphragm, and a compensating

Table 8.2 Survey of adhesives used for mounting strain gauges

Cement Type	Organic							Ceramic			
	Evaporation of solvent at room temp	Meltable thermoplastic	Chemical Reaction					Evaporation hot curing	Chemical reaction hot curing	Molten spray	Very hot sinter
			Curing at room temperature			Hot curing					
Cement base	Nitro-Cellulose	Shellac	Acrylic	Polyester	Epoxy	Epoxy	Phenolic	Silicate	Phosphate	Refractory oxide	Glass
Representative example	Baldwin SR-4	DeKhotinsky	Eastman 910 Loctite IS-12B	Philips PR 9244	Araldite (various grades)	Araldite (various grades)	Bakelite Philips PR 9246	Baldwin RX-1	Brimor Cement	Rokide A Rokide C	Hommel L-6AC
Cure time (usually higher temp. gives accelerated cure)	2–24 h	Cooling time	5 min	5–15 min	12–72 h	2–10 h	1–6 h (room temp. cure takes 72 h)	1 h	1 h	None	30 min
Cure temperature	20–65°C 70–150°F	Melts at 130°C (275°F)	Room temp.	Room temp.	20–65°C 70–150°F	100–250°C 210–500°F	100–200°C 210–400°F	110°C 220°F	300–600°C 600–1100°F	None	1000°C 1800°F
Cure pressure	0.1–0.5 kp/cm² 1.4–7 lb/in²	0.1–1 kp/cm² 1–15 lb/in²	0.1–1 kp/cm² 1–15 lb/in²	0.5–1 kp/cm² 7–15 lb/in²	0.5–1 kp/cm² 7–15 lb/in²	0.5–3 kp/cm² 7–50 lb/in²	1–5 kp/cm² 15–75 lb/in²	None	None	None	None
Breaking strain of bond *	100000 µε	20000 µε	20000 µε	20000 µε	20000 µε	20000 µε	20000 µε	5000 µε	5000 µε	20000 µε	5000 µε
Used with strain gauge types	Paper backed	All types that can take melt temperature	All except paper	All types	All types	All types that can take cure temperature	Only types with phenolic backing	Stripped, and ceramic insulated	Stripped, and ceramic insulated	Stripped, and ceramic insulated	Stripped, and ceramic insulated
Electrical insulation	Excellent	Excellent	Excellent	Excellent	Excellent	Excellent	Excellent up to 150°C (300°F)	Decreases with higher temp.	Decreases with higher temp.	Excellent but decreases at higher temp	Excellent
Moisture resistance	Fair	Good	Fair	Fair	Good	Good	Good	Poor, dissolves in water	Fair	Good	Excellent
Remarks	Soluble in MEK and acetone, not suitable for plastics soluble in these	Glue stick melts on contact with heated test object	Has poor keeping qualities. Best for short term measurements	Has poor keeping qualities. Store at low temperatures	Two component cement, some hardeners are toxic	Two component cement, some hardeners are toxic	Has good keeping qualities	Reacts adversely with some metals and plastics	Reacts adversely with some metals and plastics	Porous	Suitable for all metals that can take cure temp.

* At temperatures substantially below 0°C the adhesive qualities of most organic cements are reduced

gauge attached to an unstrained coupon of beryllium–copper placed circumferentially above the diaphragm. This had essentially a linear strain response.

(4) Miniature surface-mountable pressure transducer: this was 6.35 mm (0.25 in) in diameter and only 0.508 mm (0.02 in) thick. The transducer had as a sensing element a specially designed, four-arm, printed-circuit type strain gauge. The transducer had low hysteresis, was temperature compensated, and provided excellent response to rapidly fluctuating pressures.

8.6.1 Load cells

The compressive forces applied to components can be measured by introducing load cells into the structure to which the load is applied. The load is carried by a cylindrical column of a special steel similar to EN24. During manufacture particular attention is paid to the final stress relieving and the mechanical ageing of the load bearing column. The strain gauges are bonded to the column under very carefully controlled conditions, using a special form of epoxy resin adhesive. The complete column is coated with special forms of non-hardening compounds to aid the water proofing of the load cell.

The assembly of the finished column in a standard compression cell utilizes 'O' ring seals, thus preventing the ingress of moisture or dirt. The outer case, base and top sealing plate of the load cell is normally steel. The load cell is of very robust construction with a typical general specification such as that in Table 8.3.

Table 8.3 Typical specification of a load cell

Accuracy	generally $\pm 0.25\%$ FLR*
Linearity	$\pm 0.2\%$ FLR*
Repeatability	$\pm 0.15\%$ FLR*
Maximum operating voltage	30 V AC or DC
Output	1 mV/V
Bridge impedance	120, 240, 350, etc. ohm
Temperature range	$-10\,°C$ to $+50\,°C$
Zero stability	$\pm 0.5\%$ FLR*
Deflection	approx. 0.2032 mm (0.008 in) (dependent on device)
Overload capability	100% FLR*
Electrical connection	4 core cable or to suit

*Full load rating.

8.6.2 Proof ring load transducer

Based on the traditional proof ring design the body of this load transducer [3] is a homogeneous elastic member machined all over from a grain orientated billet of EN26X high tensile steel. This ensures high linearity and low hysteresis (Fig. 8.6).

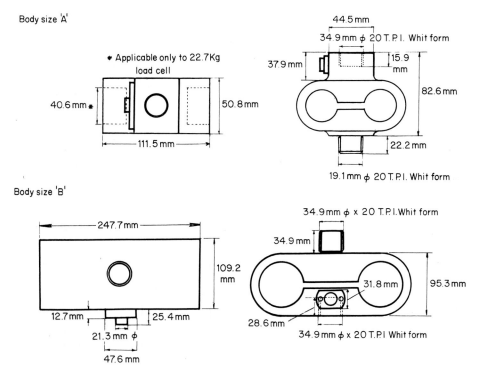

Figure 8.6 *Proof ring load transducer.*

Deformation of the ring is sensed by a miniature inductive displacement transducer with zero friction, infinite resolution and high electrical output. With a linear relationship between load and displacement it is possible to direct-read the applied load. This transducer can be used on both compressive and tensile modes under static or dynamic conditions.

8.6.3 Gauge pressure transducer

A diaphragm onto which resistors of controlled impurity are diffused on one side to form a complete Wheatstone bridge provides the pressure sensitive element of a typical Sangamo type SWC 792 transducer (Fig. 8.7). The use of a single crystal diaphragm and integrated circuit technology is well established and provides advantages of small dimensions, high linearity, freedom from creep with low hysteresis, high frequency response through low mass and minimal diaphragm deflection.

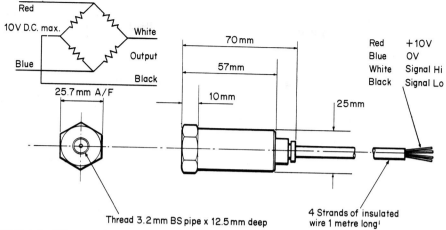

Figure 8.7 *Strain gauge pressure transducer.*

Pressures for which this transducer is calibrated range from 0 to 6.9 MN/mm² (1000 lbf/in²) with a maximum rating of 10.4 MN/mm² (1500 lbf/in²). Temperature compensation of 0.06% per °C is incorporated in the circuitry.

8.6.4 Peak strain indicator

Strain transducers developed by the British Hovercraft Corporation [4] are incorporated in the TM14A peak strain indicator. This is a self-contained unit, which displays as a steady reading on a moving coil meter. Readings are held until exceeded by a higher value or reset. Resetting is achieved manually, by depressing a panel mounted switch, or automatically by external closing contacts via a panel mounted socket. Front panel controls allow a large input zero imbalance to be nulled and overall sensitivity to be adjusted to give full scale deflection for a wide range of input levels.

The specification of this instrument is shown in Table 8.4.

Table 8.4 Specification for the TM14A peak strain indicator

Output display	Moving coil meter with mirror scale and knife edge pointer. Scale length 114 mm (4.5 in)
Peak response	Full reading of maximum load requires load duration of 1 ms
Zero unbalance	Zero control is capable of nulling bridge errors up to ± 2 mV/V
Decay time	Not worse than 1% in 10 min at 20 °C or 1% in 5 min at 30 °C
Reset	Panel switch or external contacts via socket on front panel
Dimensions	$304 \times 178 \times 178$ mm³ ($12 \times 7 \times 7$ in³) approx.
Accuracy	Better than 1.5% of full scale deflection
Bridge resistance	250–1000 ohm

8.6.5 Tourmaline underwater pressure transducers

Transducers which are suitable for the measurement of pressure waves in a liquid medium without disturbing the wave shape are built up from discs of piezoelectric tourmaline crystals [5]. These transducers are made up in single, double or quadruple element construction, each gauge pile being coated to seal the pile and maintain high insulation without detracting from the response of the gauge to pressure and frequency. A good quality antimicrophonic and anti-tuboelectric connecting cable minimizes cable signal effects. Each transducer is tested under both static and dynamic conditions by the British Calibration Service, Admiralty Weapons Research Establishment (AWRE), Foulness, Essex and a calibration certificate issued with each transducer gauge.

The general specification for each gauge is shown in Table 8.5.

Table 8.5 Specification for tourmaline underwater pressure transducer

Dynamic pressure range	0–20 000 psi 0–140 MN/m^2
Charge sensitivity	Single element, 72 pC/(MN/m^2) (0.5 pC/psi) Double element, 145 pC/(MN/m^2) (1 pC/psi) Four element, 290 pC/(MN/m^2) (2 pC/psi)
Rise time	This is dependent upon the media in which the shock wave is being measured (4μs approx. in water)
Insulation	10^{10} ohm
Capacitance (pile only)	Single element, 300 pF Double element, 600 pF Four element, 1200 pF

8.6.6 Strain gauge bridges

Very small changes in resistance are involved in the use of strain gauges. Such measurements cannot be made with a normal Wheatstone bridge with any degree of accuracy.

Typical of the accuracy needed by a strain gauge bridge: if it is required to measure a strain of 0.05% (500 microstrain) in a 100 ohm gauge with a gauge factor of 2.0, this would represent a resistance change of 0.1 ohm. If an accuracy of 1% is required, the measurement must be made to 0.001 ohm in 100 ohms. The switch contact resistance in a normal Wheatstone bridge is often greater than this.

To obtain accuracy in the special form of bridge used for strain gauge measurement a pair of ratios is used with a low-resistance slide-wire at their apex. The active and compensating gauges form the other two arms of the bridge as shown in Fig. 8.8 [6]. The slider is adjusted until balance is obtained and the percentage change of resistance is given directly by the slide wire reading. The indication is linear only for small changes but in strain measurements

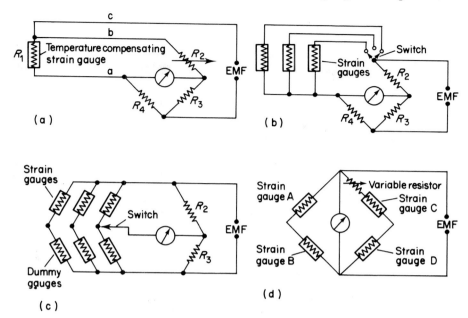

Figure 8.8 *Forms of strain gauge bridge: (a) three-lead temperature compensation; (b) multiple strain gauge circuit; (c) temperature-compensating multiple circuit; (d) four-gauge Wheatstone circuit.*

the upper limit of resistance change requred is either 0.5% or 1% and over this range the departure from linearity may be neglected, being 1/200th of the 1% reading and a correspondingly smaller fraction for smaller readings.

A second form of measuring device consists simply of a calibrated galvanometer or digital voltmeter which measures the out-of-balance voltage of the bridge which is a nearly linear function of the resistance change over a small range, inherent non-linearity being small and negligible.

8.6.7 Vibrating wire (acoustic) strain gauges

TRRL gauges were originally introduced at the Transport and Road Research Laboratory (TRRL) in 1958 as buried gauges for use in concrete roads [7]. Later, in 1968, the surface acoustic strain gauge was developed for use on steel and concrete bridge structures. Both gauges employ a single silver plated piano wire stretched between end anchorages fixed to the member under investigation. A change in the distance between the anchorages causes a corresponding strain change in the wire and a change in the frequency of vibration.

An electromagnet at the gauge centre plucks the wire and transmits the frequency of vibration to a recording device. From the change in frequency the strain change in the member can be deduced.

Usually a wire 0.254 mm (0.010 in) diameter is employed with a length of 139.7 mm (5.5 in) in the gauge although other sizes and lengths are occasionally used. The strain in the wire is given by the expression $\varepsilon = kf^2$, where f is the frequency of vibration. The constant k is the gauge factor (as discussed in Section 8.2.2) and is given approximately by the expression

$$k = \frac{4l^2\rho}{\varepsilon g} \qquad (8.2)$$

where l is the length of the vibrating wire, ρ the density of the wire, E the modulus of elasticity of the wire, and g the acceleration due to gravity ($= 9.81$ m/s^2).

In practice k is normally determined experimentally to allow for batch to batch variations in l, ρ and E. Tests on buried gauges in concrete carried out in 1968 [8] indicated that very approximately $\varepsilon = 3.0f^2 \times 10^{-9}$ for the standard wire length where f is in hertz.

A unique mechanical feature of the surface gauge is found in the stainless steel barrel which is designed to 'float' between the end blocks when the gauge is in position and serves only as a mounting for the plucking coil. Adjustment of the amount of float is made by means of the knurled tension adjuster, which can be locked by the locking ring shown in Fig. 8.9. Because of the length of the gauge it is most suited to use on full-scale structures.

In the twin-wired surface gauge a second wire has been added 6.35 mm (0.25 in) above the first and the end blocks raised the same amount. The wires are also displaced laterally by 12.7 mm (0.50 in) in order to enable the plucking coils to clear one another. Because of this the effect of a lateral bending moment (in the plane of the member surface) could influence the readings but the gauge has been specifically designed for use where this effect is small.

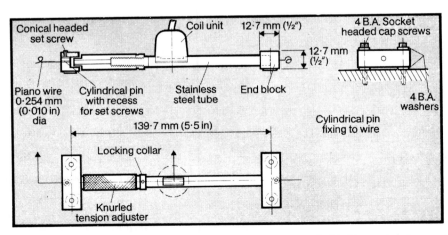

Figure 8.9 *TRRL vibrating wire (acoustic) strain gauge as embedded in steel and concrete bridge structures.*

8.6.8 Vibrating wire load cells

Compressive load cells in the form of annular rings are used to measure loads in structures such as in prestressing cables and anchor bolts. This type of load cell employs three or more vibrating wire elements to compensate for the effects of non-axial loading. The high degree of stability offered by the combination of vibrating wire element and heat treated load cell body enables small changes in load to be resolved against a background of high standing load. Temperature corrections are unnecessary since the differential temperature coefficient between the steel wire and the cell body is virtually zero.

When a load is applied to the cell the frequency of the vibrating wire changes. In the simplest system the frequency values are referred to sets of tables and to the individual calibration data supplied with each load cell to obtain the value of load. Direct reading systems are also available. Here load readings are calibrated directly in the units required on the scale of a meter. The system also incorporates an overload alarm which can be set to operate at any level of load. This system was designed for use in an elevator used for lifting small marine craft out of the water in a dockyard.

A small annular load cell measuring approximately 5 cm in diameter and 5 cm long has been introduced for measuring the tension in bolts of up to 20 mm diameter. The load cell measures loads up to 250 kN and has been used to measure the tension in bolts securing the segments of the tunnel lining in an extension of the London Underground system.

To extend the application of load cells to the field of industrial weighing, a new computing readout system has been developed. This equipment presents the load output in the required units on a digital readout system, after processing the signal from the load cells and carrying out the mathematical conversion in a microprocessor within the instrument.

8.7 Structural instrumentation

Stresses and strains of the type encountered in civil engineering structures vary slowly. High stability is needed for the monitoring of such stresses and strains.

Transducers operating on the vibrating wire principle [9] possess the required long-term stability and are therefore preferred for structural instrumentation applications. A variety of measuring devices based on this principle have been introduced in recent years. They are incorporated during the construction phase or fitted after the structure has been put into service, and are used both to confirm the assumed design data and also to give warnings of incipient failure.

Gauges are available for the measurement of surface and internal strains in concrete and steel. Gauges of this type are now being used extensively in concrete and steel box girder bridges, flyovers, motorways, strong floors, roof struts, concrete dams, pressure vessels and silos, turbine rafts, cast *in situ* piles, retaining walls, tunnel linings, and gas and fuel pipes. Load cells are also supplied, in the majority of cases being made specifically for the required application.

Vibrating wire strain gauges of various designs have been developed, some of which are described in Table 8.6.

Table 8.6 Vibrating wire strain gauges

Gauge type	Size	Applications
TSR/1 TES/2	25 mm 51 mm	Gauge lengths vary from 25 mm (1 in) to 203 mm (8 in). The shorter gauges are useful in model work; the longer ones can be used to measure strains across material interfaces such as mortar joints
TES/5.5 TSR/5.5	140 mm 140 mm	The gauge length most frequently used for surface and internal strain measurement, each is available with a temperature sensor. These guages will measure strain variations of more than 2500 microstrain. Embedment gauges of this type were installed in the columns supporting the elevated carriageway of the Adam Bridge, A27 Shoreham By-Pass, UK. These gauges are of a pattern designed by the Transport and Road Research Laboratory, England
TW/GT	140 mm	A high-temperature embedment gauge, using mineral insulated cable, allows strain measurements to be made in concrete at temperatures up to 200 °C over a range exceeding 3000 microstrain
TSRT/5.5	140 mm	To measure bending at the surface of a structure a special surface mounting gauge uses two gauges, one above the other. Being at different radii from the centre of bending their strain response differs in proportion, enabling the amount of bending to be calculated

8.7.1 Vibrating wire strain gauges applied to box girder bridges

Recommendations of the Merrison Committee Interim Report on the design and method of construction of box girder bridges have led to the use of vibrating wire gauges. These are based on a design by the Transport and Road Research Laboratory. These gauges accurately measure the strains in sections of the bridge, and the extent to which these strains vary when the bridge is hydraulically jacked up at the ends [10].

The strain distribution is measured by attaching strain gauges to predetermined points on the box girder sections. The electrical leads are then marshalled at a convenient measuring point, where they are coupled into portable automatic data logging equipment.

The gauge is fixed to the surface under investigation by means of quick setting adhesive; it can also be mounted by means of bolt attachment or arc welding to steelwork. In operation, the gauge uses a high tensile steel wire, in tension between the two end mounting blocks, to sense the variation in surface strain over the gauge length. The strain variation develops a corresponding change in tension in the wire. A plucking coil is mounted in the protective enclosing tube surrounding the wire. A current pulse fed to the coil shock-excites the wire, which then oscillates at a frequency determined by the wire tension.

Variations in strain are thus converted to changes in frequency of oscillation of the wire, observations of which are made by measuring the output from the coil, which now acts as a pickup device. A square law relationship exists between strain change and the observed frequency change (see next section). Very small strains (0.5 microstrain) can readily be measured whilst at the other extreme, the overall strain range measuring capability is about 3000 microstrain.

A portable strain unit consists basically of an electronic period counter driven by a 100 kilohertz crystal oscillator and incorporates circuits for energizing the gauges. Silicon transistors and integrated circuits are used thoughout for greater reliability and the equipment is driven from internal batteries; it is fully portable for field and site work.

8.7.2 Load/stress computer

Direct values of load or stress derived from the changes in strain registered by vibrating wire gauges are read out on the GT 1177 unit [11]. Calculations on the transducer input signal are performed in this unit by means of a microprocessor and associated circuitry within the instrument.

Vibrating wire transducers develop an output signal which is a frequency analogue of the mechanical input parameter. In most types of vibrating wire transducer the mechanical input is used to develop a strain, and the relationship between frequency and strain is a square law given by

$$\Delta\varepsilon = k(f_0{}^2 - f^2) \tag{8.3}$$

where $\Delta\varepsilon$ is the change in strain in microstrain (i.e. strain $\times$ 10^{-6}), k is the gauge or cell factor in microstrain per frequency-squared, f_0 is the initial or datum output frequency from the transducer, and f is the final output frequency from the transducer.

In many cases, improved measuring resolution is obtained if the period of wire vibration (i.e. the reciprocal of frequency) is measured. In such instances, the period of one hundred cycles (T) of wire vibration would be recorded, and in a typical measuring system the least significant digit of the display would be $10\,\mu s$ ($10^{-5}s$). The square-law relationship between strain and the displayed period T would then be:

$$\Delta\varepsilon = k \left[\left(\frac{1}{T_0} \right)^2 - \left(\frac{1}{T} \right)^2 \right] \times 10^{14} \text{ microstrain} \tag{8.4}$$

The known constants of the transducer such as k, T_0, etc. are first set by means of thumbwheel and rotary switches on the front panel of the instrument. Such zero datum readings can themselves by established by the computing readout in an initial zero load test run. The instrument then automatically plucks the gauge and processes the data in accordance with the corresponding square-law relationship.

The result is then indicated as a direct reading on a digital display in the preselected units appropriate to the measurement being made.

8.8 Interpretation of strain gauge readings

The position in which a strain gauge is placed in relation to the (local) direction of the applied strains can influence the values of the readings and the deductions (particularly in relation to stress) which are made from them.

At any one point in a material, stress may occur in three planes. At the same point, the resulting strain can be the consequence of the direct stresses, the associated shear stress and such influences as the transverse strain (Poisson's ratio).

For this reason, a recognition of the influence of these factors, particularly as related to what is known as *principal stress* is important.

8.8.1   Principal stresses: simple tension and pure shear

Simple tension acting on a bar as in Fig. 8.10, produces a complex tensile and shear stress system across a plane AC (Fig. 8.11) such that

$$\sigma_\theta = \sigma \cos^2\theta \qquad (8.5)$$

$$\tau_\theta = \tfrac{1}{2}\sigma \sin 2\theta \qquad (8.6)$$

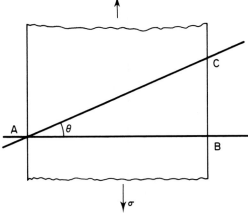

Figure 8.10 *Stresses: applied and internal (planar)*.

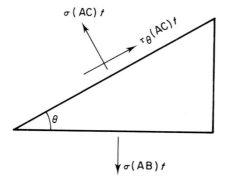

Figure 8.11 *Forces due to stress on planes for a material of thickness t.*

This shows that the shear stress τ_θ is maximum when $\theta = 45°$ (when $\tau_{\theta,max} = \frac{1}{2}\sigma$) and accounts for the critical 45° shear failure plane in ductile materials.

Pure shear stress acting on a bar as in Fig. 8.12 produces a complementary shear stress on the adjoining plane. It can be shown that with pure shear stress on two planes at right angles (Fig. 8.12(b)), the action on an element at 45° is one of equal tension and one of equal compression (Fig. 8.12(c)).

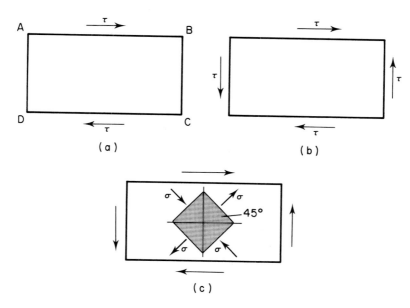

Figure 8.12 *(a) Pure shear stress acting on a bar;*
(b) complementary shear stresses;
(c) compressive and tensile stresses.

8.8.2 Principal stresses: two-dimensional stress system

If a bar is acted upon by both tensile stresses and a shear stress (Fig. 8.13), then

$$\sigma_\theta = \tfrac{1}{2}(\sigma_y + \sigma_x) + \tfrac{1}{2}(\sigma_y - \sigma_x)\cos 2\theta + \tau \sin 2\theta \qquad (8.7)$$

$$\tau_\theta = \tfrac{1}{2}(\sigma_y - \sigma_x)\sin 2\theta - \tau \cos 2\theta \qquad (8.8)$$

The normal stress σ_θ will be maximum when $\tan \theta = 2\tau/(\sigma_y - \sigma_x)$ and this is known as the *principal plane* and the maximum value of σ_θ is the *principal stress* which has a value of:

$$\sigma_{\theta,\max} = \tfrac{1}{2}(\sigma_y + \sigma_x) + \tfrac{1}{2}[(\sigma_y - \sigma_x)^2 + 4\tau^2]^{1/2} \qquad (8.9)$$

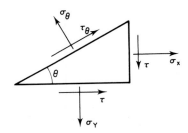

Figure 8.13 *Material under two tensile stresses and one shear stress.*

(there is a maximum value of σ_θ which is the second principal stress and represents a compressive stress maximum given by

$$\sigma_{\min} = \tfrac{1}{2}(\sigma_y + \sigma_x) - \tfrac{1}{2}[(\sigma_y - \sigma_x)^2 + 4\tau^2]^{1/2} \qquad (8.10)$$

The shear stress under these circumstances is at an angle of 45° to the principal stresses and is given by

$$\tau_{\theta,\max} = \tfrac{1}{2}[(\sigma_y - \sigma_x)^2 + 4\tau^2]^{1/2} \qquad (8.11)$$

Readers may be familiar with Mohr's circle for evaluating stresses. As an example of principal stress magnitudes, a beam under a bending load may have a bending stress of 50 N/mm² (7250) lbf/in²), shear stress 20 N/mm² (2900 lbf/in²) producing principal stresses of 57 N/mm² (8265 lbf/in²) on a plane at angle $\theta = 70° \, 40'$ and 7 N/mm² (1015 lbf/in²) on a plane at angle $\theta = 160° \, 40'$ and a maximum shear stress of 32 N/mm² (4640 lbf/in²) on a plane at angle $\theta = 25° \, 40'$.

8.9 Strain gauge rosettes

Normally, strain gauges are cemented to the surface of the part being tested (one exception is when gauges are used for measurements on concrete). Stress at a surface cannot act perpendicular to the surface plane so effectively the gauge is measuring a two-dimensional strain system. All the terms for one of the axes (for example z) can be eliminated so that the equations for principal stress can be rewritten as follows:

$$\sigma_x = \frac{E}{1 - \mu^2}(\varepsilon_x + \mu\varepsilon_y)$$

$$\sigma_y = \frac{E}{1 - \mu^2}(\varepsilon_y + \mu\varepsilon_x)$$

(8.12)

Often the directions of the principal stresses are not known in the practical strains that refer to arbitrarily positioned axes as shown below:

$$\varepsilon_\alpha = \varepsilon_x \cos^2\alpha + \varepsilon_y \sin^2\alpha + \beta_{xy} \sin \alpha \cos \alpha$$

(8.13)

The equation defines the strain ε_α at a point, where α is the angle that the strain makes with the x axis in arbitrary $x - y$ axes. The strain ε_α can be measured by a strain gauge, and when measurements are made at three different angles to obtain three values for ε_α and α, three equations can be solved simultaneously to give ε_x (the strain in direction x), ε_y (the strain in direction y), and β_{xy} (the shear strain present at the point). Further, when the values obtained are substituted into:

$$\tan 2\alpha_p = \frac{\beta_{xy}}{\varepsilon_x - \varepsilon_y}$$

(8.14)

the angle α_p that the principal planes make with the arbitrary axes can be determined. Hence, when α_p is substituted back into equation (8.13) the principal strains and shearing strain at the measuring point can be found and substituted into equation (8.12) to give the stresses acting at the point.

When the directions of the principal axes are not known three gauges can be used to give the necessary information. However, three gauges give almost three times as much labour in preparation and mounting, assuming that there is enough room for them, and a deal of extra calculation, especially when the gauges have not been mounted at an 'easy' angle from each other.

Strain rosettes avoid three complications, they are often used where two, three or four measuring grids are mounted on the same backing so that they can all be cemented at the same time onto the test specimen in one easy operation. Strain rosettes have standard, accurately determined 45°, 60°, or 90° angles between the different grids to help simplify calculations. Charts or calculators are often available for use with standard rosette types, so that strain and the direction of the principal planes can either be read directly, or found with a very minimum of simple arithmetic. Fig. 8.14 shows some of the more common strain rosette arrangements.

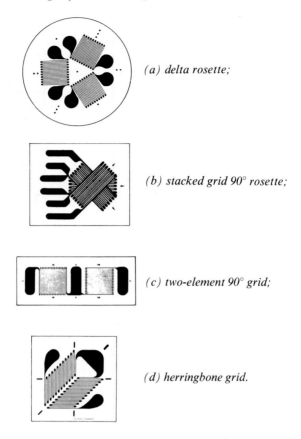

(a) delta rosette;

(b) stacked grid 90° rosette;

(c) two-element 90° grid;

(d) herringbone grid.

Figure 8.14 *Common strain rosette arrangements.*

The delta rosette in Fig. 8.14(a) is a very common type used to determine the directions of the principal axes, and the principal strains acting on them. A rosette for similar applications is shown in Fig. 8.14(b), but in this case the gauge grids have been stacked in a sandwich arrangement to save space, and to give a closer approximation to a point strain measurement. When this type of rosette is used with higher excitation voltages, heat dissipation can be a problem because of the close proximity of the grids to each other.

The two-element gauge shown in Fig. 8.14(c) has its grids arranged at 90° to each other. This arrangement can be used to augment the measured change in resistance, i.e. give a larger gauge factor, when the two grids are connected into adjacent arms of a measuring bridge.

Fig. 8.14(d) shows a two-guage rosette with grids arranged in herringbone pattern. This type of gauge, and similar arrangements incorporating four grids at 90° to each other are frequently employed for measuring torsional strains on axles and shafts.

8.10 Strain gauge sandwich

Bending strains on plates and panels can be measured by using a 'Flexagage' of sandwich construction manufactured by the Budd Company [12].

Two measuring grids are mounted, one on each surface of a thick backing piece with an accurately known thickness of material between them. When the assembly is cemented to the test panel one gauge is in contact with the panel surface and measures the surface strain. From the measurement obtained by the other grid, it is possible to deduce the position of the neutral axis, the amount of bending, and the linear strain present. Fig. 8.15 illustrates the principle.

A simple geometry gives the position of the neutral axis, while the difference between the strains projected onto the two surfaces of the panel is a measure of the linear strain present. The algebraic average of the projected surface strains gives the bending strain at the panel surface.

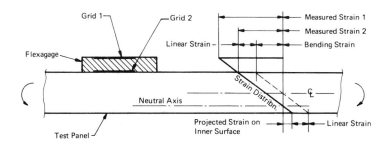

Figure 8.15 *Principle of 'Flexagage' sandwich construction.*

8.11 Thermoelastic stress analysis

Thermoelastic stress analysis (TSA) is a new non-contacting method for determining the surface stress distribution in a structure subjected to dynamic loading. Work on the method, formerly known as stress pattern analysis by thermal emission (SPATE), was funded by the Admiralty Surface Weapons Establishment, Portsmouth, UK.

TSA is based on the thermodynamic property (well known in gases), which gives rise to heating or cooling under conditions of adiabatic compression or expansion. To a much lesser extent the same phenomenon occurs in solids. Since stress is not equal throughout, as is pressure in a gas, temperature changes are localized. Such temperature changes, may amount to only a few thousandths of a degree and are dependent on the level of stress at each particular point.

Resolution and stress sensitivity of an instrument based on the method are comparable with those of strain gauge techniques, but the time taken to identify and, if desired, record areas of high stress and therefore potential structural weakness is much less than with conventional methods. Additional features of instrumentation for TSA are that it can be designed to operate remotely, and can produce a rapid display of the stress distribution over a selected viewing area of the structure. These features can be used to advantage in the fast location of areas of stress concentration which may be due to faults in design, faulty manufacture or material fatigue.

8.12 Ultrasonic goniometry: surface stress measurement

Changes in ultrasonic velocity may be used to measure surface stresses, the procedures being based on either the direct measurement of ultrasonic velocity or on critical angle phenomena [13–15]. Goniometers suitable for measuring critical angle phenomena require some means of measuring reflected energy as a function of varying angles of incidence and reflection [16–21].

Surface stresses in steel objects may be measured by use of a goniometer developed by Andrews and Keightley at the British Steel Corporation Sheffield Laboratories [22]. A simple reflection principle is used in which an ultrasonic beam in water impinges at an angle to the flat surface of a piece of steel. The reflected beam at the same angle varies in intensity according to the amount of energy transmitted through the interface into longitudinal waves (and shear waves, if the angle of incidence is small enough). These relationships are illustrated in Fig. 8.16. As θ_{1L} (incident wave in liquid 1) is increased, θ_{2L} and θ_{2S} increase until firstly θ_{2L} reaches $90°$ at the first critical angle and then θ_{2S} at

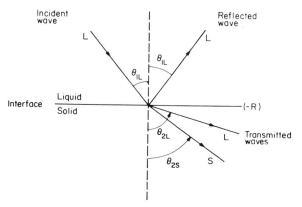

Figure 8.16 *Reflection and transmission through a plane boundary.*

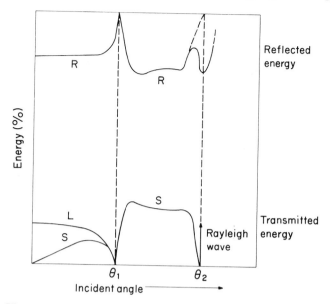

Figure 8.17 *Typical energy relations at a liquid/solid boundary.*

the second critical angle. Meanwhile the reflected longitudinal ray in the liquid makes the angle θ_{1L} with the normal. If this medium is a solid there is also a reflected shear wave at θ_{1S}.

As the angle θ_{1L} is increased, the energy reflected into these various waves follows the pattern idealized in Fig. 8.17. When the first critical angle is reached (θ_1) the longitudinal wave (L) has $\theta_{2L} = 90°$. The shear wave has zero intensity also at this angle but then reappears until it vanishes at its own critical angle $\theta_{2S} = 90°$.

The instrument is held together in a rigid framework comprising a circular base plate onto which is welded a vertical main plate and perpendicular to it at the rear, two strengthening members. The circular holes in these plates serve to reduce the overall weight. They support the top platform on which there is mounted a motor drive and gearbox to operate the vertical lead screw. There is also a revolution counter which records the angular rotation of the screw and the reading of this counter can be calibrated easily as a function of the angle of incidence in Fig. 8.18.

Two ultrasonic probes act as transmitter and receiver and have their axes along the lines of the tangents to the two circles. These transducers are in the frequency range 1–10 MHz and have diameters between 5 and 15 mm.

In operation the instrument must be immersed in water (or other liquid) sufficiently to cover the lower ends of the transducer and receiver. The lower part of the instrument thus has to be protected against corrosion.

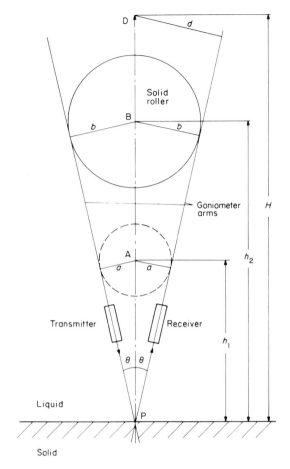

Figure 8.18 *Geometrical principles of a transportable goniometer.*

References and list of companies

1 'Strain gauge selection, criteria, procedures, recommendations', *Bulletin* TN-132, Welwyn Strain Measurement Ltd, Armstrong Road, Basingstoke, Hants, RG24 0QA, UK (Tel. 0256-62131).
2 Rogers, L., 'Applications of thermography in the steel industry', *Eurotest Technical Bulletin* E35/36 Rue du Commerce 20-22-Bte 7, B-1040 Brussels, Belgium.
3 Sangamo Weston Controls Ltd, North Bersted, Bognor Regis, Sussex, PO22 9BS, UK.
4 Experimental & Electronic Laboratories, British Hovercraft Corporation, East Cowes, Isle of Wight, UK (Tel. 0983-4121).
5 The Meclec Company, No. 7 Unit, Star Lane, Great Wakering, Essex, SS3 0PJ, UK.
6 'Methods of measurement of strain', *Bulletin* S.G.3, Tinsley Telecon Ltd, South Norwood, London SE25.

7   Potocki, F. P. (1958), 'Vibrating-wire strain gauge for long-term internal measurement in concrete', *The Engineer,* **206** (5369), 964–7.

8   Tyler, R. G. (1968), Developments in the measurement of strain and stresses in concrete bridge structures'. *Ministry of Transport, Road Research Laboratory Report,* No. 189, Crowthorne.

9   Gage Technique Ltd, PO Box 30, Trowbridge, Wilts, UK (Tel. 02214–61652).

10 (1972), 'Gauges for box garden tests', *Construction Steelwork,* January, Portal Press Ltd, 72 London Road, Croydon, Surrey.

11 Gage Technique Ltd, PO Box 30, Trowbridge, Wilts, UK (Tel. 02214–61652).

12 Vaughan, J. (1975), 'Strain measurement', October, Bruel & Kjaer DK-2850 Naerum, Denmark (Tel. 02-80-05-00).

13 Crecraft, D. L. (1967), *J. Sound Vib.,* **5**, No. 1, 173–92.

14 Elian, H. A. (1960), *Non-destructive Testing,* **18**, No. 3, 180.

15 Nononha, P. J. and Wert, J. J. (1975), *Testing and Evaluation,* **3**, No. 3 147–52.

16 Bradfield, G. *Notes on Applied Science* No. 30, National Physical Laboratory, London, HMSO.

17 Bradfield, G., (1968), *Non-destructive Testing,* **2**, 165–72.

18 Rollins, F. R., (1966), *Materials Evaluation,* **24**, 683–9.

19 Curtis, G. J., private communication.

20 Fountain, L. S. (1967), *Journal American Strain Association,* **42**, No. 1, 242–7.

21 Lambert, A. (1973), *Proc. Congress Mesucora,* 7.10–7.19.

22 Andrews, K. W. and Keightley, R. L. (1978), 'An ultrasonic goniometer for surface stress measurement', *Ultrasonics,* September, 205–9.

9
Corrosion monitoring

9.1 Introduction

Corrosion monitoring is the systematic measurement of the corrosion or degradation of a component with the aim of obtaining information for use in its control. Techniques may have two sources:

(1) Plant inspection methods;
(2) Corrosion-testing measurements.

Experience shows that attention to corrosion monitoring can have four benefits:

(1) Effective scheduling of maintenance work.
(2) Reduction of shutdowns to facilitate inspection.
(3) Prevention of shutdowns due to unforeseen deterioration.
(4) Supply data for process monitoring such as to reduce corrosive activity.

Corrosion occurs everywhere, all the time. Corrosion monitoring is important to all activities whether high or low in technology. It is therefore difficult to put a value on the benefits of monitoring prior to a failure but if the possible options are costed, then the management decision to either install or not install a monitoring scheme would be based on the best available technical and economic information. In general the cost of installing a suitable monitoring procedure is marginal compared to the total capital cost of the plant or the cost of an annual shutdown; also the effort often involved in a failure investigation would, in hindsight, have been better used at the design and construction stage in ensuring future maintainability.

9.2 Corrosion costs

Rust and other forms of corrosion destroy materials so that corrosion is a direct loss of fixed capital: in the USA it was estimated that 40% of all steel is produced to replace steel lost by corrosion.

Arguments as to the costs of corrosion do not stop at the cost of replacing the corroded article but must cover the whole life cycle cost. Thus it costs US

$12 400 a day (1978 costs) to take a typical 400 kV transmission line out of service to deal with corrosion damage. Records for a new 2000 MW power station showed that over an 18 month period corrosion deterioration cost £1.9 M.

Nationally, in the UK corrosion in 1971 was estimated to have cost £1365 M (equivalent in 1981 to £5000 M). Dr Atkinson of Queen's University (Canada) claimed in 1975 that Canada wasted $1500 M in corrosion costs.

9.3 Corrosion: electrochemical process

Excluding the deteriorating effects of high-temperature oxidation or sulphidation, fluid-wetted metals experience an electrochemical process. This process has three essential constituents:

(1) An anode, the site which loses metal.
(2) A cathode, which is not consumed.
(3) An electrolyte, the corrosive medium in which the process develops.

Effectively, the presence of an anode, a cathode and an electrolyte forms a simple electrical cell (Fig. 9.1). At the anode corroding metal passes into the electrolyte as positively charged ions. Electrons are released and participate in the cathodic reaction. A corrosion current is thereby set up.

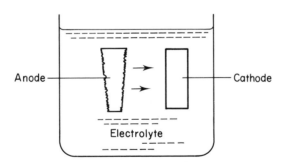

Figure 9.1 *Simple corrosion cell.*

9.3.1 Corrosion cells

The whole surface of a component may become the anode, and the surface of another component in contact with it the cathode. Usually, however, corrosion cells will be much smaller and more numerous, occurring at different points on the surface of the same piece of metal (Fig. 9.2). For example, anodes or cathodes may arise from differences in the constituent phases of the metal itself, from variations in natural or protective coatings on the metal, or from variations in the electrolyte (caused, for instance, by differential aeration).

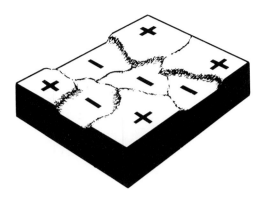

Figure 9.2 *Surface corrosion: anode/cathode areas.*

The electrolyte may be in bulk form (as when a component is immersed in or is handling a liquid during service) or it may be present only as a thin condensed or adsorbed film on the metal surface. The rate of corrosion is influenced considerably by the electrical conductivity of the electrolyte: it is lowest in poor conductors (e.g. high-purity water) and highest in good conductors (e.g. salt or acid solutions).

The reactions occurring in a typical cell are shown in Fig. 9.3. Effectively this means that a metal is reduced by corrosion to some form of oxide.

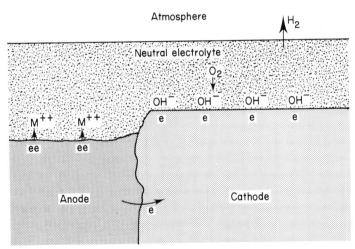

Figure 9.3 *Reactions in a typical corrosion cell.*

9.3.2 Anode: corrosion resistance

Intrinsic resistance to corrosion of a metal depends upon such factors as:

(1) Position in the electrochemical series.
(2) Physical properties of its surface film: adherence, compactness and self-healing characteristics.

Within the precedence of the electrochemical series 'base' metals corrode easily while 'noble' materials are more resistant. This is shown by Table 9.1 with magnesium as the lowest 'base' material and platinum slightly higher than gold as the highest 'noble' material. Surface films are an oxide of the metal and have a protective influence; they behave like a noble metal.

Table 9.1 Electrochemical (galvanic) series

Base	Magnesium		
	Zinc		
	Aluminium	*(commercial)*	High potential
	Cadmium		
	Duralumin	*(Al with $4\frac{1}{2}$% Cu)*	
	Mild steel		
	Cast iron		
	Stainless steel	*(type 430; 18% Cr)*	ACTIVE
	Stainless steel	*(type 304; 18% Cr 10% Ni)*	ACTIVE
	Lead-tin solders		
	Lead		
	Tin		
	Nickel		
	Brasses		
	Copper		
	Bronze		
	Monel		
	Silver solders	*(70% Ag. 30% Cu)*	
	Nickel		PASSIVE
	Stainless steel	*(type 430)*	PASSIVE
	Stainless steel	*(type 304)*	PASSIVE
	Silver		
	Titanium		
	Graphite	*(carbon) (non-metal)*	
	Gold		
Noble	Platinum		Low potential

9.4 Rust

Rusting of automobiles is one of the typically insidious consequences of the 'red marauder' – the corrosion of iron and steel to form rust. Commercially, the rusting of buried pipes, steel structures etc. is of considerable concern.

Rusting causes iron (Fe) to form a ferrous oxide such as $FeO(OH)$ or $Fe_2O_3 . H_2O$ or Fe_3O_4. Ordinary rusting is a chemical reaction called the 'oxygen-absorption corrosion cycle' which requires only water and oxygen in contact with iron to form rust. This cycle is shown in Fig. 9.4. Cancer in car

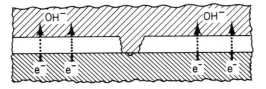

Cathode reaction $2e^- + \frac{1}{2}O_2 + H_2O \rightarrow 2OH^-$

At one or more points the electrons eventually find their way back to the electrolyte. This can happen as illustrated in the box, *It Starts with a Potential Difference*. In some cases, however, the electrons migrate directly through solid layers of coatings. At this point, electrons react with the water and oxygen to form hydroxyl ions.

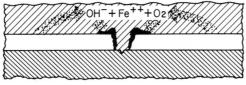

Precipitation of rust as FeO(OH), $Fe_2O_3 \cdot H_2O$, or Fe_3O_4

The circuit is completed as the ions in the electrolyte unite. Through a series of chemical reactions, rust is formed and precipitates back to the surface.

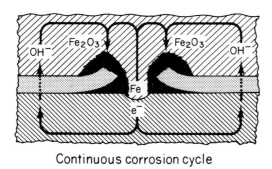

Continuous corrosion cycle

Now the cycle can continue until all the iron is consumed. In the early stages, rust spreads beneath the paint and causes it to blister. This reaction often takes place in unseen areas and is undetected until the process has worked completely through the section.

Figure 9.4 *Rusting: the oxygen-absorption corrosion cycle.*

bodies results from the presence of water either from the atmosphere or trapped in mud and dirt; heat and salt expedite its electrolytic effect.

Steel mills expend a great deal of energy driving off oxygen in the steel making process. As a result, a large amount of potential energy is stored in the steel, and the iron is thermodynamically 'anxious' to return to iron oxide. Potential effects are shown in Fig. 9.7, pp. 318/9.

9.4.1 Influence of water

All waters which come in contact with metals in industrial and other processes are ultimately derived either from sea water or rain water. They can vary greatly in chemical composition. In the case of sea water the important factors for corrosion are the degree of pollution and/or dilution caused by flow from rivers, and the dissolved-oxygen content. The composition of rain water depends on the composition of the soils through or over which it passes before draining into the river, lake or reservoir from which it is taken. For example, the calcium carbonate content of hard water can be 20 times greater than that of soft water.

Iron and carbon steel are the commonest materials of construction. When unalloyed steel corrodes, the corrosion rate is usually governed by the cathodic reaction of the corrosion process, and oxygen is an important factor. In neutral waters free from dissolved oxygen, corrosion is therefore usually negligible. The presence of dissolved oxygen in the water accelerates the cathodic reaction; and consequently the corrosion rate increases in proportion to the amount of oxygen available for diffusion to the cathode. Where oxygen diffusion is the controlling factor, the corrosion rate tends to increase also with rise in temperature. In acid waters (pH 4), corrosion can occur even in the absence of dissolved oxygen, since the evolution of hydrogen provides an alternative cathodic reaction. In most cases, the waters with which we are concerned will not be sufficiently acid for this type of corrosion reaction to occur, though the presence of dissolved CO_2 moves the pH towards the acid region (Figs 9.5 and 9.6).

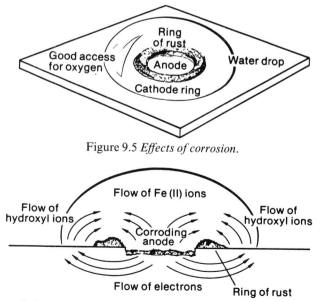

Figure 9.5 *Effects of corrosion.*

Figure 9.6 *Movement of electrons and ions during the corrosion process.*

It Starts with a Potential Difference

The rusting mechanism takes place between two points: an anode—where the rust forms—and a cathode separated from the anode by an electrical "potential" difference. But specific locations are often determined by design, environment, or internal metallurgical variations. Here are typical conditions that create the electrical potential difference between the anodic and cathodic reactions. These examples apply to unprotected as well as coated steels.

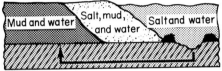

A potential difference is established between an oxygen-starved surface and a surface where oxygen is plentiful. Rusting, for example, begins at the anode formed in the centre of a droplet or a stagnant puddle of water where oxygen is scarce. Cathodic reaction takes place at the outer edges where oxygen is available.

Gradients in electrolyte composition are probably the most common cause of rust. Pipes in soil, for example, passing through strata with varying salt content, and car fenders collecting clumps of mud with salt and water of varying concentration are conditions that encourage very rapid rusting.

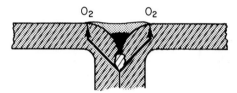

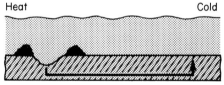

Crevice corrosion is caused by an oxygen gradient between the electrolyte at the surface and the oxygen-starved electrolyte at the bottom of the crevice. This condition is typical of weldments, sheet-metal joints, and rough surfaces where water may be trapped, and also explains why a rough microfinished surface rusts faster than a smooth surface.

Temperature graduations, variations in illumination, and even agitation of the electrolyte set up strong potentials and encourage rusting. For example, appliance parts that are partially heated and partially cooled, or exposed to stagnant water at one end and agitated water at the other are susceptible to corrosion.

Figure 9.7

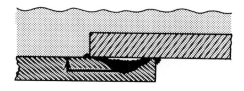

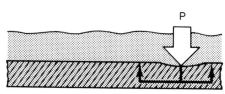

A less obvious form of crevice corrosion takes place between tightly sealed joints where the concealed metal surface is oxygen starved. Electrolyte may seep between irregularities in the mating surfaces or may be carried by moisture-bearing materials (such as wood) in contact with the steel.

Localized stresses in steel change electrical potential. Generally, a point of applied stress becomes an anode. This phenomenon is especially critical in structural parts because the metal degenerates where strength is most needed.

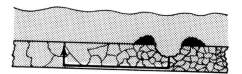

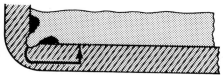

Steel can also rust all by itself—due to internal metallurgical differences. The microscopic grains in a piece of steel are comprised of a number of crystalline structures, and a potential difference can exist between them. Even variations in grain structure or heat treatment within a metal can set up potentials that drive rusting reactions.

Even localized residual stresses provide anode points: formed or bent steel parts bearing residual stresses near the point of deformation, surface-hardened or conversion-coated parts with compressive residual stresses at the surface, and even welded metal that may be stressed near the weld zones, for example. This phenomenon explains why sheet steel often rusts first at bent edges.

Figure 9.7 *Electrochemical potential effects on rust.*

Iron and steel may dissolve in neutral waters at anodes associated with defects in the surface oxide film (Fig. 9.8). The cathodic reaction can occur anywhere on the surface, and the combination of small anodes and a large cathode leads to pitting corrosion. Anodic inhibitors act by causing an insoluble compound (ferric oxide) to precipitate at the anodic sites, so stifling the the anodic reaction and inhibiting corrosion. Cathodic inhibitors have to act over the whole surface of the corroding steel and are therefore less efficient. They reduce corrosion by forming a film of high electrical resistivity over the surface, which provides a barrier to the corrosion current.

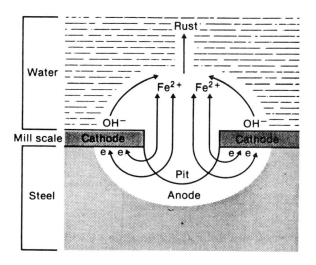

Figure 9.8 *Corrosion at surface defects: intense pitting corrosion at a break in mill scale on a steel.*

Corrosion of metals in the atmosphere includes attack on metals exposed out of doors, indoors in buildings, inside equipment and in closed packages. Corrosion depends on two factors: the contaminants present in the air (the dust content and gaseous impurities) and the moisture content of the air (Fig. 9.9).

Table 9.2 Variation of corrosion rate with atmospheric conditions

Test site	Average corrosion rate (μm/year)
Nkpoku (Nigeria)	5
Brixham	38
London (Battersea)	85
Sheffield	90
Calshot	70

(Courtesy, British Steel Corporation)

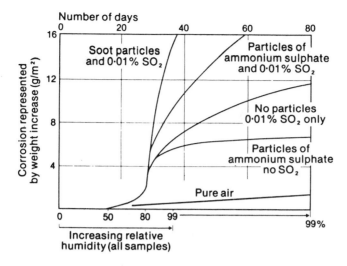

Figure 9.9 *Effects of humidity and pollution.*

Atmospheres can differ considerably in the amount and type of contaminant present, depending on whether the area is industrial, urban, rural or marine; and they often vary in this respect with the time of day or year (Table 9.2). Near the coast, sea salts will be present. In an industrial region the air will include appreciable quantities of carbon, carbon compounds and SO_2 (the waste products of burning coal, fuel and petrol and smaller amounts of H_2S, and NO_2, NH_3 and various suspended salts. The atmosphere can therefore contain both gaseous and solid contaminants. SO_2 (which is converted to sulphuric acid) is the main accelerator of atmosphere corrosion. Dust particles also play an important role, either because they are themselves corrosive and, with moisture, form a corrosion cell when they settle on a metallic surface, or because, though themselves innocuous, they can absorb SO_2 from the atmosphere and so cause corrosion in contact with a metal.

The vapours of the lower aliphatic (formic, acetic, propionic) acids evolved from certain woods, plastics, glues and paints are another source of atmospheric corrosion. In a damp atmosphere, concentrations as low as 0.1 ppm can cause some metals to corrode rapidly, but fortunately such so called 'vapour corrosion' is encountered only in confined atmospheres inside equipment and in packages. In the open, the vapours are generally dispersed by ventilation before they can cause any damage.

Relative humidity is important in considering atmospheric corrosion. Clean iron does not rust in inland and industrial atmospheres with a relative humidity of 70% or less, and, in general, for any corrodible metal there is a critical humidity below which it does not corrode. The critical humidity is determined largely by the hygroscopic nature of any solid contamination which may be present, and also by that of the corrosion product, and hence is influenced greatly by pollutants in the atmosphere. Broadly, however, the critical humidities for steel, copper, nickel and zinc can be regarded as falling within the range of 50–70% relative humidity.

Direct exposure to rain can be beneficial or detrimental, as can shelter from rain. Rain water can leach soluble inhibitors from protective coatings, and can wash away corrosion products. Parts under shelter may be shielded from coarser solid particles, but, on the other hand, could be subject to contamination by wind-blown fine particles, and to polluted condensation, with none of the beneficial effects of rain washing.

Whether condensation occurs indoors depends on such factors as the weather, temperature variations, internal heating, and, of course, any generation of moisture in the vicinity.

9.5 Influence of soil on corrosion

As with corrosion in water an electrolyte is involved when corrosion occurs in soil. An electrolyte dissolves metals under the influence of electrochemical activity (Fig. 9.10). Corrosion can differ from area to area because of differences in soil environments, and therefore, as with atmospheric corrosion, a metal might perform well in some areas but corrode elsewhere.

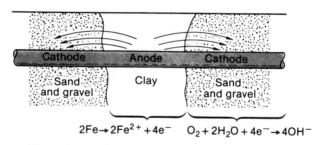

$$2Fe \rightarrow 2Fe^{2+} + 4e^- \qquad O_2 + 2H_2O + 4e^- \rightarrow 4OH^-$$

Figure 9.10 *Differential aeration corrosion in soil.*

The main factors determining whether soil conditions are conducive to corrosion are: moisture, access to oxygen (aeration), electrical conductivity (which is influenced by the presence of dissolved salts), and the pH of the soil.

The electrolyte to which a buried steel structure is exposed will normally not be sufficiently acid for corrosion to occur other than by a process requiring the presence of oxygen to maintain it. Corrosion therefore normally requires a soil pervious to air. An important exception is the damaging attack to which iron and steel are subject in neutral waterlogged clay in which access of oxygen from the air is precluded by the nature of the soil. Corrosion in this case may proceed by a cathode reaction governed by the presence of sulphate-reducing bacteria in the soil.

Much corrosion of buried metal is caused by stray electric currents, emanating, for example, from electric railways. The buried metal provides the stray current with a preferential low-resistance path, acting as a cathode where the positive current enters the metal and an anode where it leaves (Fig. 9.11). The effect is worse with direct currents (alternating currents can become rectified by oxide films on the metal surface), and corrosion can be very severe.

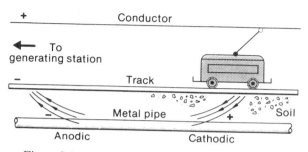

Figure 9.11 *Stray electric currents creating corrosion.*

9.6 Galvanic action

In assemblies of two dissimilar metals the one with the higher potential migrates to the other to form a deposit of its oxide on the low-potential cathode. This sacrificial corrosion continues until the metal with the higher potential is consumed. Any metal higher on the scale (at the anodic end) sacrifices itself to a metal lower on the scale when the two are in contact in the presence of an electrolyte. This is known as galvanic action.

Galvanic corrosion can be halted with a non-conductive barrier – such as a paint or a nonmetallic shim – between the dissimilar metals. In some cases, the galvanic mechanism can be used to protect steel from corrosion. Steel coated with a layer of metal higher on the galvanic scale – zinc or tin, for example – is protected because the surface metal sacrifices itself electrochemically in place of the steel. Some examples of the selective leaching of metals are given in Table 9.3.

9.7 Corrosion of metals by wood

Wood is a corrosive substance by nature. Acetic acid in wood is volatile and can corrode nearby metal without the necessity for metal-to-wood contact. Vapour corrosion may affect the contents of goods in wooden containers. Two other mechanisms are also relevant:

(1) Metal-to-wood contact of land-based structures through attack by wood acids and wood treatment chemicals.
(2) Immersion of metal-in-wood where macrogalvanic corrosion dominates.

Table 9.3 Combinations of alloys and environments subject to selective leaching, and elements removed by leaching

Alloy	Environment	Element removed
Brasses	Many waters, especially under stagnant condtions	Zinc (dezincification)
Gray iron	Soils, many waters	Iron (graphite corrosion)
Aluminum bronzes	Hydrofluoric acid, acids containing chloride ions	Aluminum
Silicon bronzes	Not reported	Silicon
Copper nickels	High heat flux and low water velocity (in refinery condenser tubes)	Nickel
Monels	Hydrofluoric and other acids	Copper in some acids, and nickel in others
Alloys of gold or platinum with nickel, copper or silver	Nitric, chromic and sulfuric acids	Nickel copper or silver (parting)
High-nickel alloys	Molten salts	Chromium, iron, molybdenum and tungsten
Cobalt-tungsten-chromium alloys	Not reported	Cobalt
Medium-carbon and high-carbon steels	Oxidizing atmospheres, hydrogen at high temperatures	Carbon (decarburization)
Iron-chromium alloys	High-temperature oxidizing atmospheres	Chromium, which forms a protective film
Nickel-molybdenum alloys	Oxygen at high temperature	Molybdenum

9.7.1 Wood constituents

Cellulose is the principal constituent of wood; it is a polysaccharide of sugar molecules. In combination with water (hydrolysis) it produces free hydroxyl radicals and acetic acid; the chemical equation is

$$\underset{\text{acetylated group}}{X - O.CO.CH_3} + \underset{\text{water}}{H_2O} = \underset{\substack{\text{free}\\\text{radical}}}{X - OH} + \underset{\text{acetic acid}}{CH_3COOH}$$

where X is the sugar unit in the chain.

In a given wood, the ratio of formation of acetic acid depends on the temperature and the moisture content of the wood, and the rate of its escape to the atmosphere depends on the geometry of the piece of wood in question. Besides acetic acid, small quantities of formic, propionic and butyric acids are present in wood, but their effects can be neglected in comparison with those of acetic acid. Fungi prevent the accumulation of acid although they may themselves produce carbon dioxide which may attack metals [1].

Timber which has been floated down river or stored in a seasoning pond may absorb salt water and produce a chlorine attack on metal.

The acidities of a number of woods are given in Table 9.4. Metals vary in their susceptibility to corrosive attack by the acetic acid emitted by wood and may be classified as in Table 9.5.

Table 9.4 Acidities of woods

Wood	Typical pH values	Vapour corrosion hazard
Oak	3.35, 3.45, 3.85, 3.9	High
Sweet chestnut	3.4, 3.45, 3.65	High
Steamed European beech	3.85, 4.2	Fairly high
Birch	4.85, 5.05, 5.35	Fairly high
Douglas fir	3.45, 3.55, 4.15, 4.2	Fairly high
Gaboon	4.2, 4.55, 5.05, 5.2	Fairly high
Teak	4.65, 5.45	Fairly high
Western red cedar	3.45	Fairly high
Parana pine	5.2 to 8.8	Moderate
Spruce	4.0, 4.45	Moderate
Elm	6.45, 7.15	Moderate
African mahogany	5.1, 5.4, 5.55, 6.65	Moderate
Walnut	4.4, 4.55, 4.85, 5.2	Moderate
Iroko	5.4, 6.2, 7.25	Moderate
Ramin	5.25, 5.35	Moderate
Obeche	4.75, 6.75	Moderate

Table 9.5 Susceptibility of metals to acetic acid

Group 1 (severe corrosion)	Cadmium
	Carbon steels
	Low alloy steels
	Lead and lead alloys
	Zinc and zinc alloys
	Magnesium and its alloys
Group 2 (moderate)	Copper and its alloys
Group 3 (very slight)	Aluminium and its low strength alloys; slightly greater for Al–Cu and Al–Zn alloys
	Nickel
Group 4 (insignificant)	Austenitic stainless steel
	Chromium
	Gold
	Molybdenum
	Silver
	Tin
	Titanium and its alloys

9.8 Forms of corrosion

The characteristics of several different forms of corrosion are as follows:

(1) Uniform – consisting of a general attack. It was suggested by Chandler [2] that the thickness reduced approximately as follows:

steel in air	25 to 100 μm/year
steel in soil	5 to 75 μm/year
steel in water	50 to 250 μm/year

(2) Galvanic – resulting from the bimetallic action on two or more metals
(3) Crevice
(4) Pitting – identifiable in that it acts in a downward direction
(5) Intermittent
(6) Inter-granular – corrosion along the grain boundaries which reduces the mechanical strength
(7) Selective leaching – mostly relevant to brasses with less than 70% copper: it is also called 'dezincification'
(8) Stress-corrosion (corrosion-fatigue) – relates to corrosive action which only occurs when the metal is under tension (not under compression)
(9) Fretting corrosion – when contacting surfaces rub under pressure: it can occur in dry air and with steels it produces the oxide Fe_2O_3
(10) Tuberculation – mounds of corrosion products form on the metal surface under special conditions, as in soft waters with a high carbonic acid content.

9.9 Examples of corrosion

9.9.1 Ship structures

Renewal of hull plates in dry dock due to corrosion loss is very rare and in fact 'general' corrosion loss of the hull structure in a typical cargo vessel is quite low.

Ultrasonic thickness surveys to Lloyd's Register requirements after 20 and 25 years are quite enlightening. For example for 'A' class vessels (built around 1947/50) the figures shown in Table 9.6 were found for exposed deck and shell plating. Thus little loss was experienced for these, admittedly well maintained,

Table 9.6 Corrosion loss for 'A' class vessels

'A' class	Number of ultrasonic readings on exposed plating	Average corrosion loss (%)	Average actual loss per plate (in)
20 year	725	4.6	0.035
25 year	806	4.9	0.037

vessels. With internal structure, however, pockets exist where local design enables stagnant, often polluted, water to lie with associated high local corrosion rates, and this causes a major part of the renewals due to corrosion.

9.9.2 Bureau Veritas

Examples of field investigations at Bureau Veritas included:

(1) Jetty at a marina collapsed after 18 months. Steel floats corroded due to over-numerous consumable anodes producing intercathodic action.
(2) Bacterial corrosion of a ship's hull due to infiltration under the paint such that sulphate-reducing bacteria react to produce H_2S and sulphides. The hull steel reduced to iron sulphide and ferrous hydroxide forming ferric hydroxide with free oxygen. Paint blisters were formed by the hydrogen sulphide.
(3) Connecting flanges of copper pipes corroded in sea water due to residual stress at the collar and differential aeration due to turbulence at the joint.
(4) Very rusted heat exchanger in a plumbing system due to insufficient water treatment and failure to provide an initial protective film inside the pipes and exchangers.
(5) Diesel engine cooling water circuit badly corroded due to excessive use of antifreeze which prevented the formation of a protective layer.
(6) Petrol pipes rusted underground with serious attack at wooden supports. With clay substrata the wooden supports restricted drainage and encouraged corrosion by differential oxygenation.

9.9.3 Oil rig structures

In the course of studies of the disaster to the Norwegian *Alexander Kielland* accomodation platform the combined effects of corrosion and fatigue have received considerable attention.

One of the main causes of the fall of the *Alexander Kielland* was the development of a crack in a weld around a hydrophone. This spread, corroded, and then spread further. The corrosive effect of welds has been known for some time and conditions in the North Sea are more severe than in any other offshore exploration area. The North Sea has unusually strong currents, high waves and great variations of oxygen content in the water. Taken together, these facts can provide a deadly mixture of stress and corrosion.

Corrosion is crucial, because its effects can turn a safe adequately designed structure into a weak one, and will worsen faults in a defective structure. Corrosion monitoring is an insurance against structural weaknesses. In the coming months, the standards of corrosion and structural stress monitoring are going to come under increasing scrutiny. Automatic instrumentation systems for examining rigs need to be used in conjunction with divers (a measure ruled out for much of the year by stormy seas).

9.9.4 Aircraft

The conjoint action of corrosion and fatigue stress produces corrosion fatigue; the conjoint action can be more damaging than corrosion or fatigue separately.

Aircraft structures experience temperatures ranging between $-45\,°C$ (at 9.6–12.8 km (6 to 8 miles) altitude to $+70\,°C$ on a tropical airfield while certain parts of engine bays, hydraulic systems, etc. may exceed $100\,°C$ under flying conditions. Humidity varies up to 100% such that flight condensation may produce several gallons of water during each flight, in addition to which there is seepage from the galley or toilet effluent. These conditions together with the presence of numerous bimetal couples produce situations highly reactive to corrosion under severe vibratory/stress reversing conditions. Hence the significance of corrosion fatigue [3].

9.9.5 Box girder bridge

Corrosion can be accelerated or caused by the growth of fungus through the film and to the substrate. Should the protective film be penetrated by elements of fungus or the film lifted, moisture can penetrate and paint lost as flakes. Additionally fungus waste products (amongst them organic acids) can be corrosive in themselves as has been discovered by aero engineers.

During 1971 it was noticed that certain parts of the interior of the box girder sections of Tinsley Viaduct (UK) were colonized by fungus growth. A full inspection of the structure revealed a very heavy growth in some areas. It was mainly confined to the floor and lower walls. In all box sections a heavy smell of fungus pervaded the air. In some areas growth was associated with faeces, urine and organic materials such as cotton wool pads, cigarette cartons, paper, etc.

The various areas of growth were sampled either by pressing nutrient media directly onto the surface of the paint for incubation on return to the laboratory, or by collection of paint flakes so that optical and scanning electron microscopy could be carried out.

More than 20 species of mould were isolated although *Penicillium canescens* Spp and *Cladosporium cladosporioides* (Fres) de Vries appeared to be dominant. The other fungi isolated included:

Alternaria sp	*Penicillium brevi compactum*
Aspergillus fumigatus	*P crustosum*
A niger	*P frequentans*

A versicolor	*Sporotrichum carnis*
Aureobasidium pullulans	*Stemphylium sp*
Botrytis cinerea	*Trichoderma viride*
Chaetomium olivaceum	
Sphaerospermum	

Two other similar bridges were examined. The environments within the bridges were evidently conducive to growth in that debris present on the floors of the sections was copiously colonized by fungi.

9.10 Corrosion monitoring trends

There are four distinct trends taking place in corrosion monitoring:

(1) The growing interest in electrochemical techniques – not just because of their dual role in corrosion monitoring and helping in identification of corrosive variables – but because of their easy interfacing with data handling systems and control centres.
(2) The concept of letting a corrosion monitor control a whole plant; in fact, it is already being done in the areas of cathodic and anodic protection.
(3) Using corrosive monitors to control additions of inhibitors to cooling water systems.
(4) Interfacing corrosion monitors with computer techniques to handle data, to operate plants closer to material limits and deliberately increase or decrease corrosion rates to reflect market demands for the products. To achieve this, better facilities are needed for handling and processing data.

9.11 Corrosion monitoring methods

Techniques available for the examination of internal corrosion [4] include the following:

(1) Coupons (weight loss)
(2) Electrical resistance (ER)
(3) Linear polarization (resistance) (LPR)
(4) Galvanic
(5) Potential
(6) Non-destructive (ultrasonics, radiography, eddy-current)
(7) Hydrogen
(8) Film resistivity (protective inhibitor film)
(9) Visual methods (miniature TV)
(10) 'Tell-tale' holes
(11) Infrared

(12) Analysis of process streams
(13) Electromechanical impedance
(14) Thin-layer activation
(15) Vibrating reed
(16) Thickness metering
(17) Acoustic emission.

In addition, measuring methods such as scrapers, feeler gauges and calipers may be used – but they can only be used when plant is shut down or by-passed.

Many methods are complementary. Unfortunately no absolute measurement of corrosion can be obtained on-line. It is unwise to rely on one method only, methods generally used are:

(1) Coupons – to provide basic data.
(2) Electronic – for constant monitoring.
(3) NDT – for checking.

Industrial situations in which these techniques can be effectively employed include:

Oil/gas production	Steel production
Water injection	Power stations
Oil refining	Desalination plant
Petrochemicals	Mining
Chemicals	Effluent lines
Pipelines	Marine
Cooling water	Breweries
Water supply	Paper mills

9.11.1 Coupons

Effectively these are samples of the material which are suspended in the working environment and subject to the same corrosive influences as the material to be monitored. They are periodically removed, weighed and replaced. It is possible they might be 'cleaned' before return to the main stream of corrosive influence.

9.11.2 Electrical resistance

This is one of the major 'instrumented' corrosion monitoring techniques. The Corrosometer ⓉⓂ uses a balanced bridge technique to measure the change in resistance of a probe as it thins away under corrosion. Details of this method were given by Bovankovich [5] of E. I. du Pont de Nemours & Co., Wilmington, Delaware, USA.

The probe, shown in Fig. 9.12 comprises an exposed U-shaped upper part which is subjected to corrosive attack and a protected internal part providing the counter balancing resistance elements. The balance readings accordingly record the effective thickness of the exposed probe and thus the implied corrosion rate in mpy (mils per year) where 1 mil = 0.001 inch.

Increased electrical resistance as the cross-sectional area of a metallic conductor decreases provides the fundamental operating principle of this instrument [6]. The heart of a corrosometer system is a probe that is introduced into a corrosive environment and with which the corrosion measurements are made. This probe functions as an *in situ* sensor that accumulates the corrosion history of the environment and displays or records the corrosion information on an instrument.

The resistance of an exposed element is compared with the resistance of a protected element. If the resistance of the exposed element is measured directly, the data will be affected by temperature as well as element thickness. By measuring the ratio of the two resistances, temperature changes that affect both elements equally are cancelled out and the data obtained are directly proportional to the thickness of the exposed element. This method is often referred to as an 'electric coupon'.

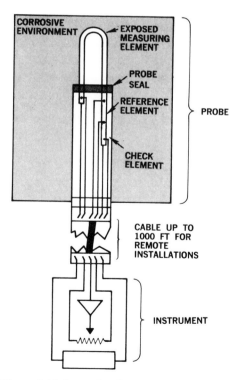

Figure 9.12 *Principle of Corrosometer® probe.*

The exposed element of the probe is called the measuring element and is available in a number of configurations and alloys. Positioned close to the measuring element, but protected against corrosive attack, are reference and check elements of the same metal as the measuring element.

To obtain valid readings, it is essential that the reference element remains protected against corrosion throughout the life of the probe. A check reading made on a probe after reading the measuring element ensures that the reference element has not been affected by corrosion and that the corrosion reading of the measuring element is valid.

The circuitry of all corrosometer instruments is designed to determine the resistance ratio of the measuring element to the reference element. The circuits are arranged so that corrosion readings are directly proportional to the amount of metal removed from the probe element. Therefore, each corrosion reading is a direct indication of the average reduction of thickness of the measuring element. The difference between successive readings indicates the amount of metal that has been removed from the measuring element during the time between readings. Readings may be accumulated periodically with a portable instrument or continuously with a recording instrument.

The importance of this particular technique is that it is the only one that may be used both in conducting and non-conducting process fluids.

Electrical resistance probes need to be used in a match with the equipment they are to monitor. For very low pressure applications such as cooling water systems the ER I plastic body probe is available. The reference element is protected by a tough silicon rubber sleeving which ensures its chemical insulation from the process fluid but provides good thermal contact to ensure its stability in varying temperature situations. For general applications at higher pressures and temperatures probes ER 70 and ER 415 are made, ER refers to electrical resistance and the figure refers to the maximum rating of the probes in bar.

The ER probes for high pressure applications are used with Casaco hollow plug fittings which enable them to be inserted into and withdrawn from the process plant whilst still under pressure at figures up to 270 bar (4000 psi). Use of ER in oil production and refining is described by Silkinson [7], Mohsen [8] and Cole [9].

In certain processes where it is common to use 'pigging out' of pipelines two methods apply, namely using either a flush mounted probe or the standard wire element probe recessed from the inside surface of the pipeline. The flush mounted probe consists of a small flat ended probe with a small corroding element. Although suitable for some applications it has limitations owing to its lower sensitivity, lower sealing integrity and ease of location of the probe tip relative to the inside surface of the pipe or vessel. Furthermore, when the 'pig' is sent down the line damage may occur to the end of the probe owing to the severe action of the pig. The standard wire element probe can be used mounted on a fitting or stub pipe or vessel. This stub pipe should be as generously sized as possible, preferably about 75 mm (3 in) diameter so that sufficient circulation is caused by eddy currents around the probe in order to obtain a representative reading.

(a) Multichannel monitoring

A six-channel corrosion data acquisition system known as 4200 Corrosometer® [10] provides direct read-out of corrosion rate and totalized metal loss. Alarm registrations can be sounded under conditions of corrosion upset; parameters can be printed out on a periodic or demand basis.

Based on the electrical resistance principle, it is compatible with all corrosometer probe configurations. The instrument reads each probe sequentially and with the use of a microprocessor calculates the current corrosion rate and totalized metal loss. The results are then displayed or printed in engineering units, either in mpy or mm/yr. Significant features are:

(1) Fast response to corrosion upsets.
(2) Read-out of corrosion rate and total metal loss.
(3) Data output in either mils per year or millimetres per year.
(4) Microprocessor flexibility and computational power.
(5) Continuous real time measurement and alarm of the corrosion process.
(6) Six-channel input capability.
(7) Compatibility with any type or combination of types or corrosometer probes.
(8) Front panel programmable functions and alarm set points.
(9) 'Human engineered' front panel for ease of operation.
(10) Data storage unaffected by power loss up to one hour.
(11) Automatic calculation of both short term and long term corrosion rates.
(12) Integral printer for hardcopy of all data and alarms.
(13) Print on demand or at programmable intervals.
(14) Key switch for administrative control of primary parameters.
(15) Auxiliary form C for contacts for external alarm annunciation.

Typical installations are in refineries, chemical plants, electrical generating plants, paper mills, pipelines, pollution control plants, ambient and process air monitoring systems.

A data acquisition system associated with this corrosion meter can survey up to fifteen probes. As shown in Fig. 9.13 up to sixteen model 4100 Corrosometers can be connected to a single data transmission line. Upon command each 4100 puts its corrosion information on the data line back to the computer. In this way, a process computer can review the staus of as many as 240 Corrosometer probes in a few seconds, for rapid, continuous access to corrosion information and early warning of upsets.

The instrument requires no calibration, adjustment or other attention from plant personnel while operating. Its acquisition and storage circuitry, controlled by a microprocessor, record new information from each probe every

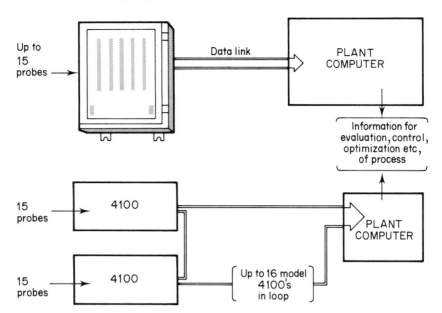

Figure 9.13 *Multichannel corrosion monitoring system: 4200 Corrosometer.*

30 minutes. A simple request from the host computer causes the 4100 to transmit currently stored data at rates up to 9600 baud, with selectable communication formats of RS232, RS422 or 20 ma. Information about metal loss is easily calculated by simple programs, with sensitivities of one millionth of an inch (0.000 025 mm) possible with presently available standard probes.

(b) Marine corrosion monitoring system

A form of Corrosometer®-designed probe Model 4016 measures corrosion in cargo holds and ballast tanks as well as in offshore rigs. It is designed for easy installation and monitors the life and effectiveness of zinc anodes as well as the protection characteristics of cathodic protection in tankers, oil rigs and underground pipelines. The system is designed to include up to twelve probes per system and provides for the storage of data on a recorder for later evaluation.

The probe is designed to withstand the rugged environment of cargo and ballast tanks in addition to the vigorous action of the sea on platform legs. It utilizes the electrical resistance technique for corrosion evaluation. As the mild steel element of the probe corrodes, the element decreases in thickness, thereby increasing in resistance. This increase in resistance is measured by the corrosometer instrument and converted to metal loss in mils per year. Each probe has a life of 0.050 in (50 mil). Corrosion rates as low as a few mils per year can be detected and quantified in a matter of days. The probe may be monitored by a portable corrosometer (CK-3) or by an automatic version.

Each probe has an integral grounding strap which connects the element of the probe electrically to the metal wall. The probe, therefore, will experience the same corrosion effects, electrical or chemical, which operate on other metal surfaces exposed to the corrosion fluid. The nature of the corrodent (liquid or gas, aqueous or non-aqueous) does not affect the probe's ability to sense corrosion. Factors which modify corrosion processes (chemical inhibitors, cathodic protection by metal anodes or impressed current, etc.) will be detected by the probe and may be evaluated accurately and easily followed so that corrective action can be taken if needed.

Where intrinsic safety may be needed, a Magna Safety Barrier may be attached. One is required for each probe. The barriers must not be more than 60 m (200 ft) from the probe installation. The extension cable from the barrier to the instrument can be up to (300 m) 1000 ft.

9.11.3 Sensors and systems

Until 1979 corrosion monitoring equipment was marketed by Nalfloc, a partially owned subsidiary of ICI Ltd. The simplest corrosion monitor used with the electrical resistance range of probes is the Sensors and Systems Corrosion Monitor RB Mark II. It is BASEEFA approved and may be used with any of the currently available resistance element probes. This has a high sensitivity when used in conjunction with the standard resistance element probe, so that changes in corrosion rate are soon apparent. It is normal to plot these readings on a graph so that the trend in corrosion rate may be clearly seen.

For fully automatic corrosion monitoring, the Corrosion Monitor RB Mark III electronically balances the bridge of the probe and when used with barrier devices can be operated with probes mounted in a hazardous area. It may be used as a single channel device and provides digital read-out of the corrosion reading with BCD and analogue outputs, the latter being suitable for use directly with recorders. It is normally used in conjunction with an autoscanning and channel switching unit whereby up to 48 probes may be monitored continuously and automatically.

The automatic switching unit operates with the RB Mark III corrosion monitor to select probes in rotation and also provides storage for the readings taken from each probe. This storage then may be interrogated independently from a computer of data logger or may be used in conjunction with a multipoint recorder.

9.11.4 Linear polarization resistance probe

As shown in Fig. 9.14 the working principle is that when a test electrode is inserted in the medium near a reference electrode, the incremental change of

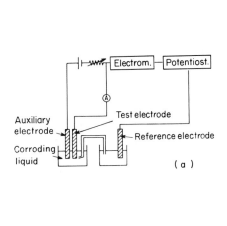

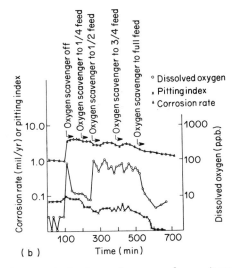

Figure 9.14 *(a) Schematic arrangement of a linear polarization resistance probe monitor;
(b) typical corrosion measurements.*

current with voltage is an index of the instantaneous corrosion rate. With only two electrodes a correction is needed for solution resistance; this does not apply when a third electrode is used [11–13].

A 'pitting index' may be obtained from the difference of reading between two identical electrodes. The general corrosion rate, the 'pitting index' or a combination of both can be utilized for alarm and/or control.

Fast response and instantaneous read-out of corrosion rate are available from this technique but it only operates in liquids. It can be used to indicate pitting or localized corrosion rates and 90% of its applications are in cooling water and waterflood situations.

Associated Octel has been using the technique in hazardous areas. The research and development department used it on a low carbon steel reflux condenser as a 'once-through water system'; it is prone to tube bundle corrosion.

Mild steel electrodes were sited in PTFE holders in the upper water outlet pipe of the reflux condenser for sixteen days. It was discovered that the rate of corrosion and scale build-up of mild steel was far too high. Examination of the test electrodes showed that the attack was not uniform and pitting was likely.

Another test involved a jacket cleaning programme to improve water flow and heat transfer properties of autoclaves. An old autoclave was fitted with electrodes to measure corrosion rates during acid cleaning. Rates were acceptable on low carbon steel at 75 °C – under 0.025 mm/day – but increased markedly at the cleaning temperature of 80 °C. The cause was ferritic ions in solution arising from dissolved rust. The inhibitor could not prevent ferritic ions attacking the steel and to boost the corrosion rate even further they would form a complex with the inhibitor to form an acid attack.

9.11.5 Corrator(TM) – corrosion rate/pitting tendency meter

By using the polarization resistance method the instantaneous corrosion rate and pitting tendency in a conductive liquid can be monitored and incorporated into an inhibitor injection control system as explained by Britton [14]. (The Corrator(TM) is a product of the Magna Corporation [6].)

Corrosion occurs at microscopic sites on a metal surface in contact with a conductive liquid, and because of variations in alloy composition, metallurgical structure and electrolyte concentration these microscopic areas have differing electrical potentials. While dissolution of the surface tends to proceed more rapidly in the anodic regions, both cathodic and anodic reactions occur on the surface, and the general rate of metal loss is a function of both types of reaction.

A measurable parameter directly related to the rate of reaction (and hence rate of corrosion) at these sites is the resistance to the flow of an electrical current from the conductive liquid into the corroding metal surface. The Corrator makes use of this principle to obtain a signal which may be read directly as corrosion rate in units of mpy or mpd.

When a test electrode (Fig. 9.15) is polarized from its free corrosion potential by means of a small applied voltage, V, either the anodic or cathodic characteristic of the corrosion reaction involved will be enhanced, depending upon the

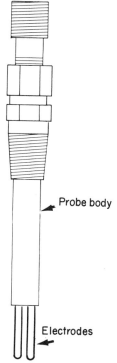

Figure 9.15 *Corrator(TM) corrosion rate probe.*

polarity of V. It is known that these anodic and cathodic characteristics may differ greatly from one another, and that the test currents obtained (I_a for $+V$, $-I_c$ for $-V$) must be combined in the following way to produce useful correlation with true corrosion rate:

$$\text{corrosion rate} = \frac{K}{V}\left(\frac{I_a I_c}{I_a + I_c}\right)$$

where K is a constant relating the units of corrosion rate to those of V and I. Corrosion rate is measured in mils per year, where 1 mil = 0.001 in.

This method of combining readings gives good agreement with observed metal loss for widely varying conditions and alloys.

Measurement involves a stable source of DC voltage, an instrument to measure current and a probe consisting of two identical metal electrodes mounted side by side and fabricated from the same alloy as that used in the system which is to be monitored. The probe is inserted into the process stream and allowed to come to equilibrium with the corrosive environment. A millivolt potential is applied across the two electrodes which slightly increases the anodic activity occurring on the surface of one electrode and the cathodic activity on the other causing a current I_a to flow; the voltage polarity is then reversed causing a current I_c from which the corrosion rate can be measured.

Pitting is related to corrosion rate and the Corrator℗ can therefore be used to monitor changes in this potential failure condition. Pitting occurs when there is an uneven distribution of the anode and cathode areas, or when the anode and cathode areas are not in close proximity to each other. The chemical conditions that lead to this type of corrosion tend, in the case of the Corrator℗ probe, to make one electrode more cathodic or anodic than the other. A potential difference thus exists between the two, in addition to that applied by the Corrator℗. This produces different readings when the polarity of the probe electrode is reversed. By averaging the two readings, an accurate value may be determined, and the difference between the two readings is a direct indication of the pitting tendency.

As the ratio of the pitting readings to the corrosion readings increases, the chances that a dangerous pitting condition exists within the system also increase.

For most industrial applications, the two-electrode probe is the best choice. Three-electrode probes are only required where the product of solution resistivity (ohm-cm) times the expected corrosion rate is greater than 25 000. A two-electrode probe can be used to accurately measure corrosion without correction curves in solutions where the product of the corrosion rate times solution resistivity is less than 10 000. Two-electrode probes will also measure corrosion in solutions where the product of the corrosion rate times solution resistivity is as high as 25 000; however, curves are used to correct the corrosion rate due to the effects of the higher solution resistivity. Where the corrosion rate–solution resistivity product is greater than 25 000, the three-electrode probe should be used.

Table 9.7 shows a comparison of corrosion monitoring methods.

Table 9.7 Comparison of corrosion monitoring methods

	Corrosometer	Corrator
Type of measurement	Electrical resistance	Linear polarization resistance
Units of measure	Total corrosion (10^{-6} in)	Corrosion rate in mils per year
Applicable environment	All environments that probe materials will withstand: liquid, gas, solid	Conductive liquids only
Available instrument and probes	Portable and recording instruments, fixed and retractable process probes	Portable, recording and controlling instruments – fixed and retractable process probes
Control capability	Not usable for control but has been adapted to alarm in some applications	Readily usable for control and alarm
Corrosion rate range	Depends on probe sensitivity and time between readings	1 mpy to 1000 mpy; several corrosion rate ranges
Effect of pitting	Does not directly measure pitting. Pits can affect readings	Reads pitting tendency. Pits do not affect corrosion reading
Effect of external current (cathodic or anodic protection)	May be read with or without external current applied to probe	Cannot be read with external current applied to probe
Limitations	Unable to measure pitting. Does not produce instant rate data	Limited to conductive liquids
Where to use	Use where cumulative corrosion information is important. Where readings can only be made occasionally. Where environment is not electrically conductive	Use where rapid data accumulation is required for studies of metals and inhibitors in conductive liquids. Where control or immediate indication of rate increase is required in conductive liquids

9.11.6 Potential monitoring

Use is made of the relationship between corrosion potential of a metal or alloy and its corrosion state identified from the corrosion rate of the material. The system uses plant equipment, such as a reaction vessel, ringlet coil or pipeline as the monitor, instead of external probes. For instance, an all-tantalum thermopocket is placed inside a rubber-lined vessel to potential monitor an Incaloy 825 ringlet coil in a sulphuric/organic acid system.

In the presence of NaNO, as an oxidizing inhibitor to maintain passivity of the Incaloy 825, the tantalum thermopocket, which is electrically insulated from the vessel, acts as a stable redox electrode. ICI Mond Division's engineers readily observed the change from passive to active corrosion by a change in potential, and again, the change back to passive corrosion was observed as the coil cooled.

The potential monitoring method is also used for monitoring pitting and stress corrosion cracking such as in Vetrecoke CO_2 removal units. Or, one can monitor presence of species producing corrosion using silver/silver chloride electrodes to monitor chlorides or cyanides or sulphide ions using ion-selective electrodes.

Another application is in monitoring of breakdown in materials slightly inferior in corrosion resistance to material in use under given conditions. For example comparing TP304 against TP316, or comparing titanium 130 against titanium 260.

9.11.7 Pipeline 'holiday' detector

Anticorrosive coatings on steel substrates can be tested for the presence of 'holidays' – defective coatings which could affect the anticorrosion protection. The portable detector consists of a lightweight unit, a battery power pack in belt form and a separate charger unit. It is portable, offering a wide range of voltages, together with audio and visual identification of holidays. Two types of probe head are available; brush and rolling spring types.

The equipment is operated by brushing the probe over the coated substrate. A high-pitched audible warning unit operates immediately a holiday is detected, irrespective of the output voltage selected. One spark is sufficient to operate the audible unit. The audible warning operates for one second duration and a built-in overload trip operates if the holiday or short circuit is not cleared within ten seconds.

9.11.8 Cathodic protection assessment

British Gas have developed a method of assessing the extent to which pipe coatings may have disbonded. Coal tar or bitumen are applied hot to a pipe – tapes may be wrapped round cold – and this protection is normally supported by cathodic protection. Holes in the coating might cause lack of adhesion and give rise to corrosion.

To test the integrity of the protective coating a ten-station cathodic disbonding unit is used. It consists of an adjusted stabilized 10 volt DC power supply capable of delivering up to 400 mA as shown in Fig. 9.16.

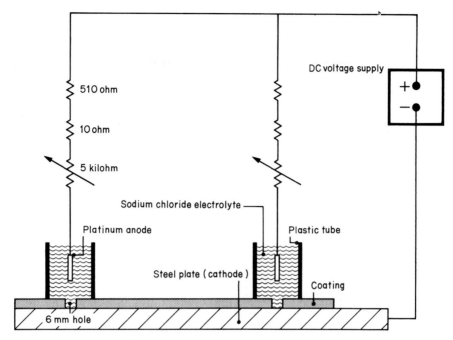

Figure 9.16 *Pipeline holiday detection in use.*

9.11.9 NDT/crack integrity appraised corrosion monitoring

Inspection methods which are particularly effective in corrosion monitoring may themselves be assessed as shown in Table 9.8.

9.12 Thickness metering

Oxide thickness measurement of nuclear reactors has been developed for in-core use [16].

A corrosion monitor uses successive pulses of laser light to drill a hole through an oxide layer, and a weaker pulse to monitor the reflectivity of the hole. A significant increase in reflected light occurs when the underlying metal is reached. If the average depth increase from each laser pulse is known by calibration then a value of the total thickness of oxide is derived.

Plate thickness metering can be based on the range of techniques for surface coating measurement which include the following:

(a) chemical dissolution
(b) 'strip and weight' (weight difference technique)

Table 9.8 Evaluation of inspection methods

Corrosion type	Inspection method				
	Radiography	Ultrasonic	Eddy current	Magnetic particle	Stress wave (acoustic) emisson
Hidden wastage of unknown mechanism	d	a	a		
General wastage	a	d	a		
Pitting corrosion	d	b	c		
Intergranular corrosion	b	d	c	b	b
Dezincification	b	d			
Corrosion fatigue	d	c	c	c	a
Stress corrosion cracking	c	d	c	c	c
Hydrogen embrittlement cracking	b	d	b	c	c

a = limited; b = can detect; c = good; d = best method.

(c) mechanical displacement
(d) microscopy (including interferometry)
(e) coulometric
(f) thermoelectric
(g) X-ray fluorescence
(h) beta – backscatter and transmission
(i) gamma absorption
(j) HF and LF eddy current
(k) ultrasonic
(l) electromagnetic induction

Tables 9.9 and 9.10 give a summary of corrosion monitoring methods [17] and characteristics of corrosion monitoring techniques [17] respectively.

Table 9.9 Summary of corrosion monitoring methods [17]

Method	Measures or detects	Applications
Linear polarisation (Polarisation resistance)	Corrosion rate is measured by the electrochemical polarisation resistance method with two or three electrode probes.	For most engineering metals and alloys in process fluids of suitable conductivity. Frequently used.
Electrical resistance	Integrated metal loss is measured by the resistance change of a corroding metal element. Corrosion rates can be calculated.	Suitable for most engineering metals and alloys in both liquid and vapour. Measurement is independent of process stream conductance. Frequently used.
Potential monitoring	Potential change of monitored metal or alloy (preferably plant) with respect to a reference electrode.	To indicate the corrosion state of the plant e.g., active, passive, pitting, stress-corrosion-cracking as characterised by a specific potential region. Directly measures plant behaviour. Moderate use.
Corrosion coupon testing	Average corrosion rate over a known exposure period by weight loss or weight gain.	Very satisfactory when corrosion is a steady rate. Useful in hazardous areas where electrical devices prohibited. Moderately cheap corrosion monitoring. Indicates corrosion type. Very frequently used.
Analytical	Concentration of the corroded metal ions or concentration of inhibitor.	Can be used to identify specifically corroding equipment. Only moderate use.
Analytical	pH of process stream.	To monitor change of pH, such as in effluent streams where acidic conditions can cause severe corrosion. Very frequent use.
Analytical	Oxygen concentration in process stream.	Control of oxygen level, usually through oxygen getters to mitigate corrosion. Moderate use.
Ultrasonics	Thickness of metal and presence of cracks, pits, etc, by changes in response to ultrasonic waves.	Generally used as an inspection tool for metal thickness or crack detection. Widely used in this capacity.
Eddy current testing	Uses a magnetic probe to scan surface.	Detects surface defects such as cracks and pits. Widely used.
Infra-red imaging (Thermography)	Spot surface temperatures or surface temperature pattern as indicator of physical state of object.	Refractory and insulation inspection, furnace tube temperature measurement, flow problem detection and electrical 'hot' spots. Not widely used.
Acoustic emission	(a) Leaks, collapse of cavitation bubbles, vibration level in equipment. (b) Cracks by detection of the sound emitted during their propagation.	Leak detection and the possibility of cavitation erosion, frettage and corrosion fatigue. Stress corrosion cracking and fatigue cracking in vessels and lines. Strictly not a monitoring tool and currently still a new technique. Not widely used.

continued on p. 344

Table 9.9 continued

Zero resistance	Galvanic current between dissimilar metal electrodes in suitable electrolyte.	To indicate polarity and magnitude of bi-metallic corrosion. Indicates dew point conditioning for atmospheric corrosion and a sensitive detector for break through of linings etc, by corrodents. Infrequent use.
Hydrogen sensing	Hydrogen probe used to measure hydrogen gas liberated by corrosion.	Typically in petrochemical industry and involving mild steel corrosion in sulphide, cyanide and other 'poisons'. Frequent use in specific applications.
Sentinel holes	Indicates when corrosion allowance has been consumed.	In specific equipment, particularly pipe bends where erosion-corrosion can produce erratic thinning and catastrophic failure can be prevented. Infrequent use.

Table 9.10 Characteristics of corrosion monitoring techniques [17]

Technique	Time for individual measurement	Type of information	Speed of response to change	Relation to plant	Possible environments	Type of corrosion	Ease of interpretation	Technological culture needed
Electrical resistance	Instantaneous	Integrated corrosion	Moderate	Probe	Any	General	Normally easy	Relatively simple
Polarisation resistance	Instantaneous	Rate	Fast	Probe	Electrolyte	General	Normally easy	Relatively simple
Potential measurement	Instantaneous	Corrosion state and indirect indication of rate	Fast	Probe or plant in general	Electrolyte	General or localised	Normally relatively easy but needs knowledge of corrosion. May need expert	Relatively simple
Galvanic measurements (zero-resistance ammeter)	Instantaneous	Corrosion state and indication of galvanic effects	Fast	Probe or occasionally plant in general	Electrolyte	General or unfavourable conditions localised	Normally relatively easy but needs knowledge of corrosion	Relatively simple
Analytical methods	Normally fairly fast	Corrosion state, total corrosion in system, item corroding	Normally fairly fast	Plant in general	Any	General	Relatively easy but needs knowledge of plant	Moderate to demanding
Acoustic emission	Instantaneous	Crack propagation, cavitation and leak detection	Fast	Plant in general	Any cavitation	Cracking, cavitation and leak detection	Normally easy	Crack propagation specialised, otherwise relatively simple
Thermography	Relatively fast	Distribution of attack	Poor	Localised on plant	Any. Must be warm or sub-ambient	Localised	Easy	Specialised and difficult
Optical aids (closed circuit TV, light tubes, etc)	Fast when access available, otherwise slow	Distribution of attack	Poor	Localised on plant	Any	Localised	Easy	Relatively simple
Visual, with aid of gauges	Slow. Requires entry on shut down	Distribution of attack, indication of rate	Poor	Accessible surfaces	Any	General or localised	Easy	Relatively simple but experience needed

Table 9.10 continued

Corrosion coupons	Long duration of exposure	Average corrosion rate and form	Poor	Probe	Any	General or localised	Easy	Simple
Ultrasonics	Fairly fast	Remaining thickness or presence of cracks and pits	Fairly poor	Localised on plant	Any	General or localised	Easy	Simple
Hydrogen probe	Fast or instantaneous	Total corrosion	Fairly poor	Localised on plant or probe	Non-oxidising electrolyte or hot gases	General	Easy	Simple
Sentinel holes	Slow	Go/no go on remaining thickness	Poor	Localised on plant	Any, gas or vapour preferred	General	Easy	Relatively simple

References and list of companies

1   Clarke, S. G. and Longhurst, E. E. (1961), 'The corrosion of metals by acid vapours from wood', *J. Appl. Chem.* **11**, 425.
2   Chandler, K. (1976), 'Corrosion control', *Engineering,* March, i–vi.
3   Forsyth, P. J. E. (1980), 'The fatigue performance of service aircraft and the relevance of laboratory data', *J. Soc. Env. Engng,* March, 3–10.
4   Britton, C. F. (1979), 'Monitoring of internal corrosion in offshore installations', *Marine Eng. Rev.,* January, 8–9.
5   Bovankovich, J. C. (1973), 'On-line corrosion monitoring', *Materials Performance,* **12**, No. 6.
6   Magna Corporation, 11808 South Bloomfield Avenue, Santa Fe Springs, California 90670, USA (Tel. (213) 863-4781; Telex 65-7443).
7   Wilkinson, L. (1977), *Monitoring Corrosion in the Oil Industry,* Society of Chemical Industry, London.
8   Mohsen, E. (1978), Paper 132, *Corrosion 78,* NACE, Houston.
9   Cole, E. L. (1978), Paper 120, *Corrosion 78,* NACE, Houston.
10   '4200 and 4100 Corrosometer', Rohrback Instruments, 11861 East Telegraph Road, Santa Fe Springs, Ca 90670, USA. (Tel. (213) 695-0421; Telex 67-7663); also 5A Oxford Road, Reading, Berks, RG1 7QG, UK.
11   Hines, J. G. *et al., Controlling Corrosion No. 6: Monitoring,* Department of Industry, England.
12   Britton, C. F. (1976) 'Corrosion', in *Corrosion Monitoring and Chemical Plant,* edited by L. L. Shreir, Newnes-Butterworth, London.
13   Rowlands, J. C. and Moreland, P. J. (1977), *'The Instrumentation and Use of Linear Polarization Measurements for Corrosion Monitoring,* Society of Chemical Industry, London.
14   Britton, C. F. (1974), 'Methods and equipment for electrochemical measurements and control in the field of corrosion', *Proc. Int. Scandinavian Congress on Chemical Engineering,* Copenhagen, January.
15   Metrotect Ltd, PO Box 1, Cleckheaton, West Yorks, BD19 3UF, UK (Tel. Cleckheaton 4222; Telex 517406).
16   Tombs, F. (1979), '25 years of nuclear power', *Atom,* No. 272, June, 146–9.
17   Department of Industry, Committee on Corrosion, Report 6, 123 Victoria Street, London SW1E 6RB.

10

Structural data assessment: stresses, stress concentration

10.1 Introduction

Structural integrity decisions must relate to an appreciation of the type and intensity of stress which is being applied. Highly dependable theories for the load-bearing ability of components under either elastic or plastic strain conditions have been evolved. Use of such theories to calculate critical stresses can be of considerable assistance when assessing the loads being applied, the residual strength of the structure, the time-to-failure and the closedown decision.

Materials at first react to an applied load by changing shape elastically. After a determinable stress the material yields and strain has a non-linear and irreversible relationship with the applied load.

The applied load may itself set up both tensile and shear 'principal stresses' within the material such that limiting values may depend upon whether the material is strong in tension (compression) or strong in shear. A defect in a ductile non-strain hardening material may produce such a stress distribution that the elastic limit might be exceeded.

Design criteria are based on:

(1) Keeping loads on a member below the elastic (yield) load so that no large-scale deflection occurs.
(2) Keeping applied loads below the ultimate load to prevent mechanical instability such as buckling or necking from taking place.
(3) Relying on elastic/triaxial constraints to preserve the structural integrity of a member or framework as a whole.

Failure of a methane storage tank at Cleveland, Ohio, in 1944 caused the deaths of 128 persons and $7 000 000 damage [1]. Investigations showed it had

346

operated at twice the loads allowed by building codes and had been made from materials which had not been properly heat treated.

Of the 4694 Liberty ships built during World War II, 1289 reported serious structural failures of which 233 caused the ship to be lost or scrapped – one ship broke in half and sank at the pier while being fitted out, the bending moment load at the time being one-half the design value. Fractures at low stresses can occur near defective welds or when design notches or cracks are present in the structures.

When conditions are right for an unstable fracture the handbook values of the 'ultimate' tensile strength or yield strength are inadequate. Brittle fracture mechanics are appropriate.

10.2 Strength of materials

10.2.1 Stress

Strength properties of materials are used to find whether components can carry the loads demanded of them without excessive deformation or failure.

These load-carrying abilities are normally characterized in terms of stress, that is the amount of load carried by a given area. Therefore stress is quoted in pressure units, force per area (for example $Pa = N/m^2$, kg/mm^2, lbf/in^2).

Similarly, other important criteria for a material, such as the limit of proportionality, the ultimate strength, and the breaking strength, are also usually given in terms of stress. For a ductile material the relationship between stress and strain is shown in Fig. 10.1. In the elastic extension part of this curve a modulus of elasticity called Young's modulus (E) is defined as the ratio of stress/strain. As strain is dimensionless, E has stress dimensions (force per area). Typical properties for a general purpose construction steel are shown in Table 10.1.

Table 10.1 Typical 'strength' properties of a general purpose construction steel

	N/mm^2	lbf/in^2	ton/in^2	kgf/mm^2
Ultimate tensile stress	478	69 310	30.9	48.8
Yield point (proof stress)	271	39 200	17.5	27.6
Elastic limit	260	37 000	16.8	26.5
Limit of proportionality	255	37 000	16.5	26.0
Young's modulus	267×10^3	30×10^6	1339	27.2×10^3
Strain at yield		-0.00124 (no dimension)		

Under shear stress conditions similar relationships apply but are of different magnitudes. Thus a modulus of rigidity (G) can be defined as shear stress/shear strain. Table 10.2 gives typical values of G for a number of materials.

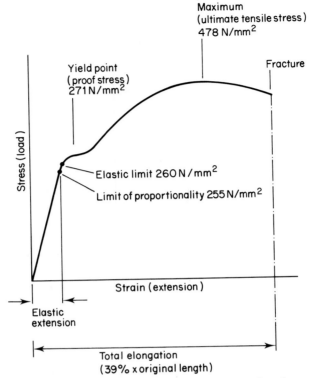

Figure 10.1 *Typical load extension curve with characteristic values for a mild steel.*

Table 10.2 Shear modulus for various materials

Material	N/mm²	lbf/in²	ton/in²	kgf/mm²
Brass	38 000	5.5 × 10⁶	2460	388
Bronze	45 000	6.5 × 10⁶	2900	460
Cast iron	41 000	5.9 × 10⁶	2600	419
Duralumin	26 000	3.8 × 10⁶	1150	266
Monel metal	70 000	10.1 × 10⁶	4500	715
Mild steel	80 000	11.6 × 10⁶	5200	818
Nickel-chrome steel	82 000	11.9 × 10⁶	5300	837
Timber	1 000	0.15 × 10⁶	670	102

10.2.2 Strain: Poisson effect

If the Young's modulus (E) of a material is known and the strain (ε) can be measured it is simple to calculate the stress (σ):

$$E = \frac{\sigma}{\varepsilon} \tag{10.1}$$

and so

$$\sigma = E\varepsilon$$

This is true when the direction of the stress is known, which is the case when a simple strut is exposed to tension. However, in all other cases there is a further complication due to transverse stress and strain (Fig. 10.2).

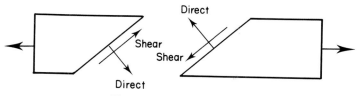

Figure 10.2 *Internal stresses.*

Fig. 10.3 shows the exaggerated deformation of a strut under simple tension loading which causes an increase in length and a corresponding decrease in cross section. This is known as the Poisson effect. It means that if strain is measured in either of the planes perpendicular to the applied load, a negative stain with lesser magnitude will be detected. The magnitude of this lesser strain depends upon Poisson's ratio (μ), a constant that varies from material to material. It is usually about 0.3, so that the strain measured perpendicular to loading will be approximately -0.3 times the strain measured parallel to the loading.

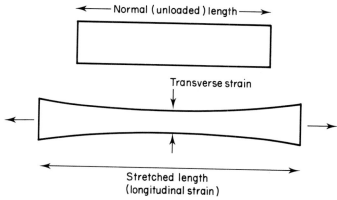

Figure 10.3 *Poisson Effect.*

10.3 Examples of stresses in structures

The stresses set up to support each particular load, vary with each particular structure.

10.3.1 Pillars, columns, struts

If the support under compressive load is short or 'stubby' the column will

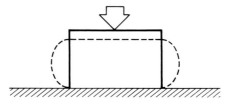

Figure 10.4 *Compressed column.*

simply squash (crush) down in the manner of Fig. 10.4. If the support is long, it will bend and stresses will be set up according to the 'end fixations' and the second moment of area of the cross section; typical 'buckling' conditions are shown in Fig. 10.5.

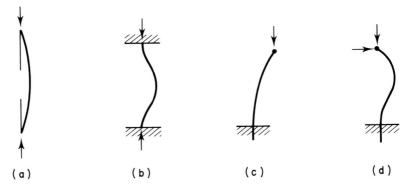

(a)   (b)   (c)   (d)

Figure 10.5 *Buckling deformation for various end-fixations: (a) hinged; (b) fixed; (c) part-fixed; and (d) laterally restrained.*

The length of a strut influences its failure stress (or the load it can support). On the basis of 'pure' theory there is a critical length for which the failure will result from squashing (crush) beyond which failure results from stresses set up by buckling. The general relationship is shown in Fig. 10.6.

10.3.2 Pipes, cylinders, spheres

Internal pressure in a vessel sets up three different stresses:

(1) Circumferential (or hoop) stress;
(2) Longitudinal stress;
(3) Radial stress.

(a) Thin pipe or cylinder

If plate thickness is less than 1/20 of the internal diameter it is reasonable to

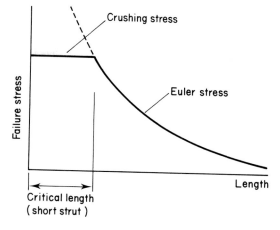

Figure 10.6 *Effect of length on failure stress (direct load).*

regard both hoop and longitudinal stresses as constant over the section; from this it can be demonstrated (Fig. 10.7) that the circumferential stress is twice the longitudinal stress – which explains why pipes and cylinders tend to rupture as shown in Fig. 10.8.

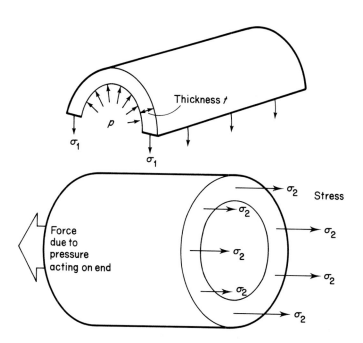

Figure 10.7 *Circumferential and longitudinal stresses in a thin cylinder.*

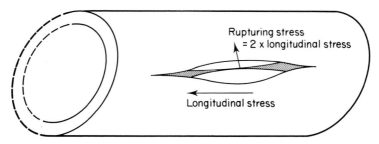

Figure 10.8 *Typical cylinder/pipe failure due to rupture.*

(b) Thick cylinder

Thick cylinders (where plate thickness is more than 1/20 of the internal diameter) cannot assume a constant stress distribution over the section. As a consequence, some parts may be overstressed and distorted into the plastic, permanent-set condition. Thus a hydraulic ram 60 mm OD, 13.5 mm ID under a pressure of 40 N/mm² is subject to 60 N/mm² maximum tensile stress and 50 N/mm² maximum shear stress.

10.3.3 Rotating discs

Centrifugal stresses incorporating Poisson effects result from rotation. Thus a steel turbine rotor disc (made from material of density $\rho = 7700 \, \text{kg/m}^3$), 0.6 m diameter and a tip thickness of 9.5 mm keyed to a 50 mm diameter shaft and rotated at 10 000 rpm would be subjected to a centrifugal stress of 200 N/mm². This is simply the stress imposed by rotation and does not include the higher stresses from loading on turbine blades. The need for awareness of such stresses is clear when it was shown (Table 10.1) that a general construction steel (at normal ambient temperature) has an ultimate tensile stress of 478 N/mm².

Temperature differentials across a disc cause expansion effects. Thermal stresses will be superimposed on centrifugal stresses when a rotating disc has a temperature difference between the inner and outer radii – a characteristic situation with turbine rotors. Typically, a steel disc 250 mm OD, 50 mm ID running at 10 000 rpm with a linear temperature differential of 45 °C will experience a maximum principal stress of 171 N/mm² (formerly, when 'cold' $\sigma = 110 \, \text{N/mm}^2$, a 64% increase).

10.3.4 Bending: beams

Loads carried by beams which are supported, cause moments (force × perpendicular distance) to be applied which make the beam bend and deflect. Measurement of deflection is one way of monitoring the applied load and also the stresses in the beam.

A characteristic of beams under load is that one side of the centre of area (neutral plane) is under tensile stress while the other side is under compressive stress as shown in Fig. 10.9. An I-beam such as that shown in Fig. 10.10, simply supported over a span of 2 m and carrying a central load of 500 kg could have maximum tensile and compressive stresses of 346 N/mm^2 (50 000 lbf/in^2).

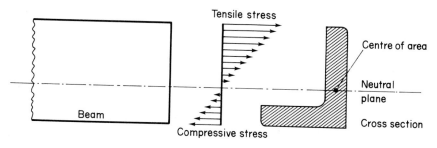

Figure 10.9 *Stress distribution across a beam.*

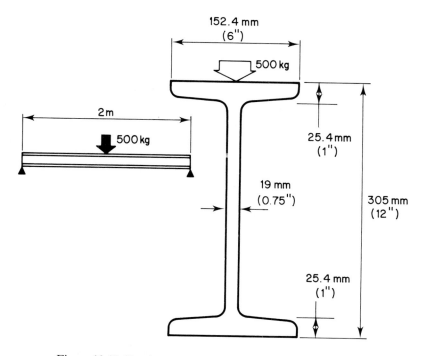

Figure 10.10 *Simply supported I-beam carrying a central load.*

10.3.5 Cantilever plates: eccentric loads

The consequences of eccentric loading of pillars is important, particularly with masonry pillars. Combined direct compressive and bending (tensile and compressive) stresses are set up in eccentrically loaded pillars and because masonry is only strong under compression any resulting tensile stresses can have a serious effect on the strength.

Cantilever plates (particularly as represented by the teeth of gear wheels) have been studied by many stress analysts [2–7]. Fig. 10.11(b) and (c) show the directional and magnitude changes of principal stress as evaluated by Hagemiers and North [8].

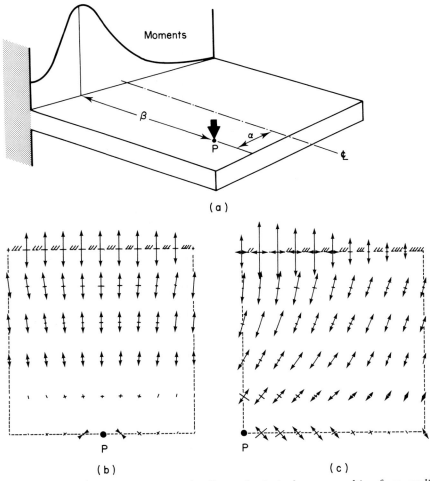

Figure 10.11 *Directional and magnitude effects of principal stress resulting from applied load: (a) eccentric point load-cantilever; (b) centrally loaded; and (c) corner eccentric loading.*

10.3.6 Torsion

Aeroplane structures, box bridges, ship's hulls, and tower blocks all experience torsional effects and suffer the resulting consequences of shear stresses. Under simple torsional conditions the effects of torque applied to shafts of uniform circular section can be related to the resulting angle of twist (torsional flexibility) and shear stress in terms of shaft diameter, polar second moment of area and shear stress.

Rectangular sections under torsion pose some further problems for evaluation. Warping when twisted causes the problem with non-circular shafts as shown in Fig. 10.12. It can be shown mathematically that the stress distribution varies from the centre as shown in Fig. 10.13.

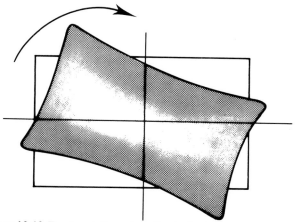

Figure 10.12 *Torsional distortion of a shaft of rectangular section.*

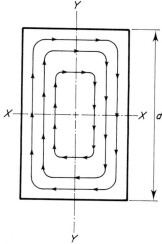

Figure 10.13 *Torsional shear stress in a shaft of rectangular section.*

10.4 Failure

10.4.1 Through elasticity to plasticity hypothesis

When a component of a structure fails, the applied load will firstly apply stresses within the elastic range, then, as either the load increases and/or the inherent strength of the material decreases, so will the component experience stresses beyond its elastic limit. A steel structure does not fail as soon as the edge stress at any cross section reaches the yield point, it continues to bear load while a control core of the material is in the elastic state.

Yield first occurs at the extreme fibres of the weakest section. These fibres are then considered to be in a plastic state. Further increased load produces considerable deflection or strain together with a redistribution of stress.

With mild steel increased strain may occur without the stress exceeding the yield point – plastic strain at the yield stress may be over 10 times the elastic strain. Resistance is not possible without excessive strain once the whole cross section of a structure becomes plastic. Thus a 'plastic hinge' develops. This effect is shown in Fig. 10.14.

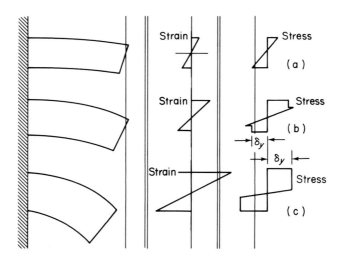

Figure 10.14 *Progress from an elastic through to a fully plastic load-bearing state: (a) elastic state; (b) partially plastic state; (c) fully plastic state;* δ_y = *lower yield stress.*

According to the design integrity of the structure (e.g. whether comprising a series of simple support beams, built-in beams of rigid frames, etc.) so the existence of one or more plastic hinges will presage collapse.

The load to produce complete collapse is called the 'collapse load'.

10.4.2 Through instability

Medieval masonry buildings depended essentially upon the shaping and arrangement of consistent stone blocks, so that each block remained in position under the combined action of its own weight and of the forces applied to it by adjoining blocks. Mortar between such blocks, unlike modern cements, was little more than an even bedding adjoining blocks; the size and weight of the structures rarely produced crushing forces.

Disturbing forces due to the weight of men or to wind forces, tended to be small compared with the gravity forces acting on the masonry itself. If a completed masonry structure could successfully 'stand up', it was likely to do so for a long time (provided there was not too much foundation settlement).

For a uniform rectangular block ABCD (Fig. 10.15) resting on a rigid horizontal surface MN with its centre of gravity at G, and subjected to a lateral force P at the level of G, resisted against sliding by a frictional force acting along CD, it was shown by Pugsley [9] after consideration of the Moseley theory [10] that no tilting will result until

$$P = \frac{a}{h} W \qquad (10.2)$$

when rotation about D will just start. As soon as this happens, the arm h of the force P will increase to $h \cos \theta + a \sin \theta$, where θ is the amount of rotation

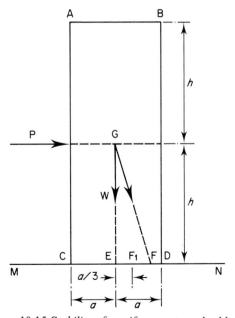

Figure 10.15 *Stability of a uniform rectangular block.*

about D, and the arm of the restoring force W will decrease to $a \cos \theta - h \sin \theta$. Thus, for equilibrium at any stage as θ grows, P must be steadily reduced from the value in (10.2) to

$$P = \frac{a - h \tan \theta}{h + a \tan \theta} W \tag{10.3}$$

This of course falls to zero when θ becomes such that

$$\tan \theta = \frac{a}{h} \tag{10.4}$$

and the conformation has reached the stage illustrated in Fig. 10.16.

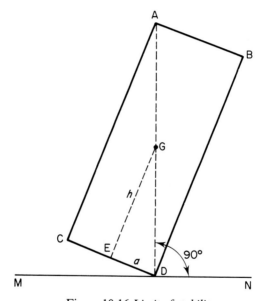

Figure 10.16 *Limit of stability.*

It was shown by Navier [11] and Rankine [12] that critical stability occurs when no tension arises anywhere across the joint. The limiting value for P is given by

$$P = \frac{1}{3} \frac{aW}{h} \tag{10.5}$$

which is one third of the overturning load of equation (10.2). Thus as a result of this no-tension condition, engineers have come to adopt a working limit for P which is one third of that starting failure by overturning. In other words, they have adopted a 'load factor' of 3.

Voussoir arch

An arch of separate stone blocks is capable of carrying some superimposed loads. The stability of such an arch was appraised by Pippard who showed that failure depended upon the fixity at the abutments.

Joints 'open' by small relative rotations of one block to another thereby forming local 'pins' or 'hinges'. One or two of these hinges can arise (as by abutment settlement, for example) with no overall danger to the arch, but as soon as four such hinges have formed the arch becomes a three-bar mechanism and collapses. Fig. 10.17 shows the various stages by which such hinges develop. It will be seen that as the concentrated load P increases, joints at positions 1, 2, 3 and 4 open progressively, the first crack appearing at a load equal to 0.44 of the failing load.

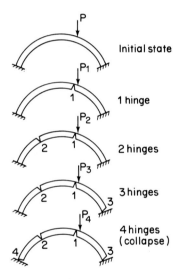

Figure 10.17 *Stages in the stability of a Voussoir arch.*

10.4.3 Rigid structures

Complete collapse is preceded by some non-catastrophic happening. There are degrees of suddenness in the mode of collapse; in the one case, the resistance to load starts to fall away more or less linearly; in the other, the 'static' resistance suddenly falls to zero, the structure having become a mechanism. In both cases the 'prelude to failure' is a minor happening – a crack; in both cases, if the applied force at failure is steadily maintained, collapse is sudden and involves the whole structure.

10.5 False brinneling: Hertz stresses

Indentation stresses which may have caused surface flow with soft materials may be associated with quite high tensile and shear stresses below the surface.

This effect was first examined by Hertz [14] who derived equations for the magnitude and type of stresses imposed on and below a surface during compression. A typical profile for pressure distribution assuming the contact surface to be a flat narrow rectangle is shown in Fig. 10.18.

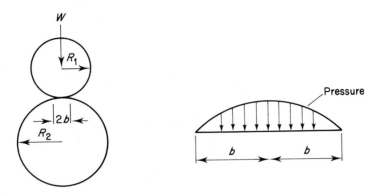

Figure 10.18 *Hertzian stress profile – two cylinders.*

Timoshenko [15] showed (Fig. 10.19) that the compressive stresses decreased with depth but that while the shear stress was zero at the surface layer it increased to a maximum value of $0.304b$ at a depth of $0.78b$.

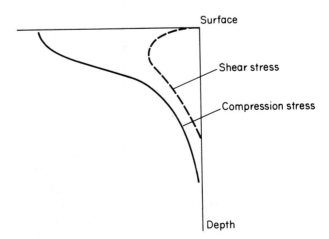

Figure 10.19 *Sub-surface stress distribution under high contact pressure (Timoshenko).*

10.6 Impact dynamics

Stress waves are set up by collisions so that the effect of impact can be the application of repeated heavy blows at rapid, microsecond intervals. The ability to analyse the dynamics of such collisions is illustrated by the following, bearing in mind that for steel with a linear elastic strain wave velocity of 5182 m/s (17 000 ft/s) the limiting velocity before the impacted material is strained beyond the elastic limit is 5.2 m/s (17 ft/s).

10.6.1 Compressive impact stresses

Metal bars which come into collision are subject to longitudinal elastic stress waves. Three impact situations are shown in Fig. 10.20:

(1) Opposing bars which become stationary at impact;
(2) A bar which strikes an infinite, rigid, unyielding mass;
(3) A bar which is struck on one end and moves off.

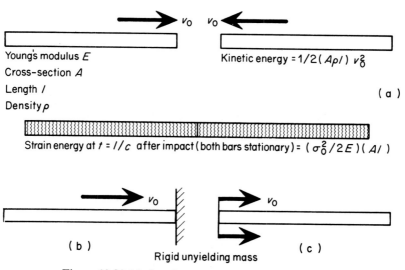

Figure 10.20 *Modes of stress wave generation in bars.*

Theoretical analysis [16] based on the equality of pre-contact kinetic energy to post-contact strain energy produces the following equation for (elastic) contact stress (σ):

$$\sigma = (E\rho)^{1/2} v_0 \qquad (10.6)$$

where E is Young's modulus, ρ density, and v_0 the contact velocity.

10.6.2 Fracture under stress waves

Materials such as rock or concrete which are strong in compression but weak in tension experience scabbing or spalling at free surfaces under impact loading. This is due to the reflection of incident compressive impulses arising from explosive or high speed impact [17]. By assuming the pulse generated by an explosion to have a triangular (sawtooth) profile, Zukas [18] derived expressions for the various stresses and showed that reflection occurs at the free surface with a maximum tensile stress.

There is a difference between static and dynamic fracture. In static fracture one fracture or crack forms and propagates through the material to separate it into two parts. Fragmentation occurs by repeated branching of the original crack. Dynamic fracture, on the other hand, consists of four basic stages:

(1) Rapid nucleation of microfractures at a large number of locations in the material.
(2) Growth of the fracture nuclei in a rather symmetric manner.
(3) Coalescence of adjacent microfractures.
(4) Spallation or fragmentation by formation of one or more continuous fracture surfaces through the material.

10.6.3 Spall under impact

Several types of spall failure can be identified:
(1) Ductile spall – cracking detectable only under microscopic examination.
(2) Phase transformation spall – occurs in materials involving a pressure-induced change.
(3) Ultimate spall – characterized by disintegration of part of the material.

Both ultimate and phase-transformation spall occur at relatively high stress levels. Ductile spall occurs at low pressures and involves the same basic mechanisms (crack nucleation and propagation) that are associated with lower strain-rate failures. It is the most extensively studied of the three types. Three levels of ductile spall have been recognized and classified according to the level of damage. The incipient spall threshold is defined as that combination of stress amplitude and pulse duration below which no damage is detected in the impacted specimen, which, after sectioning, polishing and etching, is viewed at about $100 \times$ magnification.

10.6.4 Impact deformation: cone collapse

A striking example of the capability for computational mathematics to analyse a dynamic phenomenon is given in Fig. 10.21 derived by Hallquist[19]. This steel

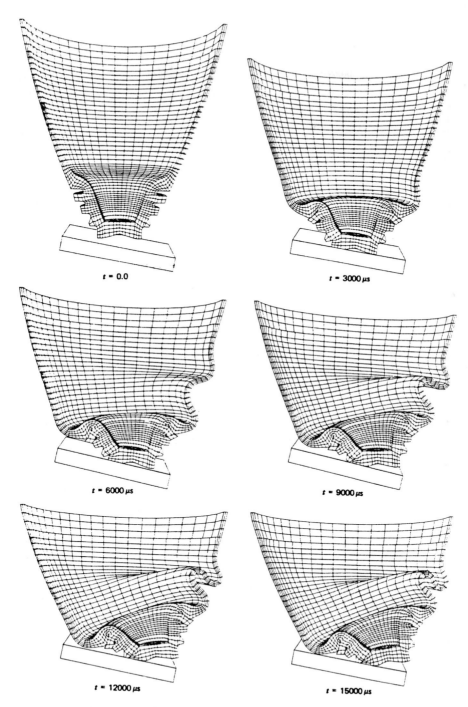

$t = 0.0$

$t = 3000\,\mu s$

$t = 6000\,\mu s$

$t = 9000\,\mu s$

$t = 12000\,\mu s$

$t = 15000\,\mu s$

Figure 10.21 *Deformation history for nose-cone impact [19].*

nose cone was designed to limit the resultant force transmitted to the aft part of a space vehicle.

Computation was performed with the DYNA 3D finite-element computer code. Deformation shapes taken at intervals of 3000 s showed that the peak deformation would be reached at 15 ms after which it would rebound – an effect which was confirmed by experiment.

10.6.5 Impact loading: penetration

During the demolition of buildings made of prestressed concrete, scabs or spall fragments may form as a consequence of rapid unloading. With the reflection of compressive stress waves from a free surface, it has been shown [20] that such debris may acquire speeds of the order of 100 m/s.

Failure due to impact penetration can occur in several ways as shown in Fig. 10.22 [21] with the following features:

(1) Spalling, tensile failure as a result of the reflection of the initial compressive wave from the rear surface of a finite-thickness plate, is a commonplace occurrence under explosive and intense impact loads, especially for materials stronger in compression than in tension.

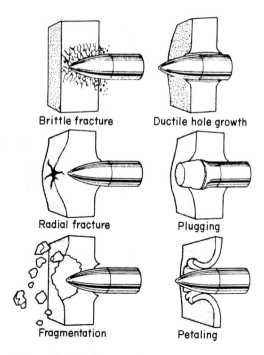

Figure 10.22 *Failure modes in impacted plates [21].*

(2) Scabbing is similar in appearance, but here fracture is produced by large deformations and its surface is determined by local inhomogeneities and anisotropies.

(3) Fracture, as a result of an initial stress wave exceeding a material's ultimate strength, can occur in weak, low-density targets while radial cracking is common in materials such as ceramics where the tensile strength is considerably lower than the compressive strength.

(4) Plugging in which impact is made by a blunt or hemispherical-nosed striker at a velocity close to the ballistic limit (the minimum velocity required for performation) of a finite-thickness target results in the formation of a nearly cylindrical slug of approximately the same diameter as the striker. The target material is constrained by the geometry and motion of the projectile to move forward in the direction of penetration. Separation of the plug from the target may occur by a conventional fracture mode, that is, void formation and growth in shear.

(5) Adiabatic shearing is a plug-separation mechanism which is characterized by the formation of narrow bands of intense shear. It is generally accepted that the adiabatic-shear instability develops at a site of stress concentrations in an otherwise uniformly straining solid. The work of plastic deformation is converted almost entirely into heat which, because of the localized high deformation rates, is unable to propagate a significant distance away from the plastic zone; shear strain rates within adiabatic-shear bands of up to $10^7 \, \text{s}^{-1}$ and temperature within the band up to $105 \, ^\circ\text{C}$ can occur.

(6) Petalling which is produced by high radial and circumferential tensile stresses after passage of the initial stress wave. The intense stress fields occur near the tip of the projectile.

Bending moments created by the forward motion of the plate material pushed by the striker cause the characteristic deformation pattern. It is most frequently observed in thin plates struck by ogival or conical bullets at relatively low-impact velocities or by blunt projectiles near the ballistic limit. Petalling is accompanied by large plastic flows and permanent flexure. Eventually, the tensile strength of the plate material is reached and a star-shaped crack develops around the tip of the projectile. The sectors so formed are then pushed back by the continuing motion of the projectile, forming petals.

10.7 Finite element analysis

10.7.1 Principles

The finite element method of analysis is a modelling technique whereby a load-bearing structure is considered as a notional assembly of discrete (finite) elements. These are derived by laying imaginary grid lines over surfaces and through solid bodies to produce subdivisions which have simple shapes.

Each grid line intersection is called a node. Elements themselves are described by the numbers and coordinates of the relevant nodes. It is assumed that the load-deformation characteristics of each element can be defined. By assembling the elements together, the characteristics of the complete structure are reproduced. Matrix algebra can be used to describe these characteristics.

Characteristics are expressed either as a flexibility matrix or a stiffness matrix. These are in terms of nodal displacements corresponding to unit nodal forces, or vice versa.

Displacement can be automated for the computer. After data defining the structural model (in terms of the constituent elements, their connectivity), and the loading and boundary (support conditions) have been read into the computer, the analysis proceeds automatically.

Fig. 10.23 for a multi-tubular joint was derived by finite element analysis to assess the stress distribution of loads applied to such a joint.

Finite element analysis is performed by many design consultants, computer engineers, Universities and Polytechnics. The following notes are based on a few, highly respected, sources who made case studies available.

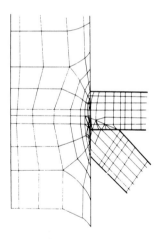

Figure 10.23 *Detailed stress analysis of a tubular joint described by curved shell elements.*

10.7.2 Offshore structure procedure

Programs ASASFLOAT and ASASLAUNCH [22] were developed to compute stresses and strains involved during the floating and launching of a jacket structure for an offshore rig. They incorporate the following:

(1) Three-dimensional launch trajectory.
(2) Jacket member load output.
(3) Small and large angle stability reports for structures.

(4) Jacket upending through controlled flooding.
(5) Program restart facility to aid parametric sensitivity studies and efficient design modifications.
(6) Ability to perform a detailed jacket stress analysis due to basic ASAS compatibility.

ASASFLOAT computes the position-dependent hydrostatic loading on a free-floating structure. The program produces a hydrostatic model from a range of elements. The initial position of the jacket must be defined relative to the water plane. The program then proceeds with the analysis. An 'out of balance hydrostatic force report' is calculated for the jacket in some arbitrary position. Through the 'stability report', the jacket is moved vertically until the displaced mass of water is equal to the mass of the structure.

Fixed increments may be applied to the initial position of the jacket. When applied in the 'out of balance' mode, the jacket may be rotated about the initial position of the centre of mass, or moved vertically at a fixed inclination. When applied in the 'stability' mode, the jacket may be rotated through a prescribed increment and then moved vertically until the displaced mass of water is equal to the mass of the structure (rotation at constant volumetric displacement).

A further command, 'equilibrium only', instructs the program to move the jacket from the initial position to a position of hydrostatic equilibrium. This position is not necessarily stable. The jacket stability can be assessed from the 'small angle stability properties' which are printed for the equilibrium position. Large angle stability characteristics for the jacket may be output by the use of rotational increments in the 'stability' mode of operation.

ASASLAUNCH (Fig. 10.24) computes the time-dependent trajectory of two structures: the jacket and the barge. The program requires hydrostatic/dynamic models for each structure, together with a description of the track

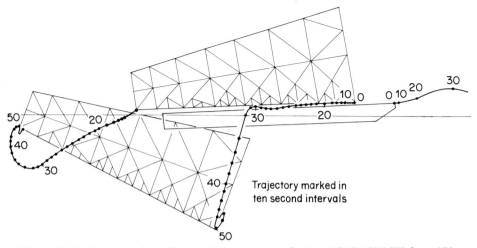

Figure 10.24 *Computer launching: trajectory appraisal using ASALAUNCH for a 170 m jacket including the effect of barge motion.*

system used for launching. One track system is located on the jacket; the other track system is defined on the barge deck, together with a rocker arm at the stern.

Initially, the jacket is positioned above the barge with the two track systems coincident. A winch force is applied to the jacket to draw it along the barge tracks and onto the rocker arm. Once the rocker arm begins to rotate or the jacket is sliding faster than the 'cut off' velocity, the winch force is released. The analysis can then proceed until the jacket comes to rest in the water.

The analysis of the jacket/barge trajectory is fully three-dimensional. The launch track system is designed to restrain the jacket from slipping sideways or yawning across the barge tracks. However, the jacket is free to roll onto one track. While the jacket remains in contact with the barge, the six degrees of freedom reaction at the rocker pin may be output, together with the component forces on the two structures.

The nodal forces derived at any one time step may be transferred as a load case for the stress analysis of the jacket structure using ASAS. The static and dynamic loads for each element in the jacket are reduced to their corresponding nodal forces. These forces are then applied to the ASAS stress model from the nodes defined in the ASASLAUNCH model.

10.7.3 Structural deformations and vibrations

In deep water North Sea platforms, a sway period of less than four seconds would guarantee acceptable levels of dynamic activity in response to storm waves. Primary bending moment periods may be calculated through ASAS. At the same time natural frequencies and mode shapes can be determined for frequency response analysis.

To find natural frequencies for large platforms (500 joints, 1000 members) is expensive in computer time, even with the most efficient techniques for eigenvector extraction. ASAS-G obtains mode shapes and frequencies by subspace iteration in an extremely accurate and economic facility for the user, results are shown in Figs 10.25, 10.26.

Features of ASAS-G natural frequency analysis include:

(1) The subspace iteration technique enables the eigenvectors to be determined for all the degrees of freedom present in the original system without the need of condensation (or eigenvalue economization). This ensures a more accurate solution and reliable stress recovery in further analysis and is recommended for all fatigue work or analyses where detailed response stresses are of paramount importance.

(2) Condensation methods may also be specified whereby the several thousand degrees of freedom normally present in an offshore structure will be reduced to one hundred or so 'master freedoms' for the purposes of dynamic

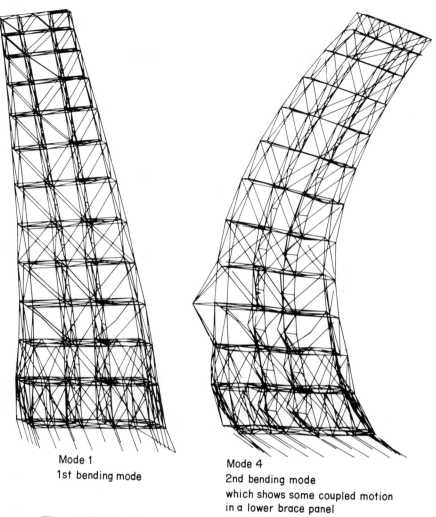

Mode 1
1st bending mode

Mode 4
2nd bending mode
which shows some coupled motion
in a lower brace panel

Figure 10.25 *Two bending modes for a jacket in 250 m water depth.*

analysis. Such methods are adequate if only deflections or overall dynamic effects are required.

(3) During execution, the program gives a good, early estimation of the lowest sway frequency and the running may be halted at any point for consideration to be given to the frequency values and subsequently restarted.

Computations which can be made include:

(1) Consolidation of all components of mass and rotational inertia to node points (program ASASMASS). Added mass is calculated and combined with structural mass and directly input non-structural mass.

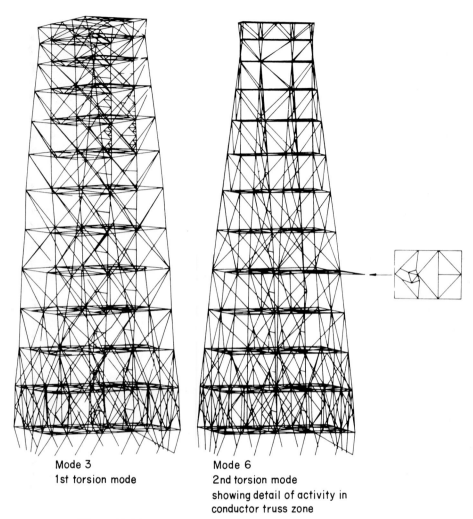

Mode 3
1st torsion mode

Mode 6
2nd torsion mode
showing detail of activity in
conductor truss zone

Figure 10.26 *Two torsional modes of a jacket platform as in Fig. 10.25.*

(2) Natural frequencies and mode shapes (program ASAS-G DYNAMICS). The program assembles a mass and stiffness matrix and produces a conventional normal mode solution. Following the necessary matrix manipulation, the resulting normalized eigenvectors are loaded on to backing file for subsequent use.

10.7.4   Some users

SESAM, a program system from Det Norska Veritas, was used by the following:

(a) Phillips Petroleum Company

A project to establish a structural contingency system for the steel platforms at the Ekofisk field. 26 major platforms, 25 tripods and 23 bridges are included.

Frame analysis in the SESAM-80 program system connects to pile-analysis procedures, such that jackets can be analysed directly with non-linear piles and interactive modelling through graphic terminals. Data for each structure are stored on separate magnetic tapes, making information available at short notice.

(b) Thyssen Nordseewerke GmbH

A series of analyses using SESAM-69 was carried out in connection with submarines built at Emden. Two overall analyses were performed for the aft-body and ballast compartment respectively. Each analysis comprised more than 100 000 degrees of freedom. The models consisted of eight-noded curved shell elements exclusively, by which the superelement technique was extensively utilized. A number of local analyses of different structural details of the submarine optimized the weight and geometry. Both eight-noded curved shell and twenty-noded hexahedron elements were utilized.

(c) MAN–Hochtief–Harms

Dynamic response of the MANDRILL 400 deep-water production platform for water in depths from 200 to 500 m was determined by the German group MAN–Hochtief–Harms. The design is based on the familiar principle of a photographer's tripod (Fig. 10.27) adapted to offshore technology. The scope of

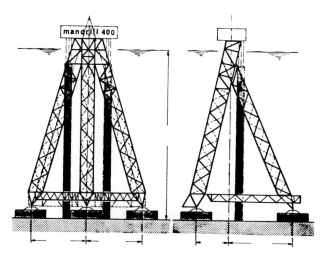

Figure 10.27 *Tripod design of MANDRILL 400.*

the analysis included evaluation of probabilistic short-term statistics and fatigue damage analysis of selected structural nodes, it was performed with SESAM-69 and related programs, using the following program modules:

(1) NV426 wave loads on fixed offshore structures.
(2) NV432 probabilistic fatigue analysis.
(3) NV337 analysis of three-dimensional frames.
(4) NV336 general superelement program.
(5) NV151 special postprocessor for NV337.

The model comprised 394 elements and 204 nodes/1224 degrees of freedom, including both riser frames.

(d) Davy McKee Minerals & Metals Ltd

Work carried out for UK process and steelmaking industries includes:

(1) Analysis of heat conduction and strength of a complex phosphoric acid reactor tank – information necessary for dimensioning the reinforcements of this concrete structure;
(2) Thermal calculations;
(3) Combined heat transfer and stress analyses;
(4) Analysis of buckling and buckling behaviour;

as applied to storage tanks, blast furnace refractories, catalytical regenerators, equipment for transport of nuclear waste.

Thermal and strength problems are incurred during the transport of molten metal. For many years refractory lined vessels have been used for the transport of molten metal, such as iron, within the steelmaking and other metals industries. Iron is carried from blast furnace to steel plant by cigar-shaped ladles mounted on railway bogies: these are generally called 'torpedo cars'.

When a new design of ladle was required which would allow lining faster and easier removal and renewal of the brick, a finite element analysis (Fig. 10.28), based on SESAM-69 program modules NV332, NV337, NV336 and NV340, was performed.

The program allows for the fact that the molten iron exerts a 'ferrostatic' pressure over the inside of the ladle, which varies as it is tilted. Similarly, the weight of bricks can be represented as a pressure distribution. Within the program NV332, such distributions are conveniently specified and node forces automatically calculated by the data-generator.

The relative expansion of the refractory lining is calculated manually, allowing for the zones of 'cracking' and 'crushing' of the bricks after the heating up period has ended. This is also represented as an internal pressure. Temperature distributions from around the shell are derived from thermographic photographs of operating cars, to find 'hot spots' due to refractory wear.

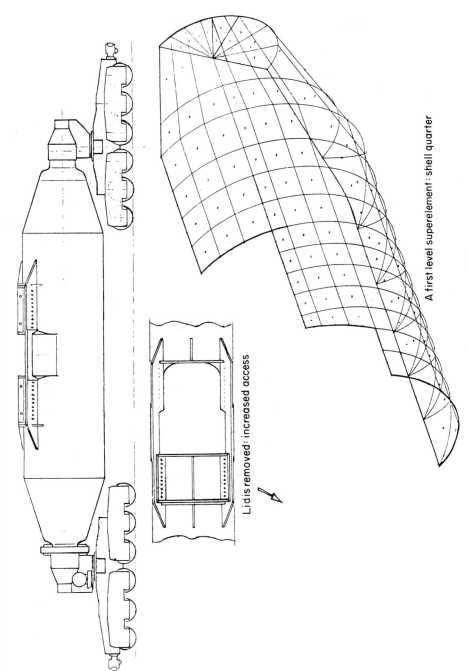

Lid is removed: increased access

A first level superelement: shell quarter

Figure 10.28 *Molten-metal carrying 'torpedo' car.*

Shunting and tilting loads are derived from mechanical calculations on the inertias and motor powers of ladle, locomotive and tilt drive.

10.7.5 Other users

(a) Saab Scania AB

A complete elastic structural analysis of the SAAB-99 car was carried out using the superelement technique applied to each part of the body. The assembly of superelements was thus directly parallel to the assembly of the metal pieces. Shell elements were used throughout, the only exception being the A-beams at the windscreen where beam elements were used. Doors, windows, bonnet and luggage boot were not included in the model, as for the present analysis these parts were not considered to take any loads.

ISO-stress plots of von Mises equivalent stresses were used. More than 500 plots were taken out for the five load cases. Output tables covered nodal displacements, main stresses, reaction forces, stress components and force-distribution, giving 7000 pages. Microfiche output was used, giving less than 50 fiches.

(b) National Industri, Drammen, Norway

A three-dimensional finite element analysis has been performed in order to evaluate the electric field pattern and potential gradient between outer coil and yoke in a transformer. The object of the investigation was to determine the influence of the yoke on the maximal electrical gradient between the coils.

Electric potentials in the modelled volume, the potentials on the surfaces of the construction and permittivities were specified. Data were given as ratio values for the surface potentials on the inner and outer coil, core and yoke, together with permittivity values for the paper insulation and for oil.

Various cross sections of the model were drawn on a graph plotter, completed with perspective plot of the various first level superelements and their assembly on higher levels. Fig. 10.29 shows some of the result-plots of the calculation. The figures are cross sections of the model with 15 degrees spacing, starting in the plane described by the centre line of the core and yoke. This shows that the equipotential curves turn outwards under the yoke, and are rising as the influence of the yoke disappears.

10.8 Stress concentration

10.8.1 Experimental evaluation

Two methods of examining stress concentrations may be used to produce experimental data: brittle lacquers and optical/photoelastic examination.

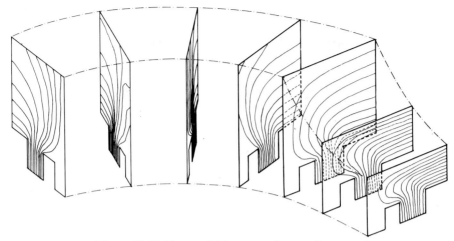

Figure 10.29 *Electric field pattern of a transformer coil.*

Brittle lacquers dry out when coated on the component specimens. Under stress, the coatings crack in a direction at right angles to the maximum tensile stress. Concentrations of stress and strain can be evaluated by comparing the crack lines with a bar, calibrated under known loads and under carefully controlled conditions of temperature and humidity.

Optical/photoelastic examination is based on double refraction exhibited by transparent materials under stress (a phenomenon first reported by D. Brewster 1818, *Phil. Trans.*, p. 156). Polarized light is passed through a plastic model of the component (Fig. 10.30). When load is applied the different stress concentrations distort the specimen such that some of the polarized light is obscured and

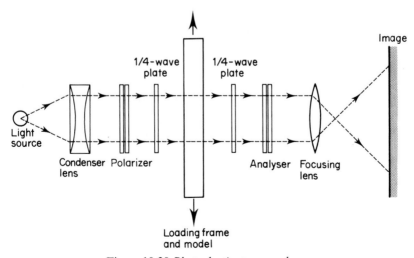

Figure 10.30 *Photoelastic stress analyser.*

part is 'interfered' into coloured fringes. The result is to form a coloured pattern on the screen (or camera plate) which corresponds to the stress distribution in the model. Such patterns (isochromatic fringes) can be used to evaluate stress concentration factors. Initially, celluloid was used for the model; more suitable proprietary materials are now available.

10.8.2 Determination of stress distributions in internal combustion engines

Local stresses in an engine component are greatly influenced by its geometrical shape. Hatton [23] reviewed the photoelastic technique as applied to internal combustion engines after indicating that the stresses within major engine components can be due to any one, or a combination of the following:

(1) Static stresses induced by clamping misalignment of mating faces, interference fits or residual casting stresses.
(2) Quasi-static stresses induced by differential thermal expansion due to temperature gradient or materials of different thermal properties.
(3) Dynamic stresses induced by combustion forces, inertia loads, or resonant vibrations.

Engineers at Perkins Engines Co Ltd, Peterborough, UK, investigated the effects of these loads by means of the following techniques:

(1) Photoelasticity.
(2) Three-dimensional araldite models.
(3) Brittle lacquers.
(4) Electrical resistance strain gauges.
(5) The finite element method and classical elastic theory.

Several complementary techniques are employed. Typically, the behaviour of connecting rods and crankshafts can be investigated (on a qualitative basis)

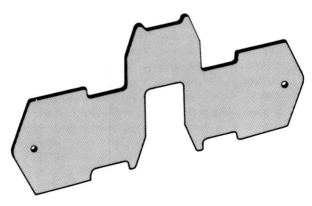

Figure 10.31 *Two-dimensional photoelastic model of a one-crank throw.*

using two-dimensional models which represent the plane of symmetry of these components. The results of such analyses have been used to give an indication of the influence of critical geometrical features on these components. Fig. 10.31 shows a two-dimensional model of one crank throw used to investigate the effect of pin and journal fillet radii size and shape on the critical stresses. The isochromatic fringe patterns displayed in Fig. 10.32 illustrate the concentration of stresses in the fillet and the effect of a poorly shaped fillet. In this case, poor shaping of the fillet will result in a 17% increase in peak fillet stress.

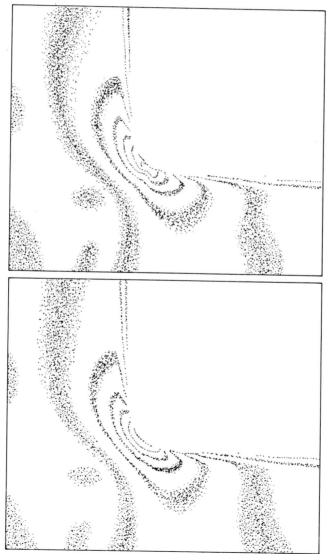

Figure 10.32 *Isochromatic fringe patterns illustrating the concentrations of stresses in the fillet.*

Two-dimensional photoelastic models are particularly useful for pilot investigations into failure mechanisms. Fig. 10.33 shows the distribution of stresses in the outside and inside surfaces of a thin wall section in a grey cast iron cylinder block. The wall blends into a panel immediately above an oilway, and fatigue failures occurred at the junction between the wall and the panel. In work to

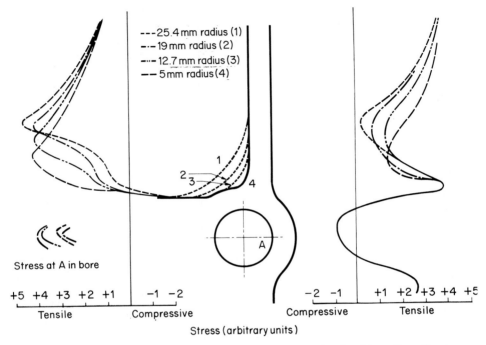

Figure 10.33 *Distribution of stresses at the outside and inside of a thin wall section in a cylinder block.*

determine the failure initiation point and the effect of modifications to the blend radius, the stress distributions obtained from a simple two-dimensional model indicated that failure would probably start from the inside surface. More significant was the fact that increasing the blend radius would not reduce the stresses, but only move the failure point further up the wall. Further investigation using strain gauges to determine the residual casting stresses and the magnitude of the dynamic strains under engine running conditions, confirmed that the failure started from the inside surface.

10.8.3 Photoelastic reflective or coating techniques

A method, similar to polarization on plastic models but which can be used on actual components *in situ,* is described by Hearn [24].

The surface of the component is coated with a layer of the photoelastic material, usually cemented with a reflective type of cement. (One which gives a silver reflective surface behind the photoelastic material). Polarized light is then reflected from the surface, and the fringe pattern is observed directly on the surface of the material by viewing through the analyser instead of by projection onto a screen.

In this case the light passes twice through the birefringent material so that the sensitivity of the technique is doubled. Relatively high strains can be experienced in the model material even under moderate loads, but in this case it is the actual strains in the component surface which are viewed, and these are generally much smaller. Most practical analysis using the coating technique is carried out within the 0–3 fringe order range.

Because the technique has been designed for use in the field (under daylight conditions if necessary) the relatively poor illumination obtained using a monochromatic source can seldom be used to obtain experimental data. Usually a more powerful white light source is used and the coloured isochromatics so obtained used for analysis. Quite often the single colour at a point of interest can be used to give a rough approximation to the strains at that point. The technique is very seldom used as a quantitative measuring tool but rather to give a general picture of the strain pattern across a wide area (Fig. 10.34).

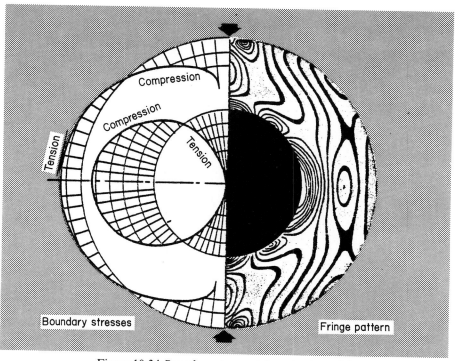

Figure 10.34 *Boundary stresses: photoelastic model.*

10.8.4   Moiré fringe patterns

The principle of the Moiré fringe technique is described in Section 5.21. It can be used for stress analysis. This is a special form of grid method of identifying strain. A grid is marked out on a material and also a transparent master grid produced. After stressing of the material, the undeformed master grid is superimposed on the deformed grid. This produces an effect such as that shown in Fig. 10.35. The resulting interference pattern (Moiré fringes) can be analysed to yield strain values by measuring the distortion differential between the deformed and undeformed coordinates.

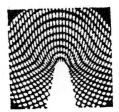

Figure 10.35 *Moiré fringe pattern: notched test piece.*

10.8.5 Strain pattern modelling

A method originated by the author many years ago [25] involved the manufacture of a sectional model as a plunger/cylinder arrangement with a clear (Perspex) front. Suitable strips of coloured clay or 'Plasticene' were packed to form the model. Application of a screw to stress the 'Plasticene' model produced deformations such as those shown in Fig. 10.36.

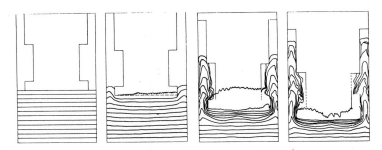

Figure 10.36 *Visualised plastic flow/strain.*

10.8.6 Magnitude of residual stresses

Variations in residual stress throughout a material are affected by depth, the process and the rate at which the process is applied. Considering some of the more usual processes:

(a) Rolling, drawing, extruding

Solid bar and strip produced by rolling, produces residual stress which depends strongly on the depth of the reduction [26]. Light rolling incurs little plastic flow which does not penetrate all the way across the section, and the surface is left in compression with the core in tension. Heavier reductions cause the entire cross section to become plastic (Fig. 10.37). It is usual with a variety of metal-forming processes, including drawing and extruding, for the residual stress to be tensile at the surface although the degree of penetration in a plastic zone depends on the ratio of the roll diameter to work piece size such that the residual stress tends to be compressive for small-diameter rolls and tensile for large-diameter rolls [27].

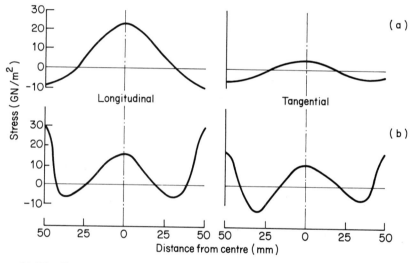

Figure 10.37 *Effect of rolling on residual stresses: (a) 0.5% reduction; (b) 1.0% reduction.*

(b) Shot peening

The favourable effects resulting from the compressive residual stress at the surface of a work piece resulting from a light rolling or drawing operation may be produced in operation from light running-in of cranks as a result of surface rolling, as shown in Fig. 10.38 [28] or by shot-peening, whereby cast iron or steel pellets are 'shot' at a surface by an airblast. The considerable surface compressive residual stress this causes is shown in Fig. 10.39 [29].

(c) Machining

Depth of cut and bluntness of the cutting tool affect the amount of surface stressing. At the very surface itself the residual stress is tensile but this becomes com-

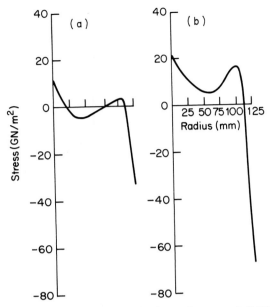

Figure 10.38 *Crank pin: effect of rolling* in situ *under a steady 8 GN/m² load.*
(a) before; (b) after.

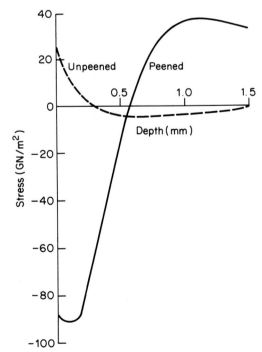

Figure 10.39 *Effect of shot-peening to produce favourable surface compressive stress.*

pressive slightly below the surface. As the metal cools thermal residual stresses develop; accordingly some allowance must be made for cutting speed and the effects of cutting lubricants.

(d) Grinding

The hardness of the metal which is being ground influences the surface residual stress. Soft steels tend to produce tensile surface stresses. Hard steels produce compressive stresses. Below the surface the general characteristic is to produce a high tensile stress which reduces to a near-zero stress with depth (Fig. 10.40).

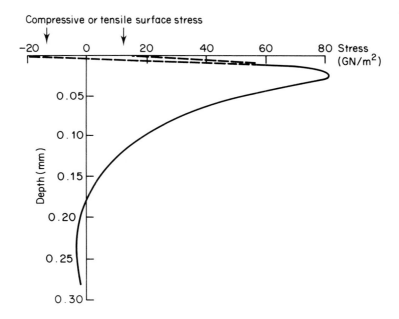

Figure 10.40 *Residual stresses produced by grinding.*

(e) Quenching

Quenching processes induce large residual stresses which give rise to cracking and distortion. Greene [30] identified residual compressive stresses ranging from 860 to 1700 MN/m² (125 000 to 250 000 lb/in²).

An investigation [31] using a deflection bending technique to derive the stress in each layer produced values for residual stress in a quenched manganese carbon steel at different depths varying from 800 MN/m² at the surface to 150 MN/m² at a depth of 0.080 mm below the surface.

10.8.7 Fasteners: limits of eccentric loading

Fasteners such as rivets derive joint strength both from the friction grip and plastic yield in the region of the joint. Some 50% of a hot-driven rivet joint's strength resides in the friction grip [32]. Bolted joints using high-strength bolts can derive more than 50% of their strength from the friction grip.

References

1   Pellini, W. S. and Puzak, P. P. (1965), NRL Report 6300, June.
2   McGregor, C. W. (1935), 'Deflections of a long helical gear tooth', *Mech. Engng,* **57**, 225–7.
3   Holl, D. L. (1937), 'Cantilever plate with concentrated edgeload', *J. Appl. Mech.* **59**, A8–10.
4   Jaramillo, T. J. (1950), 'Deflections and moments due to a concentrated load on a cantilever plate of infinite length', *J. Appl. Mech.,* **72**, 57–72, 342–3.
5   Vartak, G. W. (1956), 'Cantilever moments in plates fixed along one of the long edges and subject to concentrated loads at a point on the other long free edge', *J. Instn Engrs (India),* **37**, Part 1, November, 203–212.
6   Wellauer, E. J. and Seireg, A. (1960), 'Bending strength of gear teeth by cantilever plate theory', *J. ASME,* August, 213–20.
7   Szmelter, J., Sulikowski, T., and Lipinski, J. (1951), 'Bending of a rectangular plate clamped at one edge', *Arch. Mech. Staswanej,* **13**, No. 1, 63–74.
8   Hageniers, O. L., and North, W. P. T. (1976), 'A comprehensive study of the fixed edge bending moments of thin rectangular cantilever plates under point loading', *Trans. ASME,* **98** *Ser. B,* No. 3, August, 766–72.
9   Pugsley, A. (1966), *The Safety of Structures,* London, Edward Arnold.
10 Mosely, H. (1843), *The Mechanical Principles of Engineering and Architecture,* London, Longmans, Green.
11 Navier, C. L. M. H. (1826), *Resume des Lecons de Mechanique,* Paris.
12 Rankine, W. J. M. (1858), *A Manual of Applied Mechanics,* London, Griffin.
13 Pippard, A. J. S., Tranter, E. and Chitty, L. (1936), 'The mechanics of the voussoir arch', *Proc. Inst. Civ. Engrs,* **4**.
14 Hertz, H. (1895), *Gesammelte Werke,* Vol. 1, Leipzig.
15 Timoshenko, S. (1934), *Theory of Elasticity,* New York, McGraw-Hill.
16 Johnson, W. (1977), 'Longitudinal elastic stress waves due to impact', *Engineering,* February, 106–8.
17 Davids, N. (1960), *Stress Wave Propagation in Materials,* New York, Wiley Interscience.
18 Zukas, J. (1982), *Impact Dynamics,* New York, John Wiley, p. 38.
19 Hallquist, J. O. (1979), Lawrence Livermore Lab. Report VCID-17268 Rev. 1.
20 Johnson, W. (1979), *Mechanical Properties of Materials at High Rates of Strain: Conference Series No. 47,* Bristol, Adam Hilger/Institute of Physics.
21 Zukas, J. (1982), *Impact Dynamics,* New York, John Wiley, p. 164.
22 'A monograph on the analysis of offshore structures with ASAS/FATJACK related programs' (1979) Atkins Research and Development, Woodcote Grove, Ashley Road, Epsom, Surrey, KT18 5BW, UK (Tel. Epsom 26140).
23 Hatton, C. R. (1978), *Photoelastic Techniques in IC Engine Design,* Chartered Mechanical Engineers, September, pp. 55–9.

24 Hearn, E. J. (1968), *Stress Analysis: Design Engineering Handbooks,* Product Journals Ltd, Summit House, Glebe Way, West Wickham, Kent.
25 Collacott, R. A. (1959), 'Visualisation of plastic flow patterns', *Engineering,* 18 December.
26 Buhler, H. and Buchholtz, H. (1934), 'Effect of reduction and cross section by cold drawing on residual stresses in rods', *Arch. Eisenhuttenwesen,* **7**, 427–30.
27 Baldwin, W. M., Jr (1949), 'Residual stresses in metals', *Proc. ASTM,* **49**, 539–83.
28 Horger, O. J. (1950), 'Residual stresses', *Handbook of Experimental Stress Analysis,* M. Hetenyi (Ed.), New York, Wiley, pp. 459–578.
29 Lessells, J. M. and Broderick, R. F. (1983) 'Shot-peening as protection of surface damaged propeller blade materials', *Proc. Int. Conf. on Fatigue of Metals,* London, Inst. Mech. Engrs, 617–27.
30 Greene, O. V. (1930), 'Estimation of internal stresses in a quenched hollow cylinder of carbon steel', *Trans. Am. Soc. Steel Treating,* **18**, 369.
31 Rai, J. K., Mishea, A., and Rao, V. R. K (1978), 'Some comments on residual stresses induced during quenching', *Int. J. Prod. Res.,* **16**, No. 6, 463–7.
32 (1950), *Fastener's Data Handbook,* Industrial Fasteners Institute, p. 36.

11
Fracture mechanics

11.1 Introduction

A structural component is considered to have failed when 'permanent' deformations have occurred. The material need not have ruptured. For such conditions a number of criteria were submitted:

(1) Maximum principal stress theory (Rankine) – when the maximum principal stress reaches the value of the maximum stress at the elastic limit in simple tension [1]:
   (a) This theory is satisfactory when failure occurs under tensile cleavage fracture.
   (b) This theory is not satisfactory when applied to plastic deformations under shear components of the applied stress – when the difference between principal stresses applies.
(2) Maximum shear stress (or shear difference) theory (Guest and Tresca) – implies that failure will occur when the maximum shear stress under complex loading reaches the value of the maximum shear stress for the material in simple tension at the elastic limit [2].
(3) Strain energy (Haigh) – argues that as strains are reversible up to the elastic limit the energy absorbed by the material should be a single-valued function at failure, independent of the stress system, i.e.

$$\begin{pmatrix} \text{strain energy/unit volume} \\ \text{causing failure} \end{pmatrix} = \begin{pmatrix} \text{strain energy at the elastic} \\ \text{limit in simple tension} \end{pmatrix}.$$

(4) Shear strain energy theory (Mises and Hencky) – at failure the shear strain energy in the complex system is equal to that in simple tension [3,4].
(5) Maximum principal strain theory (St Venant) – that the maximum strain ε_1 in a complex system is related to stresses (σ_1, σ_2), modulus of elasticity (E) and Poisson's ratio (v) by

$$\varepsilon_1 = \frac{1}{E}(\sigma_1 - v\sigma_2 - v\sigma_3)$$

In practice, failure means rupture. Accordingly, the conditions preceding rupture tend to become of further interest. These conditions relate to cracks and their development.

This chapter has been introduced because it is important to establish the limits to which cracks may be allowed to grow before critical conditions arise. In other words, the aspects of fracture mechanics presented in this chapter are intended to offer persons concerned with integrity decision making some basis for their actions (or lack of action).

11.2 Flaw size significance

Course notes by Professor Garrett of the University of Witwatersrand, South Africa establish the objectives of fracture mechanics as providing answers to the questions:

(1) What is the residual (component or structure) strength as a function of crack size?
(2) What is the critical crack size at service load?
(3) How long might it take for the crack to grow from initial to critical size?
(4) What size of pre-existing flaw can be permitted? (philosophy: living with defects).
(5) How often should we inspect our structure?

Fracture mechanics is intended to quantify resistance to fast (catastrophic) failure by relating the failure stress of a component or structure containing a flaw of given size to the *size* of that flaw, through a previously determined value of material property known as *fracture toughness*. This 'triangle of integrity' links the three factors as shown in Fig. 11.1.

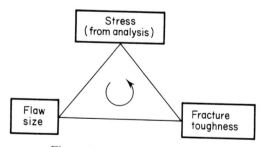

Figure 11.1 *Triangle of integrity.*

11.3 Stress concentrations

In 1913 C. E. Inglis published a paper in the *Transactions of the Institution of Naval Architects* to the effect that geometrical irregularities, such as holes, cracks and sharp corners, which had previously been ignored, may raise the local stress – often only over a very small area – very dramatically indeed. Holes and notches may cause the stress in their immediate vicinity to exceed the breaking stress of a material, even when the general level of stress in the surrounding neighbourhood is low (and the structure might appear to be perfectly safe).

Almost any hole or crack or re-entrant in an otherwise continuous solid will cause a local increase of stress. Fig. 11.2(a) shows a smooth, uniform bar or plate of material, subject to a uniform tensile stress. The lines crossing the material represent what are called 'stress trajectories'; that is to say, typical paths by which the stress is handed on from one molecule to the next. In this case they are, of course, straight parallel lines, uniformly spaced.

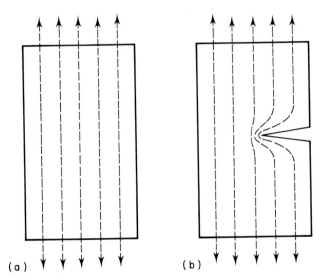

Figure 11.2. *Stress distribution in a uniformly loaded bar: (a) no crack present; (b) cracked.*

Interruption of a number of these stress trajectories by making a cut or a crack or a hole in the material, causes the forces which the trajectories represent to be balanced and react in some way. The forces have to go round the gap, and as they do so the stress trajectories are crowded together to a degree which depends chiefly upon the shape of the hole (Fig. 11.2(b)). With a long crack, crowding around the tip of the crack is often very severe. Thus in this immediate region there is more force per unit area and so the local stress is high. It is possible to calculate the increase of stress which occurs at the tip of an elliptical hole in a solid which obeys Hooke's law. Results like this apply with sufficient accuracy to openings of other shapes.

11.4 Crack characterization

Cracks can range in size from ängstrom lengths through to metres, an indication of this being shown in Fig. 11.3. It was reported [5] that the crack size at fracture

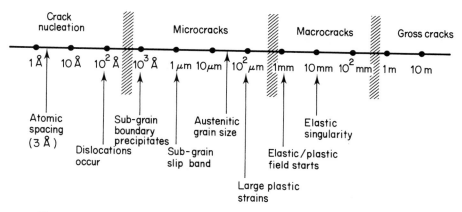

Figure 11.3 *Range of crack sizes (note that* 10^4 Å $= 1$ μm; 10^3 μm $= 1$ mm*)*.

for the 35 mm thick hot box dome of a nuclear reactor was as shown in Table 11.1. A defect reporting system is recommended as follows:

Record if a crack is
150 mm long – any depth
25 mm long – 6 mm deep
any length – 12 mm deep

Table 11.1 Crack sizes/virgin metal

	Crack depth at fracture (mm)	Uncracked depth (mm)
Dome	26	9
Skirt	50	–
Support ring	31	–

Weakening effects are not exclusively caused by holes, cracks and other deficiencies of material. One can cause stress concentrations by adding material, if this induces a sudden local increase of stiffness. Patching a thick plate of armour on the thin side of a warship will have an overall weakening effect. There is a dictum 'partial strength produces general weakness'.

This occurs because the stress trajectories are diverted just as much by an area which strains too little, such as a stiff patch, as they are by an area which strains too much, such as a hole. Anything which is elastically 'out of step' with the rest of the structure will cause a stress concentration and may therefore be dangerous.

In most engineering materials cracks grow in a direction which is perpendicular to the maximum principal applied tensile stress (in fracture mechanics terms, a Mode I crack) [6].

11.5 Crack path

The implications of stress concentrations around cracks have often been dismissed by invoking the 'ductility' of metals to imply that the overstressed metal at the tip of a crack simply flowed in a plastic way and relieved itself of any serious excess of stress. Thus, the sharp tip of the crack became 'rounded off' so that the stress concentration was reduced and safety restored. This has the merit of being at least partly true, though in many cases the stress concentration is by no means fully relieved by the ductility of the metal; local stress often remains much higher than the commonly accepted 'breaking stress'.

Even in the best engineering circles, scandals occurred from time to time. In 1928 a liner of 56 551 tons had an additional passenger lift installed. In the process rectanguar holes, with sharp corners, were cut through several of the ship's strength decks. Somewhere between New York and Southampton, when the ship was carrying nearly 3000 people, a crack started from one of these lift openings, ran to the rail, and proceeded down the side of the ship for many feet before it was stopped, fortuitiously, by running into a porthole. The liner reached Southampton safely and neither the passengers nor the press were told. Very much the same thing happened to the second largest ship in the world, the American transatlantic liner *Leviathan,* at about the same time. Again the ship got safely into port and publicity was avoided.

11.6 Critical crack failures

Cracks (either due to fatigue or initially present in the structure) may lead eventually to brittle failure. This is of a catastrophic nature.

Failure as related to a crack is assumed to be accomplished on reaching a critical length. If oil or water tightness is important then the initial crack length (e.g. a crack which can endanger this tightness) may be used as a fracture criterion. If the cracked part is expected to serve in a comparatively low temperature then the critical toughness of the material corresponding to this environment should be used as a fracture criterion. A criterion based on residual static strength will be based on a critical value after which the structure may be in jeopardy.

11.6.1 Cracks on ships

Committee XIII of the International Institute of Welding [7] reported a tanker which had 129 cracks located mostly in bulkhead stiffeners; they resulted from a purely local tank loading, not from the hogging and sagging loads induced by the sea.

The ship classification society of Japan [8] reported that a survey of the inner structural members of oil tanks in 97 tankers indicated that a considerable number of cracks occurred at welded structural discontinuities. Most of the cracking occurred when the ships were from two to four years old.

An American survey of cracks in 210 tankers exceeding 122 m (400 ft) in length and built after the Second World War found that 66 were suffering from cracks. Only one tanker of the 66 had brittle cracks, while the others had fatigue cracks.

Damage to the structural parts of a ship's hull, examined at the end of 1967, reported a large number of cracks in the bottom of a 60 000 tons (deadweight) tanker built in 1964 [9]. About 1000 cracks were found, nearly all initiated at the scallops in the web of the bottom and side longitudinals or at the foot of the tripping brackets.

11.6.2 Critical crack length

Griffith [10] showed in 1921 that Inglis's stress concentration was simply a mechanism (like a zip fastener) for converting strain energy into fracture energy, and that no failure mechanism will work unless it is continually supplied with energy. Stress concentration requires energy to keep prising the atoms of a material apart. If the strain energy dries up, the fracture process stops.

A piece of elastic material stretched and then clamped at both ends is a system for which no mechanical energy can get in or out. It is a closed system containing a defined amount of strain energy. For a crack to propagate through this stretched material the necessary work of fracture must be supplied from the energy. If we consider the specimen to be a plate of material one unit thick, then the energy bill will be WL where W is the work of fracture and L the length of crack. This is an energy debt which increases linearly as the crack length L increases.

This energy is found immediately from internal resources, i.e. somewhere in the specimen the stress must be diminished. This can occur because the crack will gape a little under stress and thus the material immediately behind the crack surfaces is relaxed (Fig. 11.4). Roughly speaking, the two triangular areas shaded in Fig. 11.4 will give up strain energy. As one might expect, whatever the length, L, of the crack, these triangles will keep roughly the same proportions, and so their areas will increase as the square of the crack length, i.e. as L^2. Thus the strain energy release will increase as L^2.

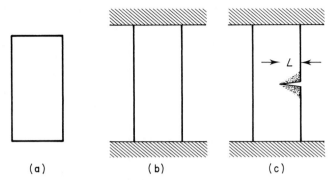

(a)   (b)   (c)

Figure 11.4 *Crack development: (a) unstrained; (b) strained/rigidly clamped; and (c) cracked – relaxation and release of strain energy.*

Thus the core of the whole Griffith principle is that, while the energy debt of the crack increases only as L, its energy credit increases as L^2. The consequences of this are shown graphically in Fig. 11.5 where OA represents the increased energy requirement as the crack extends, and is a straight line. OB represents the energy released as the crack propagates, and is a parabola. The net energy balance is the sum of these two effects and is represented by OC.

Up to the point X the whole system is consuming energy; beyond point X energy begins to be released. It follows that there is a critical crack length, which

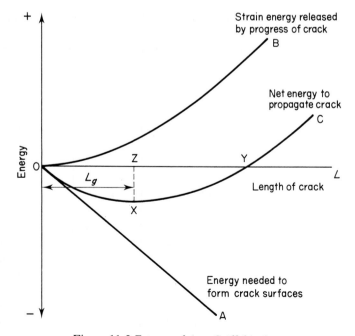

Figure 11.5 *Energy release: Griffith's theory.*

we might refer to as L_g, which is called the 'critical Griffith crack length'. Cracks shorter than this are safe and stable and will not normally extend; cracks longer than L_g are self-propagating and very dangerous. Such cracks spread faster and faster through the material and inevitably lead to an 'explosive', noisy and alarming failure. The structure will end with a bang, not a whimper. Thus even if the local stress at the crack tip is very high – possibly much higher than the 'official' tensile strength of the material – the structure is still safe and will not break so long as no crack or other opening is longer than the critical length L_g. The value of L_g is calculated from the following equation:

$$L_g = \frac{1}{\pi} \times \frac{\text{work of fracture per unit area of crack surface}}{\text{strain energy stored per unit volume of material}}$$

or

$$L_g = \frac{2WE}{\pi s^2}$$

where strain energy is $\frac{1}{2} es = s^2/2E$, since $E = s/e$, and

W = work of fracture (in J/m^2) for each surface
E = Young's modulus (in N/m^2)
s = average tensile stress in material near the crack (taking no account of stress concentrations) (in N/m^2)
L_g = critical crack length (in m)

Thus the length of a safe crack depends simply upon the ratio of the value of the work of fracture to that of the strain energy stored in the material, inversely proportional to the 'resilience'. In general, the higher the resilience the shorter the crack one can afford to put up with.

Rubber stores much strain energy but its work of fracture is quite low, hence the critical crack length L_g for stretched rubber is quite short, usually a fraction of a millimetre. Thus on sticking a pin into a blown-up balloon, it bursts with a very satisfactory pop. Although rubber is highly resilient and will stretch a long way it breaks in a brittle manner, very much like glass.

11.7 Mechanics of fracture

Fracture is an inhomogeneous process of deformation that causes regions of material to separate and load-carrying capacity to decrease to zero. High stresses which break atomic bonds are concentrated at the edges of inhomogeneities. Fractures appear as microcracks of one or two grain diameters in length; they occur typically at existing flaws or notches. Atomic fractures occur over regions whose dimensions are of the order of 10^{-8} cm. When bonds between atoms are broken across a fracture plane a new fracture plane is created. Cleavage failure (Fig. 11.6) results from the breakage of bonds perpendicular to the fracture

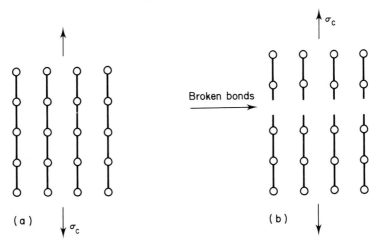

Figure 11.6 *(a) Loaded metal; (b) cleavage.*

plane; shear failure (Fig. 11.7) results from the fracture of bonds along the plane. Failure in each case occurs when the local stresses exceed the cohesive strength, i.e. at about 1/10th the elastic modulus.

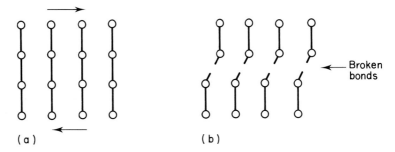

Figure 11.7 *(a) Twisted metal; (b) shear.*

11.8 Crack reaction of different metals

Griffith's analysis is for ideally brittle solids involving elastic deformation. In practice, structural metals and alloys deform plastically. The Griffith theory was based on the maximizing of stored energy plus surface energy for a crack developing in the manner of Fig. 11.8. To provide for the behaviour of plastic metals Orowan and Irwin proposed the addition of a further term to represent plastic working around a crack. This led to the introduction of fracture toughness as a fracture criterion.

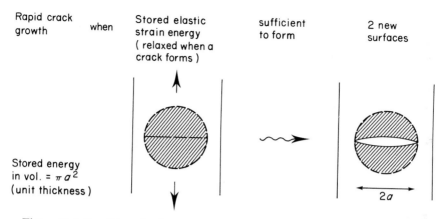

Figure 11.8 *Conditions for fracture; shaded areas represent energy affected zones.*

11.8.1 Fracture toughness

Although brittle fractures occur in glass and hardened steel (with little deformation before fracture and a macroscopically flat fracture surface) the most common break follows necking or deformation and a final cup-and-cone fracture surface. Moderately thick plates with through-the-thickness cracks behave similarly, as shown in Fig. 11.9.

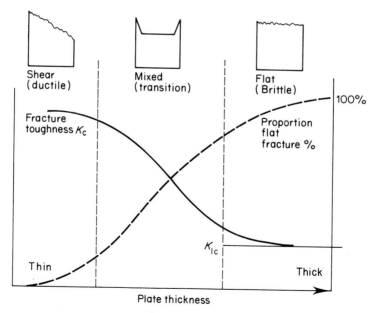

Figure 11.9 *Variation in K_c, amount of flat surface and failure mode with plate thickness.*

Fracture surface types and values of the experimentally measured fracture toughness K_c (the critical stress intensity factor at failure) vary with plate thickness. The minimum value of K_c is designated K_{Ic} and is independent of geometry for a very thick plate. K_{Ic} is designated the plain strain fracture toughness.

Reduction of fracture toughness with plate thickness was explained by Parker [11] by considering two small rectangular elements of material, one in the centre of a moderately thick plate, close to the crack tip, the other in a similar position relative to the crack, but at the free surface of the plate normal to the crack front (Fig. 11.10). As the remote loading is increased, each of these elements will fail

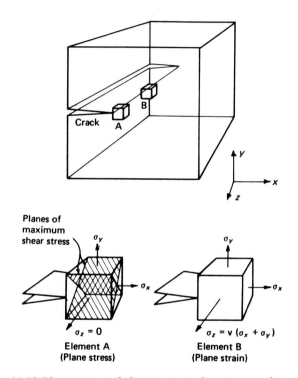

Figure 11.10 *Plane stress and plane strain in elements near the crack tip.*

at some particular load level, either by shear or by cleavage. A hydrostatic stress state ($\sigma_1 = \sigma_2 = \sigma_3$) cannot produce a ductile failure. When ductile failure of the plane stress element (A) occurs on planes at $\pm 45°$ to the crack plane, the near-hydrostatic system experienced by the plane strain element (B) will be incapable of producing ductile failure until a much greater load is applied. Element B may then fail by cleavage in the plane of a crack before it is able to achieve a critical shear stress level. Slant (shear lip) failure is thus associated with a ductile failure on inclined planes, and the flat (crack plane) failure with cleavage separation.

As the crack grows, the plastic zone ahead of the crack tip also becomes larger. Less through-the-thickness restraint occurs on internal elements, and conditions approach plane stress right through the thickness. Under these conditions the proportion of flat fracture surface reduces as the crack extends, and the proportion of slant (shear lip) increases. The resulting failure is illustrated in Fig. 11.11.

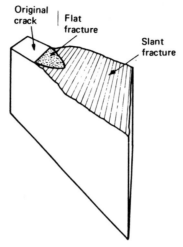

Figure 11.11 *Final fracture surface: moderately thick plate.*

11.8.2 Fracture toughness testing [12]

Single edge notched specimens fatigue pre-packed prior to testing are commonly used to evaluate K_{1c}. A specimen may be subjected to three-point bending or may be pin-loaded to produce combined bending and tension.

A clip gauge mounted across the crack-gap mouth measures the amount of crack opening and values for K_{1c} can be evaluated by formulae [13, 14].

11.8.3 Critical stress intensity factor

Fracture toughness is an experimentally-derived material constant which relates to the situation when the stress at the crack tip has a critical value such that the propagation energy is sufficient to overcome the work needed to extend the crack. Values for stress intensity factor (measured in $MN/m^{3/2}$) depend upon crack size and shape, loading system and geometry. Here the dependence on crack size and shape is illustrated. Failure occurs when the fracture toughness value is reached, which in Fig. 11.12 is at a crack size of 10 mm for a straight crack and at 23 mm for a 'thumbnail' crack.

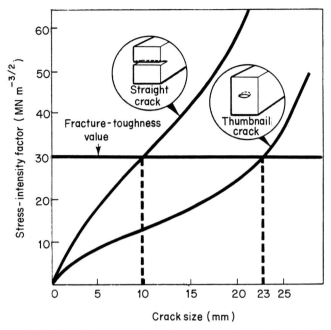

Figure 11.12 *Identifying critical crack sizes: straight and thumbnail cracks.*

For steel the fracture toughness is very dependent upon temperature, and can change by a factor of ten over a small temperature range as a result of an inherent transition from a brittle to a ductile failure mechanism. It is also markedly influenced by other metallurgical effects. Two examples of embrittlement are the lowering of toughness by diffusion of various elements into the material, and radiation damage of the material in nuclear applications.

Linear elastic fracture mechanics (the basis of this theory) is limited by the fact that it implies an infinitely large failure stress when the crack length is small. The deviation is shown in Fig. 11.13 which shows experimental values of the fracture stress for spherical steel pressure vessels with varying crack lengths. As crack length is reduced, the data deviate from the predictions of linear elastic fracture mechanics because of the increasing importance of plasticity at the crack tip. Agreement with the more sophisticated elastic–plastic fracture mechanics theory is excellent down to vanishingly small crack lengths.

Elastic–plastic fracture mechanics theory based on elaborate mathematics can be applied to situations in which failure occurs at a smaller crack size or a lower stress than does linear elastic fracture mechanics. It can be dangerously optimistic to use the linear elastic prediction if plasticity is important—as this example, for a material of fracture toughness $50 \, \mathrm{MN/m^{3/2}}$ the critical crack size falls from 23 mm to 13 mm. The difference between the two predictions is related to the size of the plastic zone, which increases as the stress intensity factor increases (Fig. 11.14).

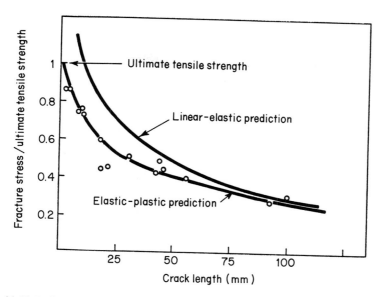

Figure 11.13 *Difference between elastic-plastic and linear elastic fracture mechanics values.*

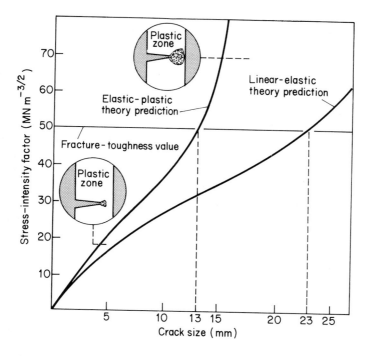

Figure 11.14 *Prediction differences: elastic-plastic theory and linear elastic theory.*

11.8.4 Failure assessment diagrams

Construction of this type of diagram forms part of the CEGB failure assessment route as described by Chell [15].

The curve relating the plastic stress intensity factor to applied load may be regarded as a failure curve for a given crack size, in the sense that all points corresponding to the critical conditions for failure lie on it. For very large toughness values, corresponding to large values of plastic stress intensity factor at failure, the corresponding failure load tends to a maximum value known as the 'plastic collapse load'.

The collapse load depends upon crack length, structural geometry, loading and the yield stress of the material. In collapse, failure occurs by a different mechanism, unrelated to fracture toughness, and is typified by gross deformation of the structure. There are thus two extremes of fracture behaviour: at low loads we can safely apply linear elastic fracture mechanics to determine the failure load; whilst at high loads, plastic limit theory gives the maximum load the structure can bear.

Both of these calculations are much easier than the elastic–plastic theory calculations necessary to determine failure at intermediate conditions. However, it is simple to construct a comprehensive failure curve from the more readily accessible extreme failure behaviour. This curve, shown in Fig. 11.15, describes all failure conditions.

In constructing the failure assessment diagram the apparent fracture toughness is used as calculated by applying the simpler linear elastic theory to the load and crack length at failure, expressed as a fraction of the true toughness value. The corresponding measured failure loads are expressed as a fraction of the plastic collapse loads. This is done for a variety of circumstances involving increasing amounts of plastic deformation.

Calculations are made for the given crack length and applied load which the structure can tolerate: (a) the stress intensity factor is determined from linear elasticity theory; (b) the plastic collapse load is calculated from plastic limit theory. With such data a point on the diagram – the 'failure-assessment point' – may be defined. If an additional specific safety factor is required, then the point must lie proportionally closer to the origin – curve CC′ is a revised curve incorporating a safety factor of two in the safe loading.

11.8.5 Fracture speed (cleavage velocity)

By introducing surface energy concepts into crack development calculations it is possible to show that the theoretical catastrophic failure of a completely brittle substance occurs at about 40% the speed of sound in the material. If steel were considered to be completely brittle (as in the cast state) the terminal fracture velocity would be 4876 m/s (16 000 ft/s) [16].

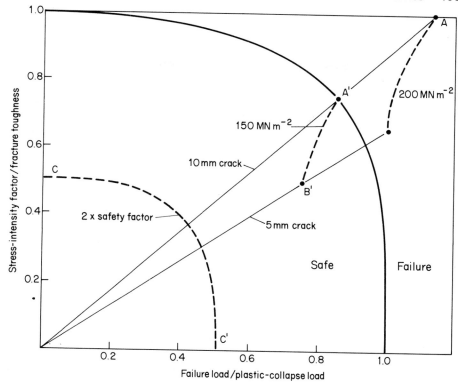

Figure 11.15 *Failure assessment diagram.*

Engineering materials usually work-harden before becoming brittle and reaching the failure stage. The very high stress intensities near the tip of a crack produce a localized plastic zone which reduces the intergranular failure.

11.9 Structural failure assessment procedure

A procedure has been codified [17] by Berkeley Nuclear Laboratories and the Central Electrical Research Laboratory (UK) for the failure assessment of engineering structures based on the criteria of Dowling and Townley [18] and Bilby, Cottrell and Swindon [19] as applied to the Dugdale model of yielding ahead of a crack tip [20].

This procedure consists of independently assessing the risk of failure: (1) under linear elastic conditions only; and (2) under plastic collapse conditions only. These two limiting criteria are then plotted as a coordinate point on a failure assessment diagram.

Linear elastic failure mechanics is well established as a means for assessing the integrity of cracked structures, it is used as the basis of Appendix A, Section XI of the ASME Boiler and Pressure Vessel Code.

Failure assessment diagram procedure as codified by Harrison, Loosemore and Milne [21] is based on the following procedure:

(1) Define the flaw shape, size and location.
(2) Define the loading conditions (relative orientation of principal stress and plane of the flaw).
(3) Determine K_r, the ratio K_1/K_{1c}, where K_1 is the linear elastic stress intensity factor and K_{1c} the fracture toughness. Values for K_1 may be obtained by using an ASME XI technique [22] or tabulated solutions by Tada *et al.* [23] or a suite of computer programs.
(4) Determine $S_r = \sigma_f/\sigma_1$ or L_f/L_u preferably maximized using conventional limit analysis, slip line theory or finite element analysis or model testing:

$$\begin{aligned} \sigma_f &= \text{failure stress} \\ \sigma_1 &= \text{collapse limit stress} \\ L_f &= \text{failure load} \\ L_u &= \text{collapse load} \end{aligned}$$

(5) Apply appropriate factors of safety to determine the assessment point K_{ra}, the factored value of K_r used on the failure assessment diagram.
(6) Determine whether this assessment point falls on the failure assessment diagram:
   (a) if it falls on/or outside – failure is predicted
   (b) if it falls inside then:
(7) Estimate the amount of flaw growth during the life of the plant or between inspections.
(8) Determine whether this amount of flaw growth has led to the flaw becoming critical according to the failure assessment diagram.

11.10 Fracture criteria: rolled and forged metals

Prediction of the workability of metals shows that for cracks at free surfaces a strain history is significant. Lee and Kuhn [24] confirmed the results of Kudo and Aoi [25] and Kobayashi [26]; they concluded that the fracture criterion is expressed by

$$\varepsilon_1 = \alpha - \tfrac{1}{2}\varepsilon_2$$

where ε_1 (tensile) and ε_2 (compressive) are the principal strains at fracture in the plane of a free surface.

Values for α range between 0.092 and 0.133 for an alluminium alloy 7075–T6.

The type of crack in an alloy can be predicted from the stress components. In Fig. 11.16 the paths for the axial stress component σ_z during upsetting are shown for various friction conditions. It has been demonstrated [27] that if the axial stress at a point on an equatorial free surface is tensile at fracture, the crack is of a normal type, and if compressive, of diagonal type.

The results shown in Fig. 11.16 predict that the cracks are of a diagonal type for all *m*-values between 0.07 and 0.75. Fig. 11.17 shows the cracks for the cases of highest and lowest friction conditions achieved in the experiments, and the observed cracks were indeed all of diagonal type.

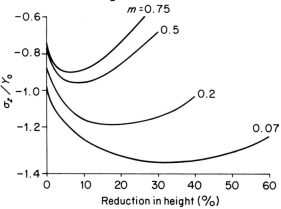

Figure 11.16 *Variations of axial stress component related to height reduction* ($Y_0 = proof$ *load*).

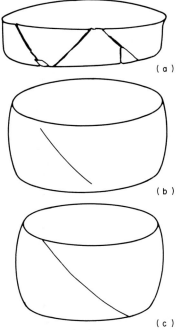

Figure 11.17 *Typical cracks in an upsetting test.*

11.11 Permissible defect sizes

Fracture mechanics applied to cracking problems in boiler downcomers showed that a defect standard cannot be generally applied but must be related to a specific application. Typical limiting sizes in terms of 'standard cracks dimensions' are given in Table 11.2.

Table 11.2 General specification of permissible defect sizes

| | Defect size $2a$ (mm) | | | |
| | Before stress relief | | After stress relief | |
	Elongated, $a > b$	Penny-shaped, $a = b$	Elongated, $a > b$	Penny-shaped, $a = b$
Embedded isolated defects	14	32	30	47
Defects within $2a$ of the surface	7	16	15	23

11.12 Fatigue life

11.12.1 Crack initiation and propagation

Fatigue life of a component or structure consists of two periods:

(1) Crack initiation. This starts with the first load cycle. It ends when a crack is detected.
(2) Crack propagation. This starts with a detectable crack. It ends when the remaining cross section can no longer withstand the loads applied.

Different portions of the total life are consumed according to the type of component or structure, thus:

(1) Massive, mildly notched components made of high strength, brittle materials, stressed by severe load spectra at high mean stresses have a relatively short crack propagation period.
(2) Built-up structures, made of low strength, tough materials, have long crack propagation periods.

If cracks or crack-like defects are present from the time of construction (which is common with welded structures) then practically the whole fatigue life consists of crack propagation alone. Failure predictions must account for both periods; a crack assumed to be present from the start would restrict the prediction to the propagation phase; on the other hand, for vital components the appearance of a crack signals imminent danger so that crack initiation life alone would be critical [28].

11.12.2 Evaluation techniques

There are four fundamental categories:

(1) Simple methods (based on constant amplitude experiments) omitting any of the complex events. Miner's rule for the crack initiation period and a simple linear addition of the crack extension due to each individual cycle using Forman's or Paris' equation for the crack propagation period are examples.

(2) Methods (based on constant amplitude tests) which account for some complex events at the notch root or crack tip. Assumptions are made for example about the size and shape of the plastic zone and its effects on fatigue life or crack propagation. For crack propagation, the Willenborg model [29] is typical. For crack initiation the ESDU method [30] is typical.

(3) Methods which try to measure what actually happens at the notch root or crack tip. For crack initiation the companion specimen method [31] is typical. For crack propagation the closure concept [32] is typical.

(4) Methods using realistic sequence test data and assumptions on how to read across from these data to the actual load sequences. The so-called 'relative' Miner rule [33] used in a similar form by the Royal Aircraft Establishment [34] and others is typical of the crack initiation period. The Wheeler model [35] is typical for estimation of the crack propagation period.

11.12.3 Damage accumulation methods

Fatigue life monitoring may be based on local elastoplastic strains [30–37] where it is assumed that the strain history at the notch root determines the life to crack initiation. Elastoplastic strain history at the notch root may be determined by means of mathematical procedures (mainly finite element calculations [38], approximation procedures [39] are experimental methods [40], i.e. measuring the actual strain by strain gauge, laser, etc. Damage is assumed to develop linearly with time (cycles).

11.12.4 T-joint welded tube life

Damage accumulation was proposed by Dover [41] so that steel offshore structures situated in the northern North Sea could be appraised by fracture mechanics then by 'hot spot' strain. Failure usually starts at the toe weld and initial defects of 0.05 to 0.25 mm are likely. Critical crack size might be half the wall thickness.

Tests made by Dover, Hibberd and Holbrook [42] involved radiological examination carried out by Brown and Root (UK) Ltd.

Strain was analysed by rosette strain gauges using Gay tracking voltmeter, Prosser RMS voltmeter with 100 s averaging period and Hewlett Packard pen recorder.

Cracks, resembling edge cracks in tension rather than a semi-elliptical surface crack, grew fast around the circumference but penetrated slowly into the metal.

11.12.5   Life estimates

Normally, a joint was considered to have failed at a crack depth equal to half the wall thickness as crack growth rate accelerated rapidly towards the end of the test. Tests carried out on 16 mm thick tube would have regarded failure to occur at 8 mm crack depth. At a crack depth of 1.65 mm after 85 380 cycles the test was terminated and the life estimated by integration. This showed that 93% of the life had already expired as only 6000 more cycles remained.

Using radial strain measurement with Miner's rule and the AWS modified fatigue life curve [43] gave the life used fraction as only 0.47.

Life estimation based on fatigue crack growth data assumes that one can use data from parent metal specimens without a weld present. Using the crack growth expression for the parent metal the first period of cycling caused the crack to extend to 0.26 mm, during the second period it increased in length to 0.39 mm. This means that the final period could be calculated to require a further 44 640 cycles and accordingly and estimated total of 124 120 cycles. These life estimates are given in Table 11.3.

Table 11.3 Life estimates

	AWS modified		BS153 modified		Fatigue crack growth analysis	
	'Hot spot' strain	Radial stress	'Hot spot' strain	Radial stress	Parent metal	*In situ*
Actual life/ estimated life	0.47	0.87	0.63	1.01	0.74	1.12

Radial stress assessment using finite element methods or an empirical method such as that of Kuang *et al.* [44] and fatigue crack growth analysis gave the best results.

11.13 Post-yield fracture mechanics

Before any fracture assessment can be carried out for either the integrity of a cracked component or the specifying of a defect acceptance standard the following information must be known:

(1) Future loads to which the component will be subjected so that the most onerous loading can be identified and used.

(2) Region or regions of the component where cracks are most likely to be found (important when defining a critical defect size for a given design stress or load).

(3) Defect shape to be analysed – conservative assumptions concerning defect size can lead to unrealistically small critical defect size.

(4) Relevant materials properties, for example fracture toughness, ultimate strength, yield stress.

Post-yield fracture mechanics theory [45] is applicable when appreciable plastic deformation is likely to occur and can be based on either: (a) a critical crack opening displacement [46]; or (b) J-integral [47].

References

1   Nadai, A. (1950), *Theory of Flow and Fracture of Solids*, McGraw Hill, New York.
2   Phillips, A. (1956), *Introduction to Plasticity*, Ronald, New York.
3   Von Mises, R. (1928), *Z. Ang. Math. Meth.*, **8**, 161.
4   Hoffman, O. and Sachs, G. (1953), *Introduction to the Theory of Plasticity for Engineers*, McGraw Hill, New York.
5   Tate, L. A. and Rigg, G. J. (1980), *Acceptance Criteria for Weld Defects in the Hotbox Dome of an Advanced Gas-cooled Reactor*, Fourth Edn, Vol. I, pp. 7-16.
6   Pook, L. P. (1982), 'Mode branch cracks and their implications for combined mode failure', Report T.803417(c), National Engineering Laboratory, East Kilbride, Scotland.
7   Report of Fatigue Failure, Committee XIII of IIW, Report IIW/IIS 14–59.
8   Yamaguchi, I. (1967), 'Fatigue failures in ship structures and their countermeasures', Contribution to ISSC 1967, Committee 3rd Report.
9   Haaland, A. (1968), 'Damages to important structural parts of the hull', Det Norske Veritas Publication No. 61, Oslo, January.
10 Griffith, A. A. (1921), 'The phenomena of rupture and flow in solids', *Phil. Trans. R. Soc. London* A, **221**, 163–97.
11 Parker, A. P. (1981), *The Mechanics of Fracture and Fatigue*, E. & FN. Spon, London.
12 Cartwright, D. J. and Rooke, D. P. (1973), 'Methods of determining stress intensity factors', Report TR 73031, RAE, Farnborough.
13 'Methods of test for plane strain fracture toughness of metallic materials' (1977), BS5447, British Standard Institution.
14 'Standard test methods for plane-strain fracture toughness of metallic materials' (1978), ANSI/ASTM E399, Annual Book of ASTM Standards part 10.
15 Chell, G. G. (1978), 'Developments in fracture mechanics and failure assessment', *CEGB Research*, January, No. 6, 36–44.
16 Teterman, A. S. and McEvily, W. (1967), *Fracture of Structural Materials*, John Wiley, New York.
17 Harrison, R. P. and Milne, I. (1977), 'A unified approach to failure assessment of engineering structures' *CEGB Report* R37.
18 Dowling, A. R. and Townley, C. H. A. (1975), 'The effect of defects on structural failure – a two criteria approach' *Int. J. Press. Vessels and Piping*, **3** (77).
19 Bilby, B. A., Cottrell, A. H. and Swinden, K. H. (1963), 'The spread of plastic yield from a notch', *Proc. R. Soc.* A, **272**, 304.

20 Dugdale, D. S. (1960), 'Yielding of steel containing slits', *J. Mech. Phys. Sol.,* **8** (100).

21 Harrison, R. P., Loosemore, K. and Milne, I. (1976), 'Assessment of the integrity of structures containing defects', CEGB Report R/H/R6.

22 'Rules for inservice inspection of nuclear power plant components', (1974), ASME Boiler and Pressure Vessel Code, Section XI Appendix A.

23 Tada, H., Paris, P. C. and Irwin, G. (1973), 'The stress analysis of cracks handbook', Del Research Corp.

24 Lee, P. W. and Kuhn, H. A. (1972), 'Fracture in Metal-Working Processes', Technical Report to American Iron and Steel Institute, April.

25 Kudo, H. and Aoi, K. (1967), 'Effect of compression test condition upon fracturing of a medium carbon steel – study on cold forgeability test; part II', *J. Japan Soc. Tech. Plasticity,* **8**, 17–27.

26 Kobayashi, S. (1970), 'Deformation characteristics and ductile fracture of 1040 steel in simple upsetting of solid cylinders and rings', *J. Engng for Industry, Trans. ASME,* **92**, 391–9.

27 Kobayashi, S., Lee, C. H. and Oh, S. I. (1973), 'Workability Theory of Materials in Deformation Processes', USAF Technical Report AFML-TR-73-192, May.

28 Schutz, W. (1972), 'The prediction of fatigue life in the crack initiation and propagation stages – a state of the art review', AGARD-CP-118 *Symposium on Random Load Fatigue,* Lyngby, Denmark.

29 Willenborg, J., Engle, R. M. and Wood, H. A. (1971) 'A crack growth retardation model using an effective stress concept', AFFDL TM-71-1FBR.

30 ESDU Item No. 71 028 (1971), 'Cumulative fatigue damage calculations (effect of correcting mean stress at stress concentrations)', The Royal Aeronautical Society, London.

31 Stadnich, S. J. and Morrow, J. D. 'Techniques for smooth specimen simulation of the fatigue behaviour of notched members', ASTM-STP 515.

32 Elber, W. (1969), 'Fatigue crack propagation under random loading: an analysis considering crack closure', *Proc. Tech. Session of the 11th ICAF-Meeting,* Stockholm.

33 Schutz, W. (1972), 'The fatigue life under three different load spectra-tests and calculations', AGARD-CP-118, *Symposium on Random Load Fatigue,* Lyngby, Denmark.

34 Maxwell, R. D. J. (1971), 'The practical implementation of fatigue requirements to military aircraft and helicopters in the United Kingdom', *Advanced Approaches to Fatigue Evaluation,* NASA SP-309.

35 Wheeler, O. E. (1970), 'Crack growth under spectrum loading', General Dynamics, Report FZM 5602.

36 Dowling, N. (1972), 'Fatigue failure predictions for complicated stress-strain histories' *J. Materials.*

37 Smith, L. R. (1965), 'A method for estimating the fatigue life of 7075-T6 aluminium alloy aircraft structures', Report No. NAEC-ASL-1096.

38 Davidge, R. W. and Massmann, J. (1975), 'Mechanical property testing of high temperature materials', AGARD R-634.

39 Stowell, E. Z. (1950), 'Stress and strain concentration at a circular hole in an infinite plate', NASA TN-2073.

40 Schutz, D. (1972), 'Durch veranderliche Betriebslasten in Kerben erzeugte Eigenspannungen und ihre Bedeutung fur die Anwendbarkeit der linearen Schadensakkumulationshypothese', LBF-Bericht No. FB-100.

41 Dover, W. D. (1976), 'Fatigue crack growth in offshore structures', *J. Soc. Env. Engng,* **15**, No. 1, 3.

42 Dover, W. D., Hibberd, R. D. and Holbrook, S. (1977), 'Fatigue crack growth in tubular welded joints', *Fracture Mechanics in Engineering Practice,* Ed. P. Stanley, Applied Science Publishers, Guildford, pp. 211–30.
43 Smedley, G. P. (1973), 'Welded steel offshore structures – the avoidance of fatigue and fracture', *Proc. Conf. Structures in the Ocean,* University of California, Berkeley, September.
44 Kuang, J. G., Potvin, A. B. and Leick, R. D. (1975), 'Stress concentration in tubular joints', *Proc. Offshore Technology Conf.,* Houston, Paper 2205.
45 Chell, G. G. (1977), 'A combined linear elastic and post yield fracture mechanics theory and its engineering applications, *Fracture Mechanics in Engineering Practice,* Ed. P. Stanley, Applied Science Publishers, Guildford, pp. 83–96.
46 Burdekin, F. M. and Dawes, M. C. (1971), 'Practical use of yielding and linear elastic fracture mechanics with particular reference to pressure vessels', *I. Mech. E. Conf. on Practical Applications Fracture Mechanics,* May.
47 Rice, J. R. (1968), Mathematical Analysis in the Mechanics of Fracture in *Fracture,* Vol. 2, Ed. H. Liebowitz, Academic Press, New York, pp. 191–331.

Further reading

'Aircraft Structural Integrity Program: airplane requirements' (1972), Military Standard MIL-STD-1530 (USAF) Sept. 1, Dayton, Ohio.
American Bureau of Shipping (1976), *Rules for Building and Classing Steel Vessels,* 45 Broad Street, New York.
American Society for Testing and Materials (1964), *Fracture Toughness Testing and its Applications,* ASTM Special Technical Publication No. 381, Philadelphia.
American Society for Testing and Materials *Standard Method of Test for Plane-Strain Fracture Toughness of Metallic Materials,* ASTM E399-74, Part 10.
American Society for Testing and Materials (1967), *Stress Corrosion Testing,* ASTM STP 425, Dec.
Bannerman, D. B. and Young, R. T. (1946), 'Some improvements resulting from studies of welded ship failures', *Welding J.* 25, No. 3, March.
Barsom, J. M. (1975), 'Development of the AASHTO fracture toughness requirements for bridge steels', *J. Engng Fracture Mechanics,* 7, 605.
Barsom, J. M. (1974), 'Fatigue behaviour of pressure-vessel steels', *WRC Bull.* 194, May.
Benham, P. P. and Warnock, F. V. (1973), *Mechanics of Solids and Structures,* Pitman, London.
Benjamin, W. D. and Steigerwald, E. A. (1967), 'An incubation time for the initiation of stress-corrosion cracking in precracked 4340 steel', *Trans. Am. Soc. Metals,* 60, No. 3, Sept.
Bennett, J. A. and Mindlin, H. (1973), 'Metallurgical aspects of the failure of the Pt. Pleasant bridge,' *J. Testing and Evaluation,* March, 152–61.
Brebia, C. A. (1978), *The Boundary Element Method for Engineers,* Pentech Press, Plymouth.
Broek, D. (1974), *Elementary Engineering Fracture Mechanics,* Noordhoff, Leiden.
Brown, W. J. (1970), 'Review of developments in plane strain fracture toughness testing', ASTM STP 463.
Clark, W. G. and Hudak, S. J. (1975), 'Variability in fatigue crack growth rate testing', *J. Testing and Evaluation,* 3 (6), 454–76.
Cudney, G. R. (1968), 'Stress histories of highway bridge', *J. Structural Division, ASCE,* 94, No. ST12, Dec., 2725–37.

Dowling, A. R. and Townley, C. H. A. (1975), 'The effect of defects on structural failure; a two-criteria approach', *Int. J. Press. Ves. & Piping*, 3.

Dugdale, D. S. (1968), *Elements of Elasticity*, Pergamon, Oxford.

Egan, G. R. (1973), 'Compatability of linear elastic and general yielding (COD) fracture mechanics', *J. Engng Fracture Mechanics*, **5**, 167–85.

Evans, P. R. V., Owen, M. B. and Hopkins, B. E. (1971), 'The effect of purity on fatigue crack growth in a high strength steel', *J. Engng Fracture Mechanics*, **3** (4), 463–73.

Frost, N. E. (1959), 'The effect of specimen width and thickness on the propagation of fatigue cracks in sheet specimens', NEL DSIR Report PM 287, East Kilbride, Glasgow.

Frost, N. E., Marsh, K. J. and Pook, L. P. (1974), *Metal Fatigue*, Clarendon Press, Oxford.

Greenberg, H. D., Wessel, E. T., and Pryle, W. H. (1970), 'Fracture toughness of turbine-generator rotor forgings', *J. Engng Fracture Mechanics*, **1**, 653–74.

Gross, J. H. (1970), 'Transition – temperature data for five structural steels', *WRC Bull.* 155, Oct.

Hall, W. J., Kihara, H., Soete, W., and Wells, A. A. (1967), *Brittle Fracture of Welded Plate*, Prentice Hall, Englewood Cliffs, NJ.

Hartbower, C. E. (1957), 'Crack initiation and propagation in the V-notch charpy impact specimen', *Welding J.* **35**, No. 11, Nov., 494.

Hearn, E. S. (1977), *Mechanics of Materials*, Pergamon, Oxford.

Irwin, G. R. (1960), *Structural Mechanics*, Pergamon, Oxford.

Johnson, W. and Mellor, P. B. (1962), *Plasticity for Mechanical Engineers*, Van Nostrand, New York.

Klippstein, K. H. and Schilling, C. G. (1975), 'Stress spectrums for short-span steel bridges', *ASTM Symp. on Fatigue Crack Growth under Spectrum Loads*, Montreal, June.

Knott, J. F. (1973), *Fundamentals and Fracture Mechanics*, Butterworths, London.

Liebowitz, H. (1968), *Fracture – An Advanced Treatise*, Academic Press, New York.

McIntyre, P. (1974), 'Advanced methods for monitoring fatigue crack growth', *J. Soc. Env. Engng*, 13–3 (62), 3–7.

McIntyre, P. (1980), *Trans. I.Mech.E.* C89/80, 217–22.

Maddox, S. J. (1974), 'Assessing the significance of flaws in welds subject to fatigue', *Welding J.*, **53**, No. 9.

Manson, S. S. (1966), *Thermal Stress and Low-Cycle Fatigue*, McGraw-Hill, New York.

Miner, M. A. (1945), 'Cumulative damage in fatigue', *J. Appl. Mech.* Sept.

Morton, J. and Ruiz, C. (1971), 'Through cracks in pipes and pipe bends' *Proc. I. Mech. E.*, **185**, No. 19, 229–41.

Nibbering, J. J. W. (1970), 'Permissible stresses and their limitations', Ship Structure Committee Report Serial No. SSC–206, US Coast Guard Headquarters, Washington DC.

Paris, P. C. and Sih, G. C. (1965), 'Stress analysis of cracks', *Symp. on Fracture Toughness Testing and its Applications*, ASTM/STP 381 pp. 30–83, Philadelphia, Pa.

Parker, E. R. (1957), *Brittle Behaviour of Engineering Structures*, Wiley, New York.

Pook, L. P. (1974), 'Fracture mechanics analysis of the fatigue behaviour of welded joints', NEL Report 561, National Engineering Laboratory, East Kilbride, Glasgow.

Pook, L. P. (1976), 'Basic statistics of fatigue crack growth', *J. Soc. Env. Engng*, December, 3–10.

Reissner, E. (1952), ' Stress–strain relations in the theory of their elastic shells', *J. Maths. Phys.*, **31**, 109–19.

Rolfe, S. T., Rhea, D. M. and Kuzmanovic, B. O. (1974), 'Fracture control guidelines for welded steel ship hulls', Report SSC-244, Washington DC, USA.

Rolfe, S. T. and Barsom, J. M. (1977), *Fracture and Fatigue Control in Structures*, Prentice Hall, Englewood Cliffs, NJ.

Rice, J. R. (1968), *Fracture*, Vol. 2, Academic Press, New York.

Rooke, D. P. and Cartwright, D. J. (1976), *Compendium of Stress Intensity Factors*, HMSO, London.

Ruiz, C. (1978), 'Ductile growth of a longitudinal flaw in a cylindrical shell under internal pressure', *Int. J. Mech. Sci.*, 76, 277–81.

Ruiz, C. and Koenigsberger, F. (1970), *Design for Strength and Production*, Macmillan, Basingstoke.

Shank, M. E. (1953), 'A critical review of brittle failure in carbon plate steel structures other than ships', Ship Structure Committee Report, Serial No. SSC-65, National Academy of Sciences – National Research Council, Washington DC, Dec. 1.

Sih, G. C. (1973), *Handbook of Stress Intensity Factors for Researchers and Engineers*, Institute of Fracture and Solid Mechanics, Lehigh University, Bethlehem, Pa, USA.

Sih, G. C. (1973), *Mechanics of Fracture*, Noordhoff, Leiden.

Sinclair, A. C. E., Formby, C. L. and Connors, D. C. (1976), *Fatigue Crack Assessment from Proof Testing and Continuous Monitoring*, Applied Science Publishers, Guildford.

Smith, R. A. (1979), *Fracture Mechanics: Current Status, Future Prospects*, Pergamon, Oxford.

Smith, R. A. (1980), 'A fracture mechanics based re-validation of air storage vessels', *I. Mech. E.*, Paper C106/80, pp. 339–44.

Smith, R. A. (1973), 'The determination of fatigue crack growth rates from experimental data', *Int. J. Fracture*, 9, 352.

Sneddon, I. N. and Lowengrub, M. (1969), *Crack Problems in the Classical Theory of Elasticity*, Wiley, New York.

Spencer, C. S. (1968), *An Introduction to Plasticity*, Chapman and Hall, London.

Sprawley, J. E. and Esgar, J. B. (1966), 'Investigation of hydrotest failure of Thiskol Chemical Corporation 260-inch diameter SL-1 motor case', NASA Report TMX-1194, Cleveland, Jan.

Stanley, P. (1977), *Fracture Mechanics in Engineering Practice*, Applied Science Publishers, Guildford.

Tada, H., Paris, P. and Irwin, G. R. (1973), *The Stress Analysis of Cracks: Handbook*, Del Research Corp., Hellertown, Pa, USA.

Tang, W. H. (1973), 'Probabilistic updating of flaw information', *J. Testing and Evaluation*, 1(6), 459–67.

Tipper, C. F. (1962), *The Brittle Fracture Story*, Cambridge University Press, Cambridge.

Timoshenko, S. P. and Goodier, J. N. (1970), *Theory of Elasticity*, McGraw-Hill, New York.

Topper, T. H., Sandor, B. I. and Morrow, J. D. (1969), 'Cumulative fatigue damage under cyclic strain control', *J. Materials*, 4, No. 1 March.

Weibull, W. (1951), 'A statistical distribution function of wide applicability', *J. Appl. Mech.*, 18.

Westergaard, H. M. (1939), 'Bearing pressures and cracks', *Trans. ASME, J. Appl. Mech.*

Wheeler, O. E. (1970), 'Spectrum loading and crack growth', General Dynamics Report FZM-5062, Fort Worth, June 30.

Wilson, W. K. (1973), 'Stress intensity factors for compact specimens used to determine fracture mechanics parameters', Research Report 73-IE7-FMPWR-R1, Westinghouse Research Laboratories, Pittsburgh, July 27.

12
Structural integrity monitoring: case studies

12.1 Introduction

Since facilities for condition monitoring became apparent in the mid-1970s [1,2] the techniques have been adopted by many organisations.

Examples range from the use of acoustic emission techniques on Brooklyn Bridge, New York to the use of load cells and data loggers by the British Ceramic Research Association, Stoke-on-Trent, UK to study the cracking of loaded brick arches.

12.2 Optical checks: Bureau Veritas

Following their use of laser optics to verify the alignment of line shaft bearings on ships, additional uses have been developed by Bureau Veritas [3]. These include:

(1) Checking the alignment of high speed line shaft bearings of land installations.
(2) Checking the deformation of a structure following a fire.
(3) Verifying the alignment of rudder stocks.
(4) Verifying the relative positions and horizontality of parts of offshore platforms.

Preventive integrity surveillance using direct sighting or laser as well as distance-measuring triangulation covers:

(1) Relative initial position of various parts of a structure.
(2) Periodical verification of relative positions which may vary due to:

(a) unstable bottom;

(b) distortion caused by impact, settlement, fire etc.

Optical methods have also been used to check low frequency vibratory displacement.

12.3 British Gas: On-line Inspection Centre

On-line inspection in British Gas started as a research project at the Engineering Research Station in the early 1970s. The objective was to evaluate methods of pipeline testing, and test the available commercial systems to recommend a 'best buy' for in-service inspection.

The engineering requirements were for accurate, reliable fault finding, a low incidence of spurious pipeline excavations and to avoid disruption of normal operations. It became apparent that there was a large discrepancy between these requirements and the performance of the systems then available, and it proved impossible to stimulate further development of these systems. In February 1974 British Gas took the decision to design and build an inspection system to meet its own operational needs. The first inspection vehicles entered service in July 1977 in 61 cm (24 in) pipelines.

The aim of the On-Line Inspection Centre is to develop and operate inspection systems to maintain the structural validity of British Gas Corporation's major asset – the pipeline. The pipeline involves an all-welded steel transmission system which consists of 660 sections of pipe totalling almost 16 000 km (10 000 miles) in length with nine diameters in the range 300 mm (12 in) nominal to 1065 mm (42 in) and with wall thicknesses ranging from 6 mm (0.25 mm) to 19 mm (0.75 in). The system works at pressures in the range nominal 7 bar (100 psi) to 70 bar (1000 psi), and the maximum length of a section is about 80 km (50 miles). The system is mainly land-based but there are now offshore lines associated with the Morecambe Bay and Rough projects.

Any on-line inspection system must detect all forms of structural weakness. To achieve optimum performance by inspection there is in general a preferred technique for non-destructively examining each type of fault. Three vehicles are used which are aligned to the principal defects occurring in the British Gas system, namely:

(1) A geometric vehicle to detect deformation in the pipe wall.
(2) A magnetic flux vehicle to detect metal loss in the pipe wall.
(3) An elastic wave vehicle to detect longitudinal cracks in the pipe wall.

British Gas use the commercially available T. D. Williamson Kaliper pig to carry out the geometry role.

The magnetic flux vehicle was developed within British Gas. At present 30.5, 61 76.2 and 91.5 cm (12, 24, 30 and 36 in) diameter magnetic vehicles are in operational service, the 45.7 and 106.5 cm (18 and 42 in) diameter are developed, and the 35.5 and 40.6 cm (14 and 16 in) are being designed. The technique and signal processing for the elastic wave vehicle have been subcontracted to the Atomic Energy Research Establishment, Harwell. These have now been incorporated into a 91.5 cm (36 in) diameter inspection vehicle, which has just entered operational service. The 61 cm (24 in) diameter version is in design and a 106.5 cm (42 in) diameter version is planned.

The mode of operation of the vehicle is shown in Fig. 12.1; it is representative of both the magnetic and elastic wave vehicles. An excitation signal is fed into the pipe wall. A defect, pipe fitting or anomaly modifies this signal and this modification is picked up by sensors and converted into an analogue electrical signal. The latter signal is recorded in a concise form and following the run is replayed and analysed to give the shape, size and position of the initial anomaly.

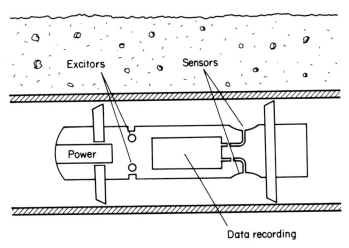

Figure 12.1 *Simplified pig operation.*

Thus the development is a mix of many disciplines:

(1) Physics to determine the excitation and allied sensors.
(2) Electronics for the data handling and processing.
(3) Mechanical engineering to provide the basic packaging and reliability.
(4) Production engineering to procure and assemble the equipment.
(5) Computing to analyse the raw data and turn them into defect data.
(6) Service operations to perform the inspection and liaise with Production and Supply Division and the Regions.

After the completion of a service operation in a pipeline the inspection tape is recovered from the vehicle and taken back to the On-Line Inspection Centre to be analysed.

The high density of data on the inspection tape is replayed and copied to several computer tapes. These tapes are then analysed in detail by three VAX 11/780 superminicomputers. Data reduction methods based on pattern recognition and signal processing techniques are used to automatically analyse and eliminate these standard features by the shape and pattern of the signals. Existing maps of the pipeline and results of previous inspection runs in the same pipeline, where available, are also used to check and eliminate data.

In the final stage a mathematical sizing model is used to get a direct estimate of the size and shape of any potential defects (Fig. 12.2). This technique is supplemented by an automatic search which compares the unknown signal with a large library of signals from known defects. The end products of the analysis are a listing of potential defects and a pipeline map listing.

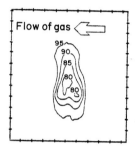

Figure 12.2 *Computer output of a typical defect.*

Attention is now being turned to a providing a commercial service to other gas companies around the world. A start was made with a very successful run for GasUnie in 1981 and other runs are planned in Europe and the USA. This is not the finish for there are still oil and product lines to tackle, as well as the whole offshore scene. The technology is available at the On-Line Inspection Centre and should repay some of British Gas's Research and Development investment.

12.4 Ultrasonic inspection: nuclear power station, Borssele, The Netherlands

During the limited time of shutdowns automatic ultrasonic inspection is performed by a MatEval system for carrying out weld inspection of the complex primary circuit of the pressurized water reactor. Later inspections have involved the main circulation pumps and the pressurizer nozzles. In practice, the inspections have been carried out by KEMA, the central laboratory for the Dutch utilities.

The basic system relies on mechanically traversing purpose-designed multicrystal ultrasonic probes along the welds. Data from the ultrasonic scans are acquired and processed by a computer and results can be output on either a teletype or a computer display terminal with a hard copy facility. Consequently there is extremely rapid acquisition of data coupled with on-line analysis so that defects can be located and sized during the down-time period. In addition, comprehensive records can be built up for comparison with the results of future inspections.

12.5 Inspection: nuclear reactors

Remotely controlled appliances developed by CEGB Berkeley Nuclear Laboratories have made it possible to perform inspection and repair tasks within the core of a Magnox power reactor. Conditions are such that if a man spent one hour near the core of a Magnox power reactor a week after it had been shut down, he would receive a radiation dose of about 50 rem - ten times that allowed to a classified radiation worker in one year [4].

The major impetus for this work was the discovery of a potential corrosion problem in certain mild-steel components in the Magnox reactors. During a routine withdrawal of steel monitoring specimens from one of the reactors at Bradwell-on-Sea in 1968, a small mild-steel bolt from a specimen holder was found broken. When recovered it was realized that the steel was corroding due to oxidation in carbon dioxide, the principal constituent of reactor coolant gas which occurs at temperatures in excess of 360 °C. Study of the bolt showed a growth of magnetite (iron oxide, Fe_3O_4).

12.5.1 Access

Access to the upper-core region is gained through one of the standpipes approximately 20 cm in diameter and 8 m long normally used for refuelling. The reactor is shut down and depressurized. Equipment is lowered from the pile cap and manoeuvred into position with a manipulator operating through an adjacent standpipe and controlled from the pile cap. Closed-circuit television is used to check the location and movement of components within the reactor vessel. The television camera is attached to a second manipulator, and steered to follow the operation of the first.

12.5.2 Manipulators

Many types of manipulator exist, and while a simple manipulator will suffice for an operation directly beneath a standpipe, a greater number of degrees of freedom is required as the location becomes more remote. This is done by increasing the number of pivoted joints and introducing swivel joints. Arms and joints must be sufficiently strong to sustain the stresses imposed by lifting operations at the extremity of its reach, yet still be capable of withdrawal into a small cross section in order to pass through the standpipe. As complexity increases, speed of operation reduces.

12.5.3 Location

Locating components in a three-dimensional space using the two-dimensional view from a closed-circuit television is inadequate. Two techniques which have been evolved are:

(1) A self-propelled vehicle which can locate components accurately with the aid of mechanical sensors. The device uses the steel beams surrounding the reactor core as a railway track, as shown in Fig. 12.3. It locates individual bolts in the beam through a simple lever and switch. Having stopped in exactly the right position, it will grind the end of the bolt perfectly flat so that an ultrasonic examination may be performed.

(2) A stereoscopic television viewing system which provides the additional depth perspective required. Two 'black-and-white' cameras view the operation from slightly different directions. Output from one is connected to the

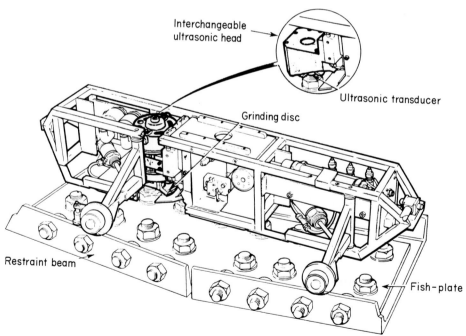

Interchangeable ultrasonic head

Ultrasonic transducer

Grinding disc

Restraint beam

Fish-plate

Figure 12.3 *Mobile ultrasonic inspection vehicle developed at Berkeley Nuclear Laboratories.*

red beam of a colour television monitor and the other to the green beam. The combined picture is effectively two views of the same objects displaced laterally and thus appears as a blurred red and green image. The operator uses a pair of spectacles having red and green filters. The two images are then presented separately to each eye to give a three-dimensional impression of the scene in front of the cameras.

12.5.4 Inspection sensors

Various techniques for crack detection, oxide thickness measurement and strain measurement suitable for in-core use are adaptations of existing techniques; others have necessitated the development of new areas of science and technology.

(a) Surface-oxide Thickness Measurement.

Three principal methods developed by the CEGB were:

(1) A laser corrosion meter (described by B. A. Tozer in *CEGB Research* No. 4, September 1976). Successive pulses of laser light are used to bore a hole through the oxide, and the reflectivity of the hole is monitored with a weaker pulse. A significant change occurs when the underlying metal is reached (as described in Section 9.12 and Fig. 9.19).
(2) An eddy-current monitor. The sensitive element is a small coil wound on a ferrite rod forming the inductance of a freely resonating electrical oscillator. When the coil is brought close to a steel component, the oscillating magnetic field causes circulating eddy currents to flow in the steel. In their turn these eddy currents produce their own magnetic field, which is opposed to the field of the coil. This effectively reduces the inductance of the coil and increases the frequency of oscillation.

  A layer of oxide interposed between the steel and the probe, changes the frequency relative to a probe in free air depending on the oxide thickness. The effect saturates if the oxide becomes so thick that the steel no longer plays any part in the interaction (Fig. 12.4).
(3) An abrading method which is able to remove a small specimen for analysis. The instrument is seated over the location to be examined with the aid of an annular vacuum seal. A jet of tiny glass beads (10–35 μm in diameter) wears a hole through the oxide, while debris and beads are removed by the vacuum system. The reflectivity of the hole is monitored by an optical system involving a light-emitting diode and photocell. A marked change occurs with the transition from black magnetite to bright metal. Abrading is then stopped, and the hole depth measured with a pneumatically operated precision linear-displacement transducer.

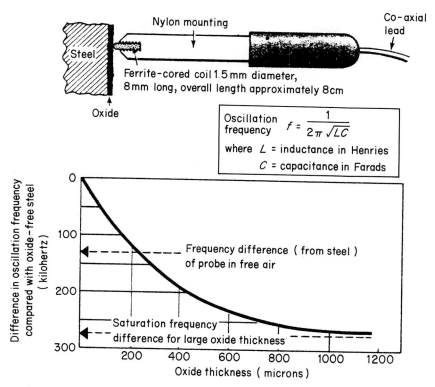

Figure 12.4 *An eddy-current probe. The upper part shows a simple hand-held probe. In the lower part the change in frequency of oscillation due to increasing oxide thickness is shown. The line represents the results of over 80 measurements on specimens from five reactors.*

In an alternative mode of operation, four abrading jets cut a small pyramid of oxide and steel from the component which is recovered by sucking it up into the vacuum system. The oxide thickness may then be measured under a microscope before subsequent chemical and metallurgical examination is made of the specimen.

(b) Interfacial-oxide measurement

Initial attempts to apply ultrasonic spectroscopy to the measurement of exposed surface layers proved fruitless. Broad-band ultrasonic pulses transmitted through a steel plate were reflected back to the ultrasonic transducer from an interfacial oxide layer. Frequency analysis of the signals was used to measure the oxide thickness.

Ultrasonic spectroscopy compares the reflected pulse with the transmitted pulse as shown in Fig. 12.5. There are 'missing' frequencies in the spectrum which are the odd harmonic frequencies related by the equation shown in

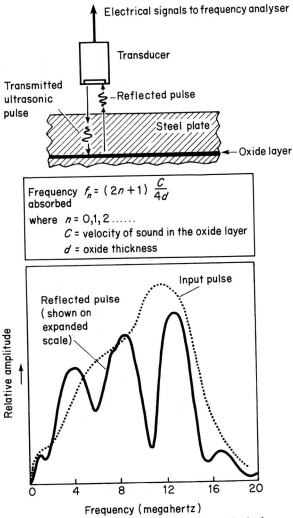

Figure 12.5 *Ultrasonic spectroscopy used for oxide thickness evaluation.*

Fig. 12.5 to the velocity of sound in the oxide (4700 m/s for magnetite) and to the thickness of the oxide layer. The pulse is greatly attenuated by its passage through the steel, and the amplitude scales for the incident and reflected pulses in the graph are not the same. The attenuation is also slightly greater at higher frequencies, but there is no difficulty in picking out the missing harmonics.

(c) Thread strain

An ultrasonic transducer applied to a bolt receives an echo from the end of the bolt. If the bolt is cracked then another echo will reach the receiving transducer somewhat earlier. Improved precision timing measurements modified the ultrasonic technique for the measurement of bolt strain (elongation). The technique is described in Section 3.11.

Fig. 12.6 presents a cut-away drawing of an instrument for measuring thread strain in 'Warwick link' bolts in the Dungeness 'A' reactors (a Warwick link is a flexible coupling which allows the graphite core a certain degree of freedom within the core-restraint structure). This machine can be handled by a simple manipulator. The transducer is coupled to the bolt head by a fluid, and the ultrasonic pulse enters the bolt at an angle after refraction at the surface. The interval between successive echoes from the threads is related to the thread pitch, and hence reveals any elongation.

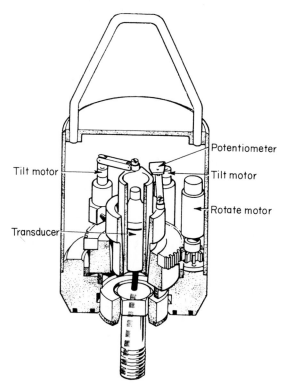

Figure 12.6 *Thread strain measurement in a 'Warwick-link' bolt.*

12.6 Offshore platforms

In UK waters it can be considered that there are two separate areas of production each with its own operation problems.

In the southern part of the North Sea drilling platforms and gas production platforms have been in operation for several years and have thus had considerable exposure to the marine environment. These structures stand in relatively shallow water and in general are built to well-tried designs which represent no special advancement of design practice over that in use elsewhere in the world.

In the northern part of the North Sea the extraction of oil rather than gas becomes dominant. Much greater sea depths, ranging from 120 m (400 ft) to 180 m (600 ft), are involved.

The history of operation of the present generation of steel structures such as those in the southern North Sea has not shown an instance of catastrophic failure due to structural factors although more minor failures have occurred in a number of cases. The justification for a monitoring system must be based on three points:

(1) The possibility that as the structures age the failures that occur may become more serious in nature.
(2) The extrapolation in design for structures emplaced in deeper water.
(3) The fact that there is no operational experience with concrete structures of this type and scale.

12.6.1 Monitoring: a review of requirements

In reviewing the needs and specifying the requirements of a monitoring system for offshore use the following considerations apply:

(1) Monitoring is a technique which must operate without interferring with normal operations; inspection implies the suspension or modification of operations.
(2) A monitoring process implies a degree of continuous surveillance but, depending on the technique selected, the actual sampling of data need not be continuous. A sampling interval of an hour is about the maximum that could be tolerated since this is compatible with the time scale of changes in weather conditions.
(3) Any monitoring system should continue to operate under storm conditions when catastrophic failure would be most likely to occur.
(4) Detection of serious loss of structural integrity should be made well before any potentially destructive storm; the structural changes which occur during a storm must be assessed by the monitoring system.
(5) The monitoring system and other associated instrumentation should be capable of a high degree of reliability. No more than an annual overhaul of the system should be envisaged.
(6) Information provided by the monitoring system must be compared with design data, as modified by the history of deterioration of the structure, before conclusions concerning the state of any structure can be made.

(7) Criteria are needed against which the data from the monitoring system can be compared and decisions can be made on appropriate action.

(8) For new structures a monitoring system should be an integral part of the design and certification requirements. By this same token, it may not be possible to monitor existing structures to an equivalent reliability standard by retrospective action.

12.6.2 Applicable monitoring techniques

There is a division between those techniques which: (a) give warning of catastrophic failure; and (b) anticipate the failure of structural members at an earlier stage.

Warning of catastrophic failure will prevent loss of life and possibly a limited recovery of valuable equipment – these are the definite techniques for structural integrity monitoring of offshore platforms.

Techniques which anticipate failure of structural members at an earlier stage allow repairs to be undertaken or a substantial recovery programme to be mounted. The economic advantages of this option are self-evident provided they can be reliably installed and operated at reasonable cost. Many potential monitoring processes are capable of satisfying both options; in general, one would expect that techniques capable of satisfying the second option would also be capable of satisfying the first.

Techniques applicable to offshore platforms are:

(1) Attitude measurement.
(2) Engineering technique.
(3) Optical and sonar techniques.
(4) Systems based on NDT techniques.
(5) Acoustic emission systems.
(6) Vibratory techniques.

12.6.3 Selection of monitoring sites

In a practical situation it is both costly and complex to provide complete coverage of a structure to determine all flaws. It is important to select those sites which are either most prone to failure or make most contribution to the integrity of the structure. These sites could then be selectively investigated. Appraisal of techniques must be made in the context of the state of current design, construction, operation and inspection practice and of current knowledge of the processes which may lead to structural failure.

In an assessment of structural criticality an important difference between steel and concrete structures (apart from their design and construction material) lies in their method of emplacement. Metal structures are typically piled into the sea floor; concrete structures normally have a large concrete raft (or rafts) which rests on a prepared area of the sea bed. The probable failure modes of these structures will be quite different.

(a) Metal structures

Any collapse of a metal structure would be likely to commence with the complete severance of one or more structural members caused by the growth of fatigue cracks. Load paths would redistribute through the structure and progressive failure of sound members would occur through overloading. Periodic inspection should identify this chain of events before the structure reaches a stage of complete collapse.

Corrosion-induced and foundation failures appear to be much less probable. The initial failures of the metal structures are likely to occur in the welded nodes of the structure.

Structural members subject to changes of buoyancy loading are particularly vulnerable; they are immersed, or uncovered, by waves and the water velocity within the waves causes significant alternating loading. Dynamic response to these forces and wave slamming must include the likely effects of marine growths, build-up of ice other non-structural loadings. Members are also subject to temperature stress variation caused by heating from the sun and cooling by immersion or evaporation. The flow of water over members by currents or by large waves causes the shedding of eddies.

Corrosion is most serious in surface water where the oxygen content is high and where local attack can be accelerated by the secretions of marine organisms. Parts of the structure in deep water arre less liable to corrosion, and the loading conditions can be calculated with greater confidence.

With modern piling practice piles do not sink axially and it is possible for their lateral support stiffness to be diminished. In extreme cases of sea bed movement lateral support might be removed for a pile depth of several tens of feet but common failure of all the piles on one leg is barely conceivable.

(b) Concrete structures

Sea water penetration through cracks in the concrete or movement of the foundation raft are thought to be major causes of failure of a concrete structure. There may also be greater errors in the calculation of structural strains exaggerated by the proximity of natural frequencies to those of larger waves. It is generally recognized that there may well be a higher probability of failure associated with the foundation raft of the structure due perhaps to uneven erosion or compaction of the sea bed which could cause very large stresses in the raft.

Periodic inspection should detect these potential failure mechanisms at an early stage.

Accidental collision by ships is an ever-present hazard. Structures which survive such collisions require the monitoring system to indicate the seriousness of the consequent damage.

12.6.4 Steel structures: critical sites

On the basis of calculations (in particular finite element stress analysis) together with some limited monitoring and operational experience, certain areas of a platform are identifiable as being of particular concern. Broadly these areas may be defined as: (a) the top 30 m (100 ft), and (b) the bottom 15 m (50 ft).

The loss of one or more member can occur without causing a dramatic loss of structural strength owing to in-built redundancy. Failures of main legs are improbable and failure to carry their normal axially compressive load is difficult to conceive, until a large proportion of bracing has failed. The proliferation of support structural members also provides effective barriers to the propagation of failure, thus giving time for the detection and repair of damage. Computer programs exist which can assess the effects of loss of structural members.

Information derived from models, structural analysis and practical experience of typical tubular steel platforms suggest that the most critical areas are the nodes or locations at which numerous braces varying from 610 to 1500 mm (2 to 5 ft) in diameter come together from various directions and are welded into stub units. Welds in these node regions are highly restrained and particularly subject to crack development if crack initiation occurs. This is despite the fact that the node welds are typically of very much better quality than the welds where the braces are joined to the stub.

Corrosion is a possible failure mechanism but it is expected to occur at a slow rate and not cause a catastrophic failure unless the corrosion site generates a crack itself. Stress corrosion may thus be important. Normal corrosion can possibly rely on inspection methods.

Critical defect size is defined (in non-destructive testing terms) as the size beyond which defects may be expected to grow under cyclic stress conditions. In marine platforms critical defect dimensions in the region of 6 mm ($\frac{1}{4}$ in) may occur.

12.6.5 Concrete structures: critical sites

A concrete offshore platform comprises a raft with one or more concrete towers to the surface and supporting a working deck. It would probably have a section thickness not less than 1 metre. Being strong in compression and reinforced to avoid tension in the concrete itself, these structures would not be expected to suffer from fatique damage and would be almost impervious to water after a minimal cement removal by leaching.

In the upper regions of concrete structures the effects of marine growth and of impact accidents remain important. Alternating wet and dry cycling is conducive to surface cracking due to dimensional changes. Such cracking can lead to corrosion of the steel reinforcement. Where this has been known to occur the subsequent steel corrosion leads to a further dimensional change, resulting in spalling, and an escalation of the problem.

Structural integrity depends upon the suitability of the sea bed surface before emplacement and on the maintenance of a stable undersurface afterwards. Scouring of the sea bed due to quite normal current or high wave reactions could undermine a structure but scouring protection would normally be provided. Experiments have also suggested the possibility that the cyclic motion of such rafts on a sandwich bed structure of sand and clay could cause collapse of the sand by densification.

12.6.6 Monitoring system reliability

Installation of a monitoring system which is itself unreliable, or which gives ambiguous data, cannot be relied on to give a warning when needed. To establish the extent of the reliability required it is first necessary to evaluate the acceptable risks of the failure to evacuate crew or failure to avoid economic loss. The formal procedure for such an exercise is established and practised by the UK Atomic Energy Systems Reliability Service and other such organizations.

When the elements of the proposed monitoring system are specified and when the reliability for each element can be derived, it is possible to calculate the overall system reliability. This procedure leads to selection between alternative monitoring methods and the optimization of the number of sensors or data acquisition points for each of the selected methods to give the required overall reliability.

12.6.7 Monitoring techniques appraisal

(a) Attitude measurement (see also Section 12.10)

Various opinions relate to the behaviour of platforms. Deck displacements on a 45 m (150 ft) high structure would be expected to have a total sway of less than 5°; a 23 cm (9 in) movement at the top of a 120 m (400 ft) to 150 m (500 ft) platform may be expected.

Typical equipment available to measure tilt, twist and displacement includes:
(1) A proprietary water levelling device which, it is claimed, can achieve a

measurement of 0.076 mm (0.003 in) difference in levels at points 30 m (100 ft) apart.

(2) An updated spirit level device developed at Langley Research Center (NASA). Light from a centrally situated lamp passes upward through a collimator to a circular spirit level and to a concave reflector. The bubble in the spirit level and the reflector make a lens system which reflects light to four photocells connected in a bridge circuit. If the surface being measured deviates from the level by plus or minus 15 seconds of arc, the bridge unbalances to give a DC output signal.

(3) Twist monitoring of a steel lattice mast structure by the Independent Television Authority. Two methods were considered:

(i) An inertial method employing various types of gyroscope; for example a stabilized platform (closed loop) and an open loop system using rate-integrating or rate gyros. Analysis showed that these inertial methods are inherently more suitable for the measurement of relatively high frequency twist.

(ii) A capacity-compass method, in which the moving vanes of a variable capacitor are mounted on the same axis as the compass needle. Twist is converted to capacitance and to a variable voltage. After temperature compensation this device is stated to be useful for deflections within a range of plus or minus 5°. This second method as developed is capable of measuring twist to an accuracy of approximately 1/16° within a range of ± 3.5°. The system responded to alternating twist if the periods were greater than approximately 3 minutes. If the upper frequency limit of 20 cycles per hour is exceeded then the open-loop gyro transducer technique or, ideally, a combined system may be preferable.

(4) A proprietary Electrolevel, used as an accelerometer to measure angular changes on tall mast structures is claimed to be sensitive to changes as small as one part in 2×10^5 (1 cm in 2 km).

(5) Useful information could be obtained by using some form of displacement gauge to measure changes between horizontal and vertical members of a space-frame structure.

(b) Witness system

Some simple 'engineering' methods which have been considered include those which only give an indication of gross failure and others which show progressive deterioration of the structure as cracks grow. Gross structural failure can be detected as members become detached, either completely or to an extent where they no longer contribute to the strength or rigidity of the structure. A system of witness wires, or pipes has been considered. Wires can transmit an electrical current and pipes contain fluid pressure when mounted inside the hollow structural members such that failure of a member (or of its junction with another structural member) would break the witness. This provides the basis of a simple technique for monitoring.

(c) Strain gauges

Strain gauge data have been obtained from several existing fixed offshore structures, mainly to obtain data for comparison with strains calculated from the basic design loading assumptions and provide:

(1) Measurements of the overall loading distribution and changes caused by movement of foundations;
(2) Measurements to confirm detail design within the structure during tow out, launch and early life;
(3) Measurements of load or strain on members in the wave slam region and subject to the accretion of marine life;
(4) Estimates of fatigue life.

The reliability of encapsulated weldable wire resistance strain gauges is acceptable for acquiring design data provided sorting tests are used to exclude weak products and providing extra gauges are used to allow for gauge failures. Some evidence indicates failure rates up to 40% in a 4-year period, so the prospect for 30 year life is not healthy. There is really no incentive to attempt to improve the reliability of resistance wire gauges, since there are positive advantages in using much larger gauge lengths, say, of 25 cm or so. Very robust inductive gauges could readily be provided which use existing technology, could be sealed reliably and would be free from short or long term drift. Strain gauges to the same reliability standard could be produced for use in concrete structures.

(d) Leak testing

Leak tightness of the hollow tubular members of platforms provides a crack-integrity monitor. In principle this is a simple method but typical difficulties are illustrated by application to the Brown and Root designed platforms for the Forties field. These may be used for storage of fuel oil so that the problem is vastly different when considering a leg full of oil. If compressed air were used as sea, a pressure of only 41 psi would achieve equal pressure at the sea bed, and would be less if the oil level were higher. If the leg were substantially empty of oil, the air pressure would require to be about 210 psi and considerable pumping energy expended as the volume is large. There are therefore serious doubts whether the technique could be successfully applied to vertical legs unless provision was made in the design to compartmentalize the tubes so that relatively small volumes can be dealt with at one time.

For monitoring bracing members, leak testing seems a more practical proposition. Individual members are smaller, and there is probably only a limited amount of intercommunication, which among other advantages makes the monitoring more selective. The pressure used could be selected to suit the depth at which any particular member is located, rather than having to use maximum (sea-bed) pressure all the time.

(e) Underwater television inspection

Gross failure such as broken or missing spars (ideally, fully identified by embossed markings) can be monitored by automatic TV tracking. Problems which have been reported include a rapid decrease in contrast ratio from 0.25 to 0.1 over a range of approximately one metre. This fact alone would result in poor visibility as experienced by underwater observers, but add to this single property those of resolution, light backscattering, absorption, and refraction, then the varying reports of divers as to the range and clarity of observations at various depths and locations may be understood more clearly.

Observers report that in the northern North Sea (e.g. the Forties field) the range of visibility can be as great as 10 or 12 m (30 or 40 ft) using artificial light, whereas in the shallower waters of the southern North Sea this range may be reduced to 1.5–2 m (5–6 ft) and in estuarial waters visibility is frequently zero.

Under conditions of still water etc., when turbidity is low, these lower ranges are reported to be extended considerably. This turbidity factor is caused by the suspension of small particles, both organic and inorganic in origin (the result of land drainage and wind action introducing particles whose sizes cary with the distance from land and the sea's activity).

It is reported that in highly active enclosed waters such as the North Sea, particle sizes range from 2 to 5 µm, whilst close inshore particle sizes in the range 6 to 17 µm may be found.

The organic matter comprises, in general, the plankton and disintegration products, having diameters below 50 µm. The overall scattering material concentration quoted for the North Sea over a year is typically 6 mg/litre.

(f) Underwater sonar viewing

Sonar (Sound Navigation and Ranging) is an imaging system. Pulses of acoustic energy are directed towards a target and small fractions of this energy reflected from the target are picked up by an acoustic transducer and converted to electrical signals for subsequent identification. Since the suspended particles are of diametrical sizes well below 50 µm and since a wavelength of 1 mm corresponds to an acoustic carrier frequency of 1.5 MHz it is clear that the size of the scatterers found in the sea is always small compared with the sonar wavelength; and further, that scattering from particles (typically found for example in the North Sea at 1 µm to 20 µm size at 0.5 to 20 mg/litre concentrations respectively) will not cause significant acoustic attenuation. The presence of suspended matter has not been reported to impair measurably the resolution capability of a sonar system.

Objections to the use of a sonar technique for monitoring a complicated space-frame structure arose from difficulty experienced with interpretation of the data when examining responses from a large static space-frame structure. A moving object may be traced comparatively easily whereas from the static situation echoes from structural components could easily be missed for very trivial reasons of transducer misalignment, geometry of the structure and shadowing effects, etc. Workers in the field believed that interpretation of the data from a complicated structure was so difficult and dubious that an unsafe condition of spar failures could well pass unnoticed, and therefore would not recommend a sonar technique in its simplest form for the purpose of monitoring for a safety factor.

Sonar devices have been proposed as a means of accurately determining, and subsequently monitoring the distance between one point of a structure and another point. Side-scan sonar is considered to be suitable as a means of detecting the sea-bed state, particularly important when considering the integrity of the concrete raft of a concrete or hybrid platform. Detection of the scouring of the bed, the movement of sand waves, movement of debris and build-up of sand and bed components, have all been clearly demonstated by proprietary versions. Such equipment, towed behind a fast-moving boat, is capable of viewing the sea-bed on either side of the apparutus to a depth in excess of 300 m (1000 ft).

(g) Ultrasonic examination (see also Section 3.13.2)

Ultrasonics may investigate very small flaws down to an area of 1 mm^2 although the probability of finding flaws on a large structure decreases with the size of the flaw. Flaws smaller than a few wavelengths in one or more dimensions interact with the ultrasonic wave. This may cause misleading estimates of flaw size. For structural integrity monitoring such small flaws are not likely to be so critical as to need to be found before they grow. Ultrasonic techniques can reliably find defects several inches long and penetrating 50% of wall thickness; these can be tolerated before repairs become urgent.

Ultrasonic torsional excitation has been suggested as another method of detection. Vibrations at a lower frequency, e.g. 100 kHz, would investigate the reaction of members to sheer stresses.

(h) Magnetic crack detection

Leakage flux used to attract magnetic particles to the region of (or into) the cracks is probably the most commonly used magnetic technique. Such techniques can be applied successfully underwater but automatic application would not appear to be attractive. An alternative but similar approach is to bring a magnetic tape into close contact with the metal surface so that the stray field imprints its own pattern onto the tape for analysis later. These techniques provide no real indication of the depth of the crack and may not distinguish between cracks and grooves which are less important structurally.

A sensitive probe has been used to scan the magnetized area to look for areas of stray flux as they project out of the steel surface, chiefly for sharp discontinuities such as cracks, which may or may not open to the surface. These can be distinguished from other defects by the rapid changes in the nearby normal component of the magnetic field. With solid metal overlying a defect the field perturbation extends and becomes less distinguishable from probe lift-off, undulations in the surface, changes in metallurgy, etc. Alternatively the base length can be made very long to reduce the importance of localized perturbations due to cracks. Such defects as areas of thin wall, perhaps due to impact, or to localized corrosion, etc. can be found. The steel that is to be examined must be magnetized to about half its saturation flux density, which may be done with a system of permanent magnets because any electrical method would involve a large volume of equipment and comsume a considerable amount of power which might endanger the safety of diving operations.

To achieve the maximum accuracy and resolution the probe must be set as close as possible to the steel it is scanning. A gap of a few millimetres would be reasonable. Because of this it appears that the heavy weld beads which are probable will prevent complete internal inspection even if access were possible. The more uniform external weld should be capable of inspection by this technique unless the difference in hardness between the parent metal and the annealed heat-affected zone causes a great change in magnetic permeability.

It seems possible that techniques could be used to separate crack signals. It may also be possible to gain some estimate of the depth of the crack from the shape of the magnetic leakage field and this could be enhanced if an AC component is introduced into the applied field, so affecting the depth of its penetration into the material. This approach could also be used if the flux variation in the applied field is monitored as a function of the field frequency since the extent of the leakage field will fall once the penetration becomes greater than the crack depth and in the first place this is directly monitored while, in the second, it will affect the magnetic resistance of the circuit.

(i) Electric current testing

Electrical resistivity is affected by the presence of cracks or corrosion pits in a conducting material. Although a simple test might be based on the application of a voltage and measurement of current, in practice such a system will not work efficiently since the resistance of the volume of material between the probes and the material (which will continuously vary due to changes in surface condition and contact pressure). A four-probe system overcomes this difficulty, two probes to complete the electrical circuit via a substantial resistance, and two to monitor voltage changes. The effect of variations in surface conditions is usually counteracted by making use of a Thornton Probe which twists on contact to penetrate any surface film.

The DC technique functions well even when access is limited to one side. It is possible to use AC currents which tend to remain closer to the surface as the frequency increases. This provides a means of determining the depth of cracks on the surface of the material facing the probes. Additionally AC currents can be introduced without probe contact and this forms the basis of eddy current inspection.

Difficulties arise in applying this technique to offshore rigs with permanently mounted detectors. Corrosion effects may mask resistance changes due to crack development. Also, sea water is an excellent conductor so that the importance of the steel as the conducting medium is reduced. Alternatively the monitoring currents may interfere with cathodic protection.

(j) Eddy current testing (see also Section 3.14)

Eddy currents do not completely penetrate a specimen but tend to be confined to the surface closest to the exciting coil. Penetration increases as the AC frequency decreases; but the response of the detector remains preferential to the volume of metal nearest to the coil. The technique is sensitive to variations in the permeability of magnetic materials and defects in the material can only be reliably detected if the near material is extremely homogeneous. Surface roughness may mask defects at depth in the material. The effects of permeability changes may be reduced if multiparameter or pulsed eddy current techniques are used to pick out the signals returning from deeper within the material. A possibly more practical alternative approach would be to saturate the steel locally with a strong static magnetic field.

Improved techniques with a better maintaining-oscillator from which small changes in the components can be extracted, remove unwanted effects. At very low frequencies (a few Hz) penetration is sufficient to detect deep cracks penetrating from the opposite side of the steel if the permeability problem can be overcome. Digital comparison techniques and filtering techniques select sharp signals from flaws, such as cracks, form slower signals due to variation in probe lift-off, etc. For offshore rigs the problem remains that with eddy current systems small areas are interrogated so that a scanning approach might need to be used.

Eddy current techniques are not currently relevant to the testing of concrete structures but direct DC conductivity measurements of the reinforcement or of special witness conductors embedded in the structure may well provide the basis of integrity monitoring.

(k) Microwave testing

Microwaves are those electromagnetic waves of frequency between 0.3 and 300 GHz being bound by the radio wave and infrared regions. Because there is a lack of penetration of electromagnetic radiation of these frequencies into metals (unlikely to exceed 1 Å) their potential role is limited to the detection of cracks penetrating the surface of the metal. Cracks on the surface re-radiate or scatter incident radiation which can thus be observed by an appropriately placed detector.

(l) Thermal scanning techniques

Thermal pulses are applied to a material by using sinusoidally varying thermal waves in a manner similar to that of electromagnetic waves in eddy current testing. Thermal properties of a material and existing flaws are established.

(m) Infrared thermal scanning

Infrared waves of wavelength between 0.75 μm and 1000 μm (1 mm) cover the next shorter range of the electromagnetic wave spectrum to microwaves. For this reason most solid materials are opaque to infrared radiation. As a reflection technique, using an emitter and receiver it is limited to a study of surface phenomena and has no advantage over microwaves. As bodies at ambient temperature are emitters of infrared radiation, variations in the emission may be detected to monitor the state of the surface and (by conduction) about the interior. Infrared techniques detect defects if the surface of a pipe is heated or cooled in such a way that a crack would break up the pattern of heat transfer. A crack shows up as a discontinuity with good resolution as only rapid changes in relative emissivity are being studied. As applied to offshore structures the technique is only applicable as a scanning process with the additional problem that the surrounding water acts as a heat sink so that heating phenomena are very transient.

(n) Radiographic techniques (see Sections 5.22, 5.23)

X-rays and similar radiographic techniques are widely used to detect flaws in concrete and steel but the use of film is ruled out for offshore rig structures. Radiation sensitive detectors do not provide the resolution of which film is capable. Techniques of neutron and proton radiography do not appear to offer any further advantage in this application.

(o) Acoustic emission (see also Chapter 4)

Overstressing and plastic deformation with the onset of microcracking involves high frequency stress waves. If a structure contains significant defects then yielding will take place at these defects whilst the bulk of the structure is at a low stress level. Acoustic emission monitoring assumes that yielding or cracking at these defects will emit acoustic signals whilst no emission will occur from the remainder of the structure. Measurement of this emission gives prior warning well in advance of failure. On a large structure triangulation techniques locate the emitting source.

Once stressed, a metal gives little or no further emission on subsequent stressing until the material is strained beyond the initial stress. This also applies to structures containing defects. If crack growth occurs, or new defects introduced between the repeat loadings, emission will take place at the same or even lower loads.

Tests have also shown that the intensity of emission is related to the stress intensity (K) or the crack opening displacement (δ) at the tip of the defect. As failure depends on attaining a critical value of K or δ the intensity of acoustic emission can in general terms be directly related to the proximity of failure.

In an oil rig the external noise level is very high in the first 16–32 m (50–100 ft) below the deck, due to wave slam, spray and rig noises, but is less at greater depths. It is a low frequency noise while the emission from defects typically ranges from 100 kHz to a few MHz.

The effective range of the instruments varies from 2 m at 1 MHz to 7 m at 100 kHz. The number of gauges required to monitor a given structure is significantly reduced if the background noise allows effective operation at the lower frequency ranges.

The technique has potential to offshore rigs in that it could be capable of measuring crack initiation, slow crack growth and the onset of failure. In the long term it may also be possible to differentiate between crack growth rates. The effectiveness of the technique depends on the characteristics of the material. Although defect locating by acoustic emission is useful the inability to provide a reproducible means of gauging the size of defects is a significant disadvantage.

The significance or severity of a defect is a function of its length, its stress field and the toughness of the material at the tip of the crack. Informed opinion suggests that in the long term acoustic emission has the potential to identify the severity of defects, i.e. the defect which is most likely to cause failure. It also has the advantage that it can scan the complete structure or part of the structure of concern without predetermining the precise area for examination.

(p) Vibration monitoring (see also Chapter 7)

Vibration monitoring techniques may utilize the vibration present in the structure, termed 'passive'; otherwise they require a specific applied loading process and are termed 'active'.

Automatic data acquisition is confined to low frequencies with an upper limit of frequency determined by the digitizing rate. Considerable economy can be made by setting limits on the levels of some signals or selection of signals below which acquisition ceases. For a running statistical analysis of some parameters and to store only a fraction of the raw data for complete analysis at base or for posterity, either a small computer is required on each platform or raw data must be transmitted directly to a shore base for processing.

Acquisition and processing must be specified in parallel with technique development to ensure that the output from these techniques can be used in practice and to ensure that the data acquisition equipment and its communication to base are adequate. Such techniques expressed in order of ascending complexity are described in the following paragraphs.

Platform structures normally respond selectively at the natural frequencies of the lowest modes, frequencies which are higher than those of the largest waves, but which must inevitably coincide with frequencies of some waves. Several modes of vibration of the whole structure respond with the division of energy between the modes dependent on the sea state. By producing frequency spectra for the wave motion, platform tilt and deflection, a regular pattern of response can be obtained from which it is possible that changes in natural frequencies could be detected well before the crew were aware, through their natural senses, that a structural change or a change in foundation stiffness had occurred.

Interpreted signals provide a measure of the health of the whole structure and its foundations. Members subject to wave slamming vibrate at their own natural frequencies following each impact. These frequencies are sensitive to extensive cracks at the end welds, therefore the frequency spectra from a strain gauge pair or an accelerometer attached to a member shows deterioration as a crack propagates.

Modal analysis and structural response data are effective if the rig is loaded passively as a result of a storm. This makes it possible to calculate:

(1) Normal mode shapes.
(2) The normal mode frequency.
(3) Damping.
(4) Excitation forces distribution.
(5) Magnitude of the exciting forces.

All these evaluations provide a direct way of checking the original stiffness matrix of the structure and of locating the region of structural defect. Changes in damping suggest foundation changes.

Active excitation through the use of a rotating mass (swept-sine), electromagnetic, mechanical hammer or a small explosive charge can be used for:

(1) Periodic monitoring.
(2) To confirm the health of deep water parts of the structure.

It is an advantage that for this test relatively few transducers would be needed (perhaps one for ten structural members). Obvious disadvantages of this method are that:

(1) Substantial installations are required to provide the excitation and to carry the loads which must be transmitted to the structures.
(2) To obtain substantial signals distinguishable from those derived by other periodic, impulsive or random sources, demands the use of the highest fidelity FFT equipment for analysis, as for passive methods.
(3) Nuisance and disturbance to the operating crew caused by the unusual noise or vibration essential to the test method.

12.6.8 Implementation of a monitoring strategy

The preceding section has drawn heavily on the report [5] prepared by the UK Atomic Energy Authority, Harwell, in the final part of which an appraisal was made of the interactions between the design function and the automatic data acquisition system on safety monitoring.

The conceptual basis is that decisions to scram, or how to avoid economic loss, must be based on a comparison of evidence of the structural state with criteria established beforehand. Evidence comes partly from NDT techniques, which will be partially automated, and from a wholly automatic data acquisition and analysis system. Criteria can be derived by reference to assumed loading data, on evidence of manufacturing quality and on the available knowledge of corrosion fatigue. Criteria can be updated to take into account the data on actual loading, deflections or strains from the data acquisition system, on actual defects from the NDT systems and from statutory inspection and on corrosion fatigue from research programmes.

12.7 Ultrasonic inspection

12.7.1 Offshore pipes and oil risers

A system has been developed for the inspection of offshore oil and gas lines, particularly for inspecting oil risers, from inside the bore.

In operation, oil flow is stopped at the well head valve and the upper end of the riser is opened to provide access. After a period to allow the oil in the riser to cool, to a point above the wax deposition temperature, the test head is inserted. The test head carries an array of probes spaced circumferentially round the bore and features an oil-tight pressure vessel which contains the multiplexing electronic circuitry. It is driven down through the oil by wheels which are in contact with the inner surface of the riser or by a small propeller.

The head is connected to the surface instrumentation by an umbilicus carrying power signal lines and the main hoisting cable. On reaching the desired depth, it is retrieved at a controlled rate, measuring the wall thickness either continuously or at predetermined intervals up the riser.

12.7.2 Pipe wall thickness

A computerized system for checking pipe wall thickness from the inside in heat exchanger equipment of petrochemical and chemical plant has been introduced by MatEval. The system consists of 15 ultrasonic probes which have been designed to deal with corroded wall conditions. Each is mounted in a spring-loaded alignment skid and the probe array is drawn through the pipe by a synchronous motor connected to a pulley and chain.

The system is controlled by a microcomputer, which initiates a question and answer dialogue with the operator, via a teletypewriter, at the start of each pipe inspection. In this way, the amount of pipe wall corrosion is established, so that pipes can be replaced at the optimum time. Systems exist for the inspection of pipes from 15 mm to 380 mm bore.

12.8 Corrosion monitoring (see also Chapter 9)

12.8.1 Oil refinery

At the Fawley (UK) refinery of Esso Petroleum a corrosion monitoring channel has been set up in one of the crude oil distillation units. An electrical resistance probe is located in the condensing train of the overhead circuit from the atmospheric distillation tower. This area is known to be a variable corrosion situation. The probe is connected to a Magna 4800 Corrosometer and the output signals are processed into a Foxborough PDP8 computer while a second PDP8 assesses the information and stores it in a file. So far, the feasibility has been proven and another channel installed in a second distillation column connected to the PDP8 through a time-sharing device. Probes are interrogated daily by the computer. The system has proven successful over 12 months and there are plans to extend it. As it stands, it is an 'open loop' system but Esso intends to make it a 'closed loop' system which requires adequate safeguards in setting basic computer operation instructions before on-line control is possible.

At ICI Mond Division (UK) a significant proportion of 'heavy ends' arise from associated plant in its chlorinated hydrocarbon solvents plant. It was decided to monitor corrosion rates of selected plant areas to establish a norm and compare monitoring performance with weight-loss coupons. Because start-up conditions produce local, high corrosion rates for short periods, mild steel and Monel electrical resistance probes were chosen.

The media were chlorinated hydrocarbons containing dissolved chlorine, hydrogen chloride, trace bromine and water at elevated temperatures and pressures. As no commercial probes were available ICI developed its own using a replaceable head and split-seal concept. Ten probes were welded to an ASA 150 flange and bolted in place. Probe bodies were of Inconel 600 and probe elements, mild steel. The probes confirmed the high corrosion rates at start up, for example in the still reflux pot.

12.8.2 Distillation column

In an early case of stress corrosion an austenitic steel distillation column was corroding, owing to carryover of chloride from a previous stage. Corrosion monitoring showed it was not possible to reduce the carryover, but that if the worst level of carryover was prevented, stress corrosion could be avoided by controlling the acidity of the process stream – a relatively easy task. Here, carbon steel remained adequate and so avoided a more expensive titanium replacement.

12.8.3 Plant piping

The choice for a part of a plant piping system was between mild steel and stainless steel. Experience with similar installations suggested that stainless steel would last indefinitely and that a mild steel system would last 3–4 years, though with occasional failures at shorter times.

An analysis showed that, if premature failures (which could result in an expensive loss of production and were potentially hazardous) could be avoided, the economic solution would be to use mild steel and replace it at a planned time within its expected life.

A mild steel system was installed, but to minimize the risk of premature failure, and to give adequate warning for planned replacement, pipe thickness was monitored ultrasonically on-line every three months. This approach proved entirely satisfactory. It is now standard practice in many of the process industries, particularly for vapour or gas lines, or lines handling organic fluids. It has also been used to monitor the effects of erosion, with or without corrosion.

12.8.4 Recirculating water system

Plant in a small factory was able to operate at increased efficiency as a consequence of corrosion monitoring.

Corrosion of the recirculating cooling water was originally monitored and controlled by adjusting the pH of the water to achieve marginally scaling conditions, i.e. to maintain a scale on the metal surfaces thick enough to confer corrosion protection, but not heavy enough to reduce thermal efficiency or cause blockage. Irregular effects resulted: heavy scale in some sections of the plant and insufficient scaling in some others (a common occurrence – balanced scaling depends upon temperatures, velocities and other variables).

Consequently it was decided to replace pH control by one of the modern synergistic corrosion inhibitor systems. This was expected to do two things:

(1) Solve the scaling and corrosion problems.
(2) Allow an increase in the heat load that could be accommodated.

Synergistic inhibitors act through the combined effect of several substances, present in accurately adjusted quantities. 'Topping up' had to be carefully controlled: adding too little inhibitor could detract from its effectiveness; too much would add to the cost, could lead to effluent problems, and could turn the water into a skin irritant. The number of individual inhibitor constituents and fluctuations in feed-water composition prevented the use of analytical monitoring. An on-line polarization monitor was introduced to monitor corrosion twice daily. This provided the process control basis on which the operator might add another briquette of inhibitor to the water.

This procedure worked satisfactorily. In 1976 the equipment cost £300 (the cost of the measurements was trivial) and the technique saved an estimated £2000 worth of inhibitor over one year. A large cooling system could well justify the use of sophisticated equipment which would measure the corrosion rate automatically and control inhibitor addition directly.

12.8.5 Pump/pipe cavitation (see also Section 6.13.2)

Monitoring for cavitation exploits the fact that, as the bubble collapses, it vibrates the component, transmitting sound or ultrasonic waves. Whether or not bubble collapse is occurring can therefore be determined by acoustic analysis, using a pressure transducer held against the outer surface to measure the frequency spectrum. If the readings indicate bubble collapse, the flow rate and pressure can then be adjusted to prevent the possibility of cavitation.

Here a technique which, on the face of it, is unrelated to corrosion has been used to provide information relevant to a corrosion problem. The presence of collapsing bubbles does not necessarily mean that corrosion/erosion damage is taking place: damage occurs only if cavitation is sufficiently severe. In this case, monitoring simply distinguishes between conditions under which a particular form of corrosion can and cannot occur.

12.8.6 Reactor/organic chemicals

Bouts of erratic corrosive attack occurred in a stainless steel reactor used to manufacture feedstock for several organic chemicals. Basic raw material was introduced and heated to the reaction temperature and other reactants were then added in sequence, the temperature being raised or lowered as required at each stage. At the end of the cycle the reactor was cleaned by washing, first with a detergent solution, then with dilute alkali, and finally with water.

The pattern of temperature changes was not identical for each batch. Attempts to limit the maximum temperature did not alleviate the corrosion and led to operational difficulties. Polarization resistance probes were introduced, and the corrosion rate monitored over a period of 14 days, during which time 36 batches were produced. Moderate corrosion occurred during the water wash, and in some batches severe corrosion followed the addition of one of the reactants (an acid).

Hydrolysis of traces of product not removed by the alkali wash was found to cause corrosion during the wash stage. A second alkali wash was incorporated in the cycle, and further monitoring established that this had the desired effect. Monitoring also linked the severe corrosion which occurred after introduction of the acid to process temperatures higher than average for this stage but less than the peak in the cycle. Increasing the degree of cooling before and during the addition, eliminated the corrosion problem – and also improved product quality. But these changes also lengthened the batch time beyond the duration of a single working shift. Further monitoring revealed that the temperature at other stages of the reaction could be raised to speed up the process without causing corrosion or impairing the quality or the efficiency of the conversion. This resulted in the cycle time being reduced from eight to six hours.

12.8.7 Reactant detected

Erratic corrosion occurred in the nickel alloy internal heating coils of a glass liner reactor which was used to produce three generally similar products in campaigns of a few weeks' duration each. But the pattern of corrosion appeared to be unrelated to the product produced.

Monitoring with polarization resistance probes and potential measurements showed that some 'acceptable' corrosion occurred throughout most of the cycle irrespective of the product being manufactured. With some batches corrosion was severe during the early stages of processing. Potential measurements indicated that the corrosion resulted from unusual oxidizing conditions. Examination of the operating records revealed a correlation between 'corrosive' batches and the use of a reactant common to all three products – but from a particular source. Oxidizing conditions were more likely to be present with this reactant than with the same material from other sources.

Thus corrosion monitoring isolated the culprit oxidizing material and its origin.

12.8.8 Distillation column

Heavy organic chemicals were produced as a result of a process stream involving:

(1) Treatment with sulphuric acid.
(2) De-ionization bed treatment to remove the acid.
(3) Separation of the lower-boiling-point components in a still.
(4) Further purification of each stream in other distillation stills.

Laboratory corrosion tests showed that carbon steel would have a limited life if used for the main still, but that austenitic stainless steel, high nickel alloys and titanium would give satisfactory service. Since the chloride content of the process stream (a dominant factor in stress corrosion) was expected to be low, and corrosion tests suggested no likelihood of stress corrosion, the still was produced from austenitic steel.

However, after a few months' operation extensive stress corrosion cracking occurred in the lower part of the still. Temporary repairs were made and consideration was given to a replacement. The alternatives appeared to be:

(1) To modify the operating conditions to allow austenitic steel to be used; or
(2) To retain the same operating conditions and use titanium or nickel based alloys.

Both of these choices were on a relatively long delivery and a carbon steel still was therefore ordered on the basis that it would last long enough (about two years) to allow work to be done from which a decision between the alternatives could be made.

Some of this work revealed small amounts of an aqueous phase in the liquor where cracking had occurred. Only about 30–100 ppm of chloride was present in the still feed, but in the aqueous phase the chloride had concentrated to more than 3000 ppm. Several sources of chloride were identified, and their elimination reduced the chloride level in the feed to 10–20 ppm; but any further reductions would have required major and expensive modifications.

Polarization resistance and potential probes were installed in the carbon steel column (this was probably the first time potential monitoring had been used to establish whether stress corrosion conditions were present) and stressed and unstressed specimens were introduced into a bypass.

Data from this corrosion monitoring showed that:

(1) If the plant were operated at the lower chloride levels, cracking would only occur when the acid content was relatively high.
(2) At the higher chloride levels the critical acid concentration for cracking was lower.
(3) If the chloride content of the feed was kept below 20 ppm stress corrosion could be prevented by controlling the acid content.

It was considered that under such control conditions an austenitic steel replacement would give satisfactory service. However, monitoring also demonstrated that the corrosion rate of the carbon steel itself could be reduced by controlling the acid content. Subsequently, it was found possible to lower the acid level to the point where the predicted life of the carbon steel was lengthened to more than ten years.

On the basis of information provided by corrosion monitoring it was decided to operate the 'temporary' carbon steel column. It was still in good condition ten years later.

Cost savings provided by this practice are based on the savings from not purchasing the new plant. Projected costs were:

Austenitic steel replacement column	£230 000
Titanium replacement column	£450 000

Actual costs were:

Monitoring costs	£6 000
Modification costs	£24 000
	£30 000

This represented a saving (ten years ago) of £200 000 to £420 000.

12.8.9 Inorganic chemical plant

A change in the source of raw material to a major inorganic chemical plant caused severe corrosion to occur. Spares for some of the affected parts were available, but there was a real possibility that if the problem was not solved quickly, production would be drastically limited for several months, – with very serious commercial results for both the company and its customers.

It was assumed that, with the old source material, corrosion had been prevented by the presence of impurities which had acted as inhibitors, and that remnants of this inhibitor had hidden the effect during the initial plant tests with the new source material and had maintained protection for a period after the change of feed stock. Plant trials were therefore conducted to find inhibitors for use with the new feed stock, monitored by probes installed in the affected units.

Electrical resistance probes were chosen because some of the corroded areas were in vapour spaces which were subject to splashing (conditions which ruled out the use of electrochemical techniques), and by polarization resistance probes exposed fully immersed in a bypass. The data obtained on instantaneous corrosion rate and total corrosion greatly simplified the assessment of inhibitor behaviour.

Three inhibitors were tested. One inhibitor had little effect at first, but the corrosion rate began to decrease after several hours and reached an acceptable level after a few days, at which point protection could be obtained with a much lower inhibitor concentration than was necessary initially.

The others had a greater initial effect, but relatively larger concentrations were needed to maintain protection.

The first inhibitor was therefore adopted as the most economic solution, and its performance is now being monitored as routine.

A large number of probes were used, with facilities for automatically scanning and recording the data, and the results were interpreted by computer. The cost was correspondingly high (about £250 000); on the other hand a remedy was found speedily and put into practice, and cuts in production were avoided. The direct cost of cuts would have been large (several million pounds) and the indirect costs to the customers and their employees even greater. The whole of this 'saving' cannot be attributed solely to corrosion monitoring, but there is no doubt that the cost of monitoring was more than justified.

12.9 Offshore structures [16, 34]:
British Gas integrity tests (see also Section 3.13.2)

British Gas has developed an ultrasonic crack detection system that can survey offshore structures 40 times quicker than conventional non-destructive testing systems. Manufactured and marketed under the name Gascosonic it aims to sell, worldwide, 30 devices at £10 000 each [6].

Norwegian certification agency Det Norske Veritas has argued that a system like Gascosonic could have prevented the Alexander Keilland rig disaster, by detecting which steel cross members had become flooded. The difficulty with conventional offshore testing techniques is that they use bulky equipment developed for use on land and then adapted for use offshore. The problem persuaded the British Gas Engineering Research Station near Newcastle (UK) to produce a robust kit for surveying its Morecambe Bay and Rough offshore platforms.

A sensing probe can be attached magnetically to any steel member of a platform or rig. It produces ultrasonic pulses which pass through the steel. If water is present, its greater acoustic conductivity enables the signal to bridge the gap and reflect off the opposite steel wall. That registers a second echo on the display, which is then compared with the display of an air-filled structure.

The greater speed of this process enabled British Gas to test every steel structure on its Rough platform in only 25 test dives over five days. Unlike existing non-destructive testing methods, Gascosonic could test every part of the platform, not just a random sample.

12.10 Underwater structures: attitude measurement
(see also Section 12.6.8)

Platform tilt and structural stresses may be determined from measurement of settlement. To do this, a reference point is established by inserting a steel tube to a depth where loading (and therefore stress) on the soil is negligible. A bullet gun is then lowered down and shots fired through the tube wall. Each bullet contains a γ ray emitter. It penetrates 300 to 400 mm beyond the wall. A γ ray log is made of the bullet positions, and this is then a reference for all subsequent loggings. The deepest bullet is considered a fixed point, so the settlement can be monitored. At least three such holes are needed if tilt measurement is required to supplement inclinometers.

If a structure moves in a lateral direction, there is a possibility of serious damage to pipelines and conductors. It is unlikely that a piled steel jacket will move laterally on the sea bed. Most gravity platforms have 'skirts' which penetrate the sea bed to resist shear and scour.

To measure such lateral movement as may take place in one method, a plastic casing is set from the platform vertically into the sea bed to a suitable depth. The casing has grooves machined on the inside, down which a slope indicator is lowered. Readings are taken at different depths. The deformation line of the casing and lateral displacement at the mudline can then be obtained. It is assumed that the base of the casing does not move.

12.11 Concrete structures

For the integrity verification of concrete structures the Construction Industry Research Association (CIRA) reported [7] that the testing of concrete underwater is at present at the research and development stage. At the current rate of progress, it is unlikely that any practical instrument or method will be in use for several years.

In the UK research has mainly been directed to the application of ultrasonic methods, and in particular the application of PUNDIT (portable ultrasonic non-destructive digital indicating tester). PUNDIT is a well established NDT method in land-based applications and, although research is being done on the use of PUNDIT underwater, it is not yet developed enough to indicate whether defects can be distinguished.

The use of a rebound hammer has been considered. However, it can only give an indication of strength, and relative readings from one survey to another. The strength is derived by empirical methods, based on known strength–hardness relationships of test specimens.

Inspection of reinforcement and prestressing is not possible using existing techniques, although work is being done on the possibility of taking potential measurements. A method called acoustic holography is also being developed by several companies. This is a technique for forming pictures by using sound pulses. CIRA consider it will be some years before a practical instrument is developed.

12.12 Rail track force monitoring

Internal rail track forces in a long length of track can be measured and monitored without having to cut the rail. The method involved was developed by British Rail [8].

One of the most important factors in monitoring the stability of continuously welded rail is the amount of longitudinal force occurring at any time in the rail. The force is influenced by a number of factors, including rail creep, track movement or even mining subsidence. Until now, no reliable means has been available for determining the residual stress other than by cutting the rail and measuring the resultant change in length. The Rail Force Transducer simplifies the procedure, enabling fast, accurate and non-destructive measurements of the longitudinal force to be made with ease.

The transducer is cylindrical in shape, 29 mm in diameter (the standard fish-bolt hole size on BR) and 28 mm deep. It is installed in a previously drilled and reamed hole 29.77 mm diameter on the neutral axis of the rail, and fixed in position using adhesive. The transducers operate on the vibrating wire principle and each device has two sensing elements built into it to cover a stress measuring range of 77 kN compressive to 300 kN tensile load.

Readings from the transducers are taken at any time required using a portable strain measuring instrument. Temperature readings are made at the same time to monitor changes in stress free temperature. A rail force value is then determined using the measuring instrument and the calibration curve. This value, together with the temperature reading, may then be referred back to the destressing temperature and the change of stress free temperature derived.

12.13 Tunnel lining stress monitoring

Access is only available from one side of tunnels. To measure the bending moment as well as direct stress on trials during tunnelling at Chinnor, Oxford-shire, Tyler [9] used TRRL surface vibrating wire strain gauges of the type first introduced in 1968 [10]. Bending moment is measured by using two wires instead of one, both incorporated in one gauge. The first gauge of this type was used in the London Underground [11].

The two wires of the 140 mm (5.5 in) surface strain gauge are standardized by setting the instrument on the tensioning plate with one end of the wires clamped, the other free and the steel barrel loose in adjustment. The other ends of the wires are tensioned in sequence, with pliers or a tensioning device, and clamped, to give a natural frequency of vibration of 800 Hz. This ensures that at this frequency the wire lengths are 139.7 ± 0.127 mm (5.5 ± 0.005 in). The wires can then be set to any desired frequency prior to sticking or bolting by using the knurled barrel adjuster.

Both wires should be set to the same frequency of vibration prior to fixing to the structure as this ensures that the end blocks remain square relative to one another. Thus when bolting to a flat surface no rotation of the blocks will occur which could inhibit free barrel movement.

If tension is expected then the wires may be set to a frequency as low as 500 Hz; if compression, as high as 1300 Hz. The strain on the wire at 500 Hz is 750×10^{-6}; at 1300 Hz it is 5070×10^{-6} giving a strain range of over 4300×10^{-6}. If a smaller strain range is expected then it is preferable to use the lower end of the range as the gauge is more sensitive there.

With the wires set at the chosen frequency and the locking rings locked on the knurled adjusters, the gauge can be glued and/or bolted to the structure. A quick setting resin glue used over the whole area of the end block is the simplest means of attachment, and is particularly suited to tension members where fatigue could arise in service around drilled and tapped holes. The arrangement is shown in Fig. 12.7. During tests carried out on the London Underground in 1970 using single wire gauges the end blocks were machined to take up the curvature but later tests proved this unnecessary, providing glueing and screwing was carried out. (The screws were only tightened sufficiently to hold the gauge lightly in position while the glue was setting, otherwise tilt and binding of the tubes in the end blocks would have occurred.)

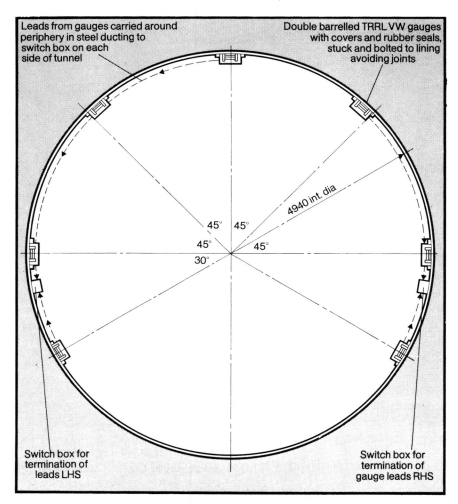

Figure 12.7 *Arrangement of strain gauges for tunnel lining stress monitoring.*

Results from gauges stuck to the lower flange of steel beams of the Loudwater Viaduct (M40), in 1968 [12] have been satisfactory in spite of traffic vibration. The overall stability of TRRL acoustic gauges from the point of view of creep in the wire has been confirmed by results obtained from the Medway Bridge.

Double wire gauges are used at Chinnor for the instrumentation of the steel lining section of an experimental tunnel. The strain readings form a small part of the data obtained during the tests. Ground movements occurring during tunnelling are logged and the performance of the steel lining, shotcrete lining and rock bolt anchors for tunnel support compared with predictions made using finite element methods.

12.14 Bridge stress monitoring (see also Section 2.4)

Displacement gauges were embedded in the Adur Bridge [13] which was opened in June 1970. Collectively, the instrumentation enables the response of the bridge to its thermal environment to be established. This information is used by the bridge designers to calculate the total movement expected, and to predict the forces in the structure. This allows the correct type of bearings and expansion joints to be specified and determines the required setting of these components during construction.

Measurements of creep and shrinkage in the concrete of the Medway Bridge (M2) Rochester, Kent, have been made since its construction in 1963. The bridge is of the cantilever type and strain gauges were embedded in one of the cantilever sections. As construction progressed, a steep increase in compressive strain was recorded, attributed to the continued construction and to prestressing later in the programme. Recordings during the next six years showed that strain is proportional to the logarithm of time.

Measurements were compared with calculations based on the German Standard 4227, 1953 'Spannbeton, Richtlinien fur Bemessung und Ausfuhrung' and were found to be in close agreement. Surface mounting strain gauges have been used extensively on a number of sites in the United Kingdom to assess the structural behaviour of steel box girder bridges in service. A programme of investigation commenced in 1971 following the failure of some bridges during the construction phase.

An important feature of tests which were carried out on bridges already in use involved the accurate measurement of strains in sections of each bridge and the extent to which these strains varied when the bridge was hydraulically lifted at its ends. This work was carried out on a box girder bridge across the M62 motorway at junction 17 near Prestwich, Lancashire. Gauges were mounted inside and outside the box girder and inside the webs.

The strain distribution was measured by attaching surface gauges (type TSR/5.5) to predetermined points on the box girder sections. The connecting leads were then taken to a convenient measuring point, where they were coupled to a portable automatic data logging system.

After releasing the bearings, the bridge was jacked up hydraulically, and strain readings were recorded. Steel strengthening plates were then welded to the box girder, and the bridge deck was lowered. A comparison of the strain readings before and after attaching the plates enabled the effects of the strengthening to be observed.

Strain gauge instrumentation was also carried out on the box girder bridge at Milford Havem, Wales, during the reconstruction period.

12.15 Structural deformation: Moiré grid technique [14]
(see also Section 5.20)

Conventional Moiré grid technique, whereby small changes in a grid pattern are recorded photographically, has been adapted so that a 35 mm camera in a modified form can be used to record the deformation and still provide an increased sensitivity.

A four-slotted mask located close to the iris, tunes the lens to resolve 300 lines/mm over the whole field of the image plane. Depth of focus remains sharp (approximately three times that achieved with the conventional circular aperture).

The camera is used with a fine grain film to record a fine pattern applied to the engineering structure. This may be random but more usually it is a regular pattern of dots or stripes running across the surface. The camera is set at a predetermined distance so that the demagnified image of the pattern appears at 300 resolved elements per mm in the image. By using electronic flash illumination, double exposure photographs are made on the same film frame before and after the object is deformed.

After processing the fine pattern detail in the image behaves as two orthogonal diffraction gratings. A beam of light directed onto the negative is diffracted into four first-order directions. By viewing the negative in one of these diffracted beams a Moiré fringe pattern, or displacement map, is seen showing dark fringes corresponding to movements of one half of a pitch in the object pattern.

From one negative two such displacement maps are generated representing either the vertical or horizontal displacements depending on whether a vertically diffracted beam or a horizontally diffracted beam is examined. Various features of material behaviour, such as cracking and slipping, are identified from these maps. Fringe values at different map positions provide strain data information by local differencing between adjacent points. Such fringe values can also be used to construct an exaggerated plot of the deformation.

12.15.1 Applications

John Laing Research and Development used Moiré grid methods to study the deformation of a 3.6 m wide brick wall supported by a new type of lightweight lintel.

The surface of the bricks was coated with a halftone paper pattern having an array of 26 dots/cm (65 dots/in) running diagonally across the paper (it was manufactured by contact copying a master halftone negative onto photographic document paper).

With a live Moiré system two separate negatives are brought in to exact register in a system similar to that used for viewing the double exposures. This generates deformation data that would normally not be available, unless many cameras are used. It can also be applied to a deformation programme over a long period which might include the removal of the camera and its replacement at a later stage. Errors in the repositioning of the camera can generally be compensated for in the analysis stage.

At the Building Research Establishment a two-camera system monitored three-dimensional motions in a concrete panel unit. In another application a 4.5 m wide quarter scale model of a multistorey building, a brick and timber panel structure, a 6 m diameter casting of a pressure vessel and a steel girder 18 m long were monitored by means of Moiré grid.

A dot pattern applied at 45° to the weld direction has been used to detect residual stress at the interface of a welded transition joint between $2\frac{1}{4}$ Cr/Mo and AISI 316 heavy section pipes. These studies for the CEGB also revealed, at a temperature of 585 °C, severe strains in the material either side of the interface.

Moiré grid used in a test over a period of 2000 h on the creep at high temperature of a small bore welded pipe has also shown a pronounced effect at the weld. Moiré technique is probably the only method of examining localized surface deformation on structures, at high temperatures, in air.

In the measurement of strain in soft materials the stencilled pattern has no influence on the material properties. It provides an alternative to strain gauges in studies of cracking thin aluminium alloy sheet and fatigue in PVC and carbon fibre composites. A *Y*-displacement fringe map from a 75 mm wide specimen of carbon fibre composite subjected to cyclic loading is shown in Fig. 12.8(a). By

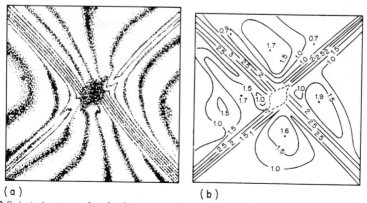

(a)   (b)

Figure 12.8 *(a) An example of a fringe map from a fatigued specimen of carbon fibre composite; (b) contours of the calculated percentage strain in the Y-direction.*

measuring the fringe spacings locally on a similar map, but at a later stage in the fatigue history, the distribution of strain has been plotted and is shown in Fig. 12.8(b). The levels and steep gradients of strain indicated are typical of the results produced by the technique.

12.16 Infrared thermography (see also Sections 2.9, 2.10)

12.16.1 LNG tank defect detection

El Paso Southern, a ship under construction at the Newport News Shipbuilding and Dry Dock Company, incorporated an LNG tank assembled inside the ship. Inspection was needed to check whether a secondary barrier located within the tank was free from defects. Visual probes, with or without image intensification were defeated by bothersome reflections from the primary stainless steel wall.

Hand-held probes for examining the surface were used with limited success. The area of defect location must be reduced to a small region by a separate method. Ultrasonic techniques using computer enhancement detect a defect remotely and establish its location by triangulation, but were unworkable in this application because normal random ship noise nulled or obscured the noise from a slight diffusion leak.

The infrared radiometric measurement was effective. Infrared scanning systems provide a wide field of view (up to 45°) and a full field of data in a short period of time (1/16 second) with a good spot resolution. The level of signal from the radiometer is a function of surface temperature and the surface emittance. The infrared scanner is able to sense temperature differences as low as 0.2 °C. If the leak causes a change greater than this level, the defect will show up on the scanner display scene as a cooler area. The extent of the defect will be more difficult to pin down with the infrared technique alone, but the speed at which the leak becomes visible at the surface is a function of flow and defect size. Consequent measurements using probe techniques can indicate the extent.

In tests carried out on the *El Paso Southern* a number of defective areas were detected by the infrared technique which were later confirmed by ultrasonic and temperature techniques. Upon physical examination of these areas, there was substantiation, and repairs were made.

12.16.2 Bridge deck delamination

Concrete bridge decks become delaminated under the influence of corrosion, particularly when de-icing salts are used for snow and ice control. In Canada this has been a costly and serious problem. Many decks experiencing distress have less than 40 mm clear cover over the steel top mat.

The detection of subsurface fracture planes, or delaminations, is necessary when determining the priority of structures for repair and for the identification of areas of unsound concrete prior to a repair contract. At the present time delaminations are detected by manual methods which are tedious and require expensive lane closures.

Infrared thermography has been found to be capable of detecting subsurface fractures on bridge decks because temperature differentials exist between sound and unsound concrete [15]. A portable instrument which electronically interprets acoustical signals generated by the instrument and reflected through the concrete may also be used [16]. This instrument, in common with the manual methods, has the disadvantage that parts of the structure must be closed to traffic while observations are made. The cost of providing traffic protection, especially in urban areas, can be very high and often causes serious delays to the road user.

It was known that substantial changes in the temperature of a bridge deck occur in the course of a day. An investigation determined different surface temperatures associated with areas of sound and unsound concrete and whether these differences could be detected and recorded using infrared thermography.

Two systems were used during the testing: the AGAtronics Thermovision 750 and the AGAtronics Thermovision Superviewer. The AGA 750 was used during testing at both ground level and from the air, while the Superviewer system was only used during airborne testing.

The 750 system was equipped with a 20° lens during the ground level testing and both the 750 and the Superviewer used a 7° lens for the airborne testing. A solar filter was used during daytime operations to filter out reflected short wavelength radiation from the Sun and to decrease the effect of shadows.

12.16.3 Location of reinforcements in concrete

Often, more than ten years lie between the planning and finishing phase of a plant. For example, in one power plant, several hundred thousand anchors were needed to be added at a later time. When drilling the required holes, the reinforcing steel bars in the concrete must not be damaged.

Hochtief [17] developed a method to locate the exact reinforcement structure in the concrete. Infrared thermography was applied in conjunction with their own technique to detect with millimetre accuracy the exact location of the reinforcement; with an ability to distinguish between the basic reinforcement, the stirrups, the hooks of shear-armature, and other metallic parts.

In this technique the reinforcement bars are heated by electromagnetic induction. After a short while of directing the electromagnetic field onto the surface of the concrete, zones of equal temperature surround the steel bars. As soon as the heat reaches the surface of the concrete, the thermal radiation can be measured. Slowly the heat on the surface grows into a heat band, where the hottest point always lies in the middle. On a thermogram of the concrete surface the heat band appears as a white line on the display screen. Each line on the thermogram represents at least one reinforcement bar in the concrete, in the shade of which there may be others. Thus, the reinforcement appears as a thermal image projected onto the surface. For follow-up work this picture is immediately marked onto the concrete.

Since March 1979 this thermographic inspection method has alraedy been successfully applied in three nuclear power plants in Germany.

12.17 Minecage monitor

Recommendations following the enquiry into the Markham Colliery disaster (1973) have led to the development of a system to monitor the minecage by providing a speed protection 'envelope' for mine winding gear. The concept is for speeds outside defined limits, or overwinding of the cage, to activate trip circuits and initiate braking. In its present form the system is essentially a monitoring device and does not control the operation of the winding mechanism.

The monitor can be considered as three sections: the sensors, electronics unit and control/display panel. The sensors employed are an optical shaft encoder mounted on the main winding shaft and magnetically encoded rope transducers.

12.17.1 Shaft encoder

This produces two pulse train outputs, one of which is fractionally 'later' than the other according to the direction of shaft rotation. Electronic pulses are produced at the rate of 1000/revoluton on each of the two ouputs. The phase relationship between the pulse trains determines the direction of movement of the cages. Speed can also be calculated by counting the pulses over a 100 ms time period. Below 4.57 m/s (15 ft/s) the equipment counts fractions as well as integral pulses enabling speeds down to 0.6 m/s (2 ft/s) to be ascertained accurately by the device.

Distance is measured by accumulating pulses from the shaft encoder over 12.5 ms time periods. This count is scaled to linear units and added to a counter (DIST 1). A reciprocal counter (DIST 2) is derived by subtracting DIST 1 from the known rope length.

Two magnetic proximity switch sensors mounted near the top of the shaft (one for each cage) provide a datum position for distance measurement. When a cage passes a switch a pulse causes the DIST 1 counter to be reset to the correct distance value. Thus the counter is not subject to accumulative error over a number of runs due to rope slip. Every 100 ms DIST 1 is scaled to the required display units and converted to binary code digits (b.c.d.) to drive the display, which may use units of feet, metres or revolutions.

The transducer is a contactless flux gate magnetometer mounted on the cage itself. This 'reads' a pattern of magnetic stripes from the fixed steel guide rope within the shaft (rather like a giant version of early wire recorders). The distance between magnetic poles on the rope is 20 cm which enables the position of the cage to be determined within 10 cm. Two read heads generate separate signals, enabling the direction of movement to be determined.

Discrepancy between drum and rope signals can be displayed to indicate the amount of slip.

The board is based on an Intel 8085 microprocessor with up to 4k system memory. Programs are held in a single plug-in PROM allowing for reprogramming in the event of changing shaft parameters.

Apart from scaling, converting and outputting the transducer signals the processor is also responsible for continuously comparing the performance of the cage against a stored speed profile. Four sets of profile coordinates are stored in memory. If the cage moves outside the profile or safety envelope (usually by 5 or 10%) an overspeed condition occurs. An overwind is when the cage starts to head upward into the winding gear and is detected by comparing DIST 1 and DIST 2 with preset distance values. Both over speed and overwind open trip relays which initiate warnings on the operator's control panel. Overspeed and overwind trips are non-latching, i.e. they are reset once the trip condition has been removed.

References and list of companies

1 Collacott, R. A. (1977), *Mechanical Fault Diagnosis and Condition Monitoring*, Chapman & Hall, ISBN 0 412 12930 2.
2 Collacott, R. A. (1979), *Vibration Monitoring and Diagnosis*, Longmans/Godwin, ISBN 0 7114 5201 6.
3 Pressicaud, J. (1983), 'Optical checks and the applied research department RD3', *Bulletin Technique Au Bureau Veritas,* 12, No. 2, 124.
4 Haines, N. F. 'Internal inspection of nuclear reactors', *CEGB Research,* October, No. 8, 3.
5 Silk, M. G., Williams, N. R., Jones, C. H., Watkins, B. and Cassie, G. E. (1975), 'The continuous monitoring of fixed offshore platforms for structural fatigue', AERE (Harwell) Report R-8055.
6 (1983), 'British Gas system finds cracks fast', *Technology,* 1, August, 19.
7 'Underwater inspection of offshore installations: guidance for designers' CIRA Underwater Engineering Group Report UR10, ISBN 0 86017 063 2, February 1978, 6 Storey's Gate, Westminster, London SW1P 3AU, UK. (Tel. 01-930 7447).
8 British Rail Technical Centre, Derby, England.
9 Tyler, R. G. (1974), 'Measuring bending moments', *Tunnels and Tunnelling,* September.
10 Tyler, R. G. (1968), 'Developments in the measurement of strain and stresses in concrete bridge structures', Ministry of Transport RRL Report No. 189, Crowthorne.
11 Thomas, H. S. H. (1960), 'The measurement of strain in tunnel linings using the vibrating wire technique', Building Research Station, Ministry of Technology, current papers, Engineering Series 34.

454   *Structural Integrity Monitoring*

12 Tyler, R. G. (1973), 'Creep and shrinkage of concrete bridge structures', *PhD. Thesis,* London University.
13 'Instrumentation of bridges' Reports LR189, 539, 641 Transport and Road Research Laboratory, Crowthorne.
14 Ferno, C. (1980), 'Moiré grid technique studies deformation in large structures', *Transducer Technology,* October, 11–13.
15 Holt, F. B. and Manning, D. G. (1978), 'Infrared thermography for the detection of delaminations in concrete bridge defects', *Proc. 4th Biennial Infrared Information Exchange,* August 22–24, St. Louis, Missouri, USA, pp. A61–A71.
16 Moore, W. M., Swift, G. and Milberger, L. J. (1973), 'An instrument for detecting delamination of concrete bridge decks', *Highway Research Record No. 451,* 44–52.
17 Hochtief AG, Abt. Qualifatskontrolle, Dr. Bernd Hillemeier, Bockenheimer Landstrasse 24, D-6000 Frankfurt/Main.

Index